D1201608

PROCESS CONTROL

PROCESS CONTROL

McGRAW-HILL SERIES IN CHEMICAL ENGINEERING

Max S. Peters, *Consulting Editor*

EDITORIAL ADVISORY BOARD

Charles F. Bonilla. Professor of Chemical Engineering, Columbia University

Cecil H. Chilton. Editor-in-Chief, *Chemical Engineering*

Sidney D. Kirkpatrick. Consulting Editor, McGraw-Hill Series in Chemical Engineering, 1929–1960

Walter E. Lobo. Consulting Chemical Engineer

Carl C. Monrad. Chairman, Department of Chemical Engineering, Carnegie Institute of Technology

Robert L. Pigford. Chairman, Department of Chemical Engineering, University of Delaware

Mott Souders. Associate Director of Research, Shell Development Company

Richard H. Wilhelm. Chairman, Department of Chemical Engineering, Princeton University

BUILDING THE LITERATURE OF A PROFESSION

Fifteen prominent chemical engineers first met in New York more than 30 years ago to plan a continuing literature for their rapidly growing profession. From industry came such pioneer practitioners as Leo H. Baekeland, Arthur D. Little, Charles L. Reese, John V. N. Dorr, M. C. Whitaker, and R. S. McBride. From the universities came such eminent educators as William H. Walker, Alfred H. White, D. D. Jackson, J. H. James, Warren K. Lewis, and Harry A. Curtis. H. C. Parmelee, then editor of *Chemical & Metallurgical Engineering*, served as chairman and was joined subsequently by S. D. Kirkpatrick as consulting editor.

After several meetings, this first Editorial Advisory Board submitted its report to the McGraw-Hill Book Company in September, 1925. In it were detailed specifications for a correlated series of more than a dozen texts and reference books which have since become the McGraw-Hill Series in Chemical Engineering.

Since its origin the Editorial Advisory Board has been benefited by the guidance and continuing interest of such other distinguished chemical engineers as Manson Benedict, John R. Callaham, Arthur W. Hixson, H. Fraser Johnstone, Webster N. Jones, Paul D. V. Manning, Albert E. Marshall, Charles M. A. Stine, Edward R. Weidlein, and Walter G. Whitman. No small measure of credit is due not only to the pioneering members of the original board but also to those engineering educators and industrialists who have succeeded them in the task of building a permanent literature for the chemical engineering profession.

PROCESS CONTROL

DISCARD

PETER HARRIOTT

Associate Professor of Chemical Engineering
Cornell University

McGRAW-HILL BOOK COMPANY
New York   San Francisco   Toronto   London

PROCESS CONTROL

Copyright © 1964 by McGraw-Hill, Inc. All Rights Reserved. Printed in the United States of America. This book, or parts thereof, may not be reproduced in any form without permission of the publishers. *Library of Congress Catalog Card Number* 63-21782

26785

II

To My Wife and Parents

Preface

This book is an introduction to the theory of automatic control and its application to the chemical process industries. Emphasis is placed on the dynamic behavior of processes and processing equipment rather than on the mechanical features of instruments and controllers. The book is intended mainly for chemical engineers, but the information on process dynamics should be useful to mechanical and electrical engineers who help design process control systems.

The book is suitable for a senior or introductory graduate course in process control or for self-study by practicing engineers. A knowledge of calculus and elementary differential equations is assumed, and the reader is expected to become familiar with the use of Laplace transforms, as most of the equations and problems use transformed notation. The Laplace transform is presented in Chap. 2, and the simple transformation rules can be grasped in one or two periods. The mathematical level of the book is suitable for a junior course, but the material on process dynamics will mean more to the student who has had courses in unit operations or in fluid dynamics, heat transfer, and mass transfer. Some knowledge of the steady-state behavior of heat exchangers, distillation columns, and reactors is assumed in discussing the dynamic behavior of these devices.

The first half of the book deals mainly with general control theory. The open-loop and closed-loop response of systems consisting of first-order and second-order elements plus dead time are discussed. Both transient response and frequency response are discussed, but frequency-response analysis is emphasized, since this seems the easiest and most useful technique for process control studies. The root-locus method is left for advanced courses, because it is fairly complex and is not practical for systems with dead time or distributed parameters.

The second part of the book gives introductory theory and practical examples dealing with the response and control of heat exchangers, flow control, level control, distillation control, and the control of temperature, composition, and pH in chemical reactors. These topics were chosen partly because of the author's interests and partly to illustrate certain types of behavior. A tubular exchanger is an example of a distributed-parameter system. Flow control is a case where the process is much faster than the measuring and controlling elements. Level control serves to illustrate the use of averaging controllers, which might be used for pressure systems as well. The response of a distillation column depends on the dynamics of fluid flow, heat transfer, and mass transfer, and one of the main problems is deciding which variables to control. A chemical reactor is particularly interesting because it is one of the few systems which can be open-loop unstable.

For a one-term course in process control, the first half of the book could be used, along with two or three topics selected from Chaps. 11 to 16. The last few topics in Chaps. 6 to 10 could be omitted or treated lightly to save time for a broader selection of applications. A weekly laboratory and computation period is recommended, and a suggested list of experiments is available from the author.

The author wishes to acknowledge the many helpful comments of his students and the technical information furnished by specialists in process control, including Leonard Barnstone, Page Buckley, Louis Bertrand, Bruce Powell, Allan Catheron, Ted Williams, and Nathaniel Nichols. Professor Thomas Weber deserves special credit for giving detailed comments on every chapter.

PETER HARRIOTT

Contents

Nomenclature

a Parameter in Eq. (11-5), constant

A Heat-transfer area, cross-section area, constant, concentration of A

A.R.   Amplitude ratio

b Bandwidth for proportional control, constant, parameter in Eq. (11-5)

B Bottoms flow rate, constant, concentration of B

c Concentration

c_p Heat capacity

C Capacity, total capacity

D Diameter, distillate flow rate

e Error, base of natural logarithms

f Frequency in cycles per unit time, friction factor

F Volumetric flow rate, feed rate

f, F Operators meaning "function of"

g Gravitational constant

g_c Newton's-law conversion factor, 32.2 ft-lb/lb force-sec

G Transfer function

h Head or depth of fluid, film coefficient of heat transfer

H Transfer function for measurement lag, liquid holdup on a distillation plate

ΔH Heat of reaction

I Interaction term in Eqs. (6-27), (6-31)

j $\sqrt{-1}$

k Reaction-rate constant, thermal conductivity, constant in flow equation

K Gain

L Time delay, load variable, inductance, length, liquid flow rate

$\mathcal{L}$ Laplace-transform operator

$\mathcal{L}^{-1}$ Inverse Laplace-transform operator

m Integer

M Mass

M_p Peak gain of closed-loop frequency response

n Integer

N Maximum slope of reaction curve, stirrer speed

p Fractional change in controller output

P Pressure, period of oscillation

Q Heat-generation rate, heat-transfer rate, volume of gas

r Chemical reaction rate, root of a polynomial, fractional change in flow, radius

R Resistance, ratio of time constants, ratio of pressure drops, reflux flow rate

s Complex variable

S Controller sensitivity, cross-section area

t Time

T Time constant, absolute temperature, tank diameter

T_D Derivative time

T_H Holdup time

T_i Integral time

T_R Reset time

U Overall heat-transfer coefficient

v Velocity

V Volume, vapor flow rate

w Width of stirrer

W Mass flow rate

x Valve-stem position, fraction converted, length, mole fraction

X Dimensionless heat-generation parameter in Eq. (15-16), valve position

y Fractional response to a step input, mole fraction

z Film thickness

Z Depth of liquid

Greek Letters

α Thermal diffusivity $k/\rho c_p$, closed-loop phase angle

β Constant

α, β, γ Dimensionless stability parameters in Eq. (15-8)

ζ Damping coefficient

θ Temperature, controlled variable

λ Heat of vaporization

μ Viscosity

ρ Density

Σ Summation

ϕ Phase angle

ω Frequency in radians per unit time

ω_n Natural frequency

Subscripts

ad adiabatic

av average

A, B Chemical substances

B Bottom product

c Set point, control point, controller

D Distillate product

e Effective

f Fluid

F Feed

i Input

j Jacket

L Load, liquid

m Measured
o outside, outlet
0   Initial
p Process
s Steam, stirrer
ss Steady state
sh Shell
T Total, tank
u Ultimate
v Valve
w Wall

In most cases h, p, x, θ, and other symbols for process variables stand for the fractional change in that variable. The average or normal value is denoted by $\bar{h}$, $\bar{p}$, etc.

PROCESS CONTROL

PROCESS CONTROL

1 INTRODUCTION

I-I. HISTORICAL REVIEW

For many years process control was an art rather than a science. Design engineers calculated equipment sizes to give a certain steady-state performance, but the control systems were chosen by rule of thumb rather than by dynamic analysis. Usually the instruments could be adjusted to give results as good as or better than manual control, and this was considered adequate. When control was poor, the instruments, sensing devices, or the process itself was changed by trial-and-error methods until a satisfactory solution was found.

Theoretical papers on process control started to appear about 1930. Grebe, Boundy, and Cermak (1) discussed some difficult pH control problems and showed the advantage of using controllers with derivative action. Ivanoff (2) introduced the concept of potential deviation and potential correction as an aid in quantitative evaluation of control systems. Callender, Hartree, Porter, and Stevenson (3, 4) showed the effect of time delay on the stability and speed of response of control systems.

Meanwhile work was progressing at a somewhat faster pace in related fields. Minorsky (5) considered the use of proportional, derivative, and second-derivative controllers for steering ships and showed how the stability could be determined from the differential equations. Hazen (6) introduced the term "servomechanisms" for position control devices and discussed the design of control systems capable of closely following a changing set point. From work on feedback amplifiers, Nyquist (7) developed a general and relatively simple procedure for determining the stability of feedback systems from the response of the open-loop system. A great deal of work was done during World War II to develop servomechanisms for directing ships, airplanes, guns, and radar antennas. After the war, several texts appeared incorporating these advances, and courses in servomechanisms became a standard part of electrical engineering training.

Although the basic principles of feedback control can be applied to chemical processes as well as to amplifiers or servomechanisms, chemical engineers have been slow to use the wealth of control literature from other fields in the design of process control systems. The unfamiliar terminology is one reason for the delay, but there are also basic differences between

1

chemical processes and servomechanisms which have delayed the development of process control as a science.

Process control systems usually operate with a constant set point, and large-capacity elements help to minimize the effect of disturbances, whereas they would tend to slow the response of servomechanisms. Time delay or transportation lag is frequently a major factor in process control; yet it is hardly mentioned in many servomechanism texts. In process control systems, interacting first-order elements and distributed resistances are much more frequent than second-order elements, just the opposite of the situation in the control of machinery. These differences make many of the published examples of servomechanism design of little use to those interested in process control.

In spite of the time delays, the large time constants, and the nonlinear elements, it is fairly easy to achieve reasonably good control of most chemical processes. Exact control is not always needed; in fact 5 to 10 per cent changes in level or pressure or 2°F changes in temperature have hardly any noticeable effect in many cases. Furthermore, the set point is usually constant, and some of the inputs are regulated by other controllers, and so the main control system has to compensate only for small load changes. The lengthy analysis needed for accurate design cannot be justified for such cases, and the control system is selected by using rules of thumb or short-cut design procedures.

Finally, for complex processes that do need close control, the weakest part of the control-system design is usually the dynamic data for the process. The response of the controller, valve, and measuring device may be known to within 5 to 10 per cent, but the error in the predicted dynamic behavior of the process itself is usually two to three times as great. The lack of accurate information on process dynamics is still a major factor limiting the use of process control theory.

As more studies of equipment dynamics become available, dynamic analysis will be more widely used in studying existing equipment and in designing controls for new processes. The economic justification will come mainly from improvements in productivity or product quality, improvements which might seem small on a percentage basis but which could save thousands of dollars per year because of the large quantities produced. Reduction of manpower requirements was one of the early justifications for automatic control, but there is not much room left for further economies here, since many plants are already operating with the minimum force needed to cope with emergencies. More critical examination of control schemes at the design stage can also reduce the investment in instruments, which may be 10 per cent of the cost of the plant. Many existing plants are overinstrumented and are really controlled by only a few automatic controllers plus manual adjustment of other signals. Dynamic analysis can show what controllers are really needed and can lead to quantitative comparisons of proposed control schemes.

1-2. BLOCK DIAGRAMS

The first step in the analysis of many engineering problems is to draw a physical diagram showing the flows of material or heat into the system or the forces acting on various parts of the system.   For control problems it is also helpful to use a block diagram to show the functional relationship among the parts of the system.   Each part is represented by a block with one nput and one output.   The block contains the mathematical relationship between input and output, which is called the "transfer function," or a general symbol to denote the transfer function.   The input and output variables are considered as signals, and the blocks are connected by arrows to show the flow of information in the system.

The addition or comparison of signals is shown by a circle or summation point with the signs written just outside the circle.   Symbols to denote addition, Σ, multiplication, $\times$, or division, $\div$, can be used inside the circle; if no symbol is used, addition is assumed.

Block diagrams for a simple control system are shown in Fig. 1-2. The top diagram shows all the major parts of the system and the corresponding signals.   Going around the loop, the signal from the temperature controller is an air pressure p, the value output is the stem position x, the jacket output is the steam temperature θ_s, the kettle signal is the temperature θ, and the signal to the controller is the measured temperature θ_m.   The values of θ and θ_m differ when θ is changing, since the temperature bulb does not have an immediate response.   The controller acts on the error e, which is the difference between the set point or desired temperature θ_c and the measured temperature.   Note that the signals are generally not the same as the physical flows in the system, though sometimes the valve output is considered to be the flow through the valve rather than the valve position.

Blocks may be combined or rearranged to simplify the algebraic manipulation of signals.   The valve, jacket, and kettle can be considered the process and combined in one block, which has the transfer function G_p. A further step combines the transfer function of controller and process to give one block relating e and θ.   If the measurement lag is very small so that θ and θ_m are practically the same, we can eliminate the measurement block to get the simplest diagram of a feedback control system, which has only one block with the combined transfer function G.

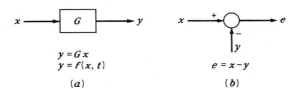

$$y = Gx$$
$$y = f(x, t)$$

$$(a)$$

$$e = x - y$$

$$(b)$$

FIGURE 1-1   Block-diagram components.   (*a*) Dynamic relationships;   (*b*) comparison of signals.

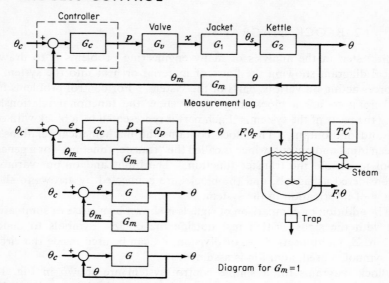

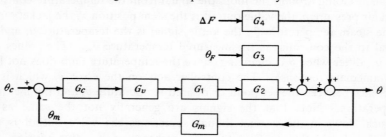

FIGURE 1-2   Block diagrams for a temperature control system.

FIGURE 1-3   Representation of load variables in a feedback control system.

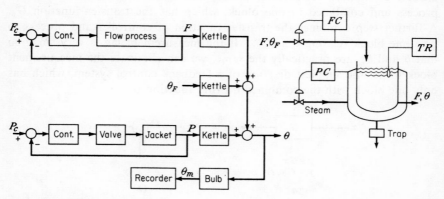

FIGURE 1-4   Open-loop control system

The block diagrams for complex control systems have multiple feedback loops as well as several inputs. Transformation theorems for simplifying complex block diagrams are presented in several texts (8, 9), and a few of the obvious transformations are used in later examples.

Each block in the diagram has only one input signal, but an element such as the kettle in Fig. 1-2 may have several inputs. The kettle temperature is affected by changes in feed rate and feed temperature as well as by the steam temperature, which is the main input variable. The uncontrolled inputs are load variables and are shown as separate branches on the block diagram. The effects of changes in load variables and changes in the controlled variable are added to get the total change in output, which is justifiable if the changes are small.

1-3. CLOSED-LOOP AND OPEN-LOOP CONTROL SYSTEMS

Figures 1-2 and 1-3 show why this type of control can be called either "closed-loop" control or "feedback" control. With feedback control the controlled variable is measured and compared with a reference input, and the difference is the signal to the control elements. An alternative method of control would be to regulate all the inputs to the process by the use of separate controllers, metering pumps, or manually adjusted valves. If the main variable is not used to adjust any of the inputs automatically, the system is an open-loop control system.

Figure 1-4 shows one scheme for open-loop control of the kettle used in the previous case. The steam pressure in the jacket is kept constant by a pressure control system, and the liquid flow is regulated by a flow controller. Although there are two feedback loops in the diagram, this is an open-loop system, since the kettle temperature, which is the main variable, is not used to adjust any of the inputs. In practice, an intermittent feedback might be provided by an operator who would adjust the process inputs when the kettle temperature became too high or too low.

The main advantage of feedback control is shown by Figs. 1-3 and 1-4. The feedback control system corrects for changes in all inputs with only one controller. With an open-loop system, every input variable must be kept constant to keep the output constant, and some inputs, such as ambient temperature or impurity concentration, are not easily measured or controlled.

Open-loop control is used for some simple processes where close control is not needed or where the input variables do not change appreciably. At the other extreme, open-loop control may be used for complex processes where feedback control is not good enough. Feedback control cannot give close control when there is a large time delay in measuring the process output, as when the main variable is a chemical composition, a molecular weight, or the quality of the product from later processing steps. To achieve close control in such cases, it may be necessary to control separately all the flows, temperatures, pressures, etc., that influence the process.

The term "feedforward control" has been applied to open-loop systems such as that shown in Fig. 1-4, but it usually refers to systems where one input variable is measured and used to adjust another input variable. For example, the feed temperature could be measured and used to adjust the steam pressure or the set point of a steam-pressure controller. The use of feedforward combined with feedback is considered in Chap. 8.

This book deals primarily with feedback control systems, and feedback should be assumed whenever the terms "control" and "control system" are used.

1-4. BASIC CONTROL ACTIONS; ON-OFF CONTROL

The on-off controller is the cheapest and most widely used type. It is the kind used in domestic heating systems, refrigerators, and water tanks. When the measured variable is below the set point, the controller is on and the output signal is the maximum value. When the measured variable is above the set point, the controller is off and the output is zero. Because of mechanical friction or arcing at electrical contacts, the controller actually goes on slightly below the set point and goes off slightly above the set point. The interval, or differential gap, may be deliberately increased to give decreased frequency of operation and reduced wear.

The performance of an on-off temperature controller with a small differential is shown in Fig. 1-5. The rate of temperature change is assumed to be more rapid with the heater on than with the heater off, and the heater is on less than half the time. If there were no lags in the transfer of heat, the process output would be a triangular wave with an amplitude equal to the differential and the frequency could be easily calculated from the differential

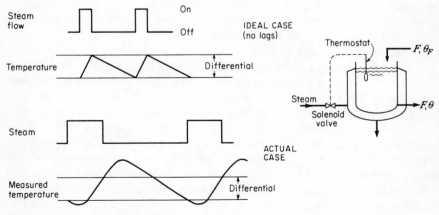

FIGURE 1-5   On-off control of kettle temperature.

and the rates of temperature increase and decrease.   The process actually shows overshoot and undershoot, which may be as great as or greater than the differential gap of the controller.

The overshoot is caused primarily by the heat capacity of the wall. After the valve closes, the hot wall continues to supply heat to the fluid in the kettle until the wall is cooled to the process temperature.   After the valve opens again, it takes time to reheat the wall and meanwhile the temperature drops below the cut-in point.   The magnitude and frequency of the oscillations can be estimated by assuming that the output is a sine wave and using the frequency-response analysis discussed in later chapters.

1-5. PROPORTIONAL CONTROL

The cycling inherent with on-off control would be objectionable for most processes.   To get steady operation when disturbances are absent, the controlled variable must be a continuous function of the error.   With proportional control, the most widely used type, the controller output is a linear function of the error signal.   The controller gain is the fractional change in output divided by the fractional change in input.

$$p = K_c e \qquad (1\text{-}1)$$

where p = fractional change in controller output
$\quad\ e$ = fractional change in error
$\quad K_c$ = controller gain
(The equations for controllers, and other process elements as well, are written in terms of changes from the initial values to simplify later mathematical treatment.)

The control action can also be expressed by the proportional bandwidth b.   The bandwidth is the error needed to cause a 100 per cent change in controller output, and it is usually expressed as a percentage of the chart width.   A bandwidth of 50 per cent means that the controller output would go from 0 to 1 for an error equal to 50 per cent of the chart width or from, say, 0.5 to 0.6 for an error of 5 per cent.

$$b = \frac{1}{K_c} \times 100 \qquad (1\text{-}2)$$

Some pneumatic controllers are calibrated in sensitivity units, or pounds per square inch per inch of pen travel.   For a standard controller with a 3- to 15-psi range and a 4-in. chart, the gain and sensitivity are related by Eq. (1-3),

$$S = 3K_c \qquad \text{psi/in.} \qquad (1\text{-}3)$$

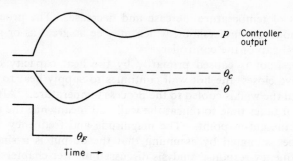

FIGURE 1-6   Response of proportional control system to
step change in load.

Figure 1-6 shows a typical response curve for a kettle with proportional control of temperature. A sudden drop in feed temperature causes a gradual drop in kettle temperature, which leads to a greater flow of steam through the control valve. The temperature eventually becomes constant at a value slightly below the set point, and this steady-state error is called the "offset." An offset is inevitable, since more steam is needed to heat the colder feed and the controller will admit more steam only if the temperature is below the set point.

Example 1-1

Water is heated continuously from about 70 to 110°F in a steam-jacketed kettle. A proportional controller is used to regulate the steam flow, and because critical flow occurs through the valve, the steam flow is directly proportional to the

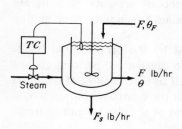

FIGURE E1-1a   Diagram for Example 1-1.

valve opening. The valve has equal-percentage trim, and a change in valve position equal to 1 per cent of the full stroke changes the valve area by 4 per cent of the original area. The temperature controller has a scale from 60 to 150°F.

Calculate the offset for a 5°F change in feed temperature for controller gains of 1 and 5.

Since we are concerned only with the steady-state changes, we can use a block diagram that shows only the steady-state characteristic or gain of each element. A consistent set of units must be used in defining the gains so that all units introduced cancel in following the signals around the loop. In this text, the individual gains are usually made dimensionless, and the numbers used appear in the definitions of two gains.

$$K_c = \frac{\text{fractional change in pressure}}{\text{fractional change in measured temperature}} = \frac{\Delta \text{ psi}/12}{\Delta\theta_m/90°F}$$

$$K_v = \frac{\text{fractional change in steam flow}}{\text{fractional change in pressure}} = \frac{\Delta F_s/\bar{F}_s}{\Delta \text{ psi}/12}$$

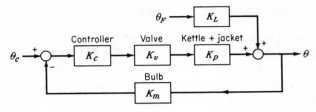

FIGURE E1-1b Block diagram for Example 1-1.

(where $\bar{F}_s$ = normal or initial value of steam flow)

$$K_p = \frac{\text{fractional change in temperature}}{\text{fractional change in steam flow}} = \frac{\Delta\theta/90°F}{\Delta F_s/\bar{F}_s}$$

$$K_m = \frac{\text{change in measured temperature}}{\text{change in actual temperature}} = \frac{\Delta\theta_m}{\Delta\theta}$$

The measurement gain is 1.0, and so the diagram can be simplified as follows: By algebraic manipulation

$$\Delta\theta = Ke + K_L \Delta\theta_F$$

Since θ_c is constant, $e = -\Delta\theta$ and

$$\Delta\theta(1 + K) = K_L \Delta\theta_F$$

The load gain K_L is the change in process temperature for unit change in feed temperature at a given steam rate. For this example

$$K_L = 1.0$$

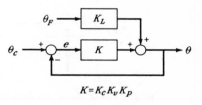

$K = K_c K_v K_p$

FIGURE E1-1c Block diagram for Example 1-1.

All the steam entering the jacket condenses; so a 1 per cent increase in steam flow is a 1 per cent increase in heat-transfer rate. Since the normal steam rate gives a 40°F temperature rise, a 1 per cent increase will give a 40.4°F rise, or an increase of 0.4°F. This increase is divided by the temperature range of the chart to get the process gain.

$$K_p = \frac{0.4°F/90°F}{0.01} = 0.445$$

$$K_v = 4.0$$

(from problem statement).

For $K_c = 1$, $K = 4(0.445) = 1.78$, and

$$\Delta\theta = \left(\frac{K_L}{1 + K}\right)\Delta\theta_F = \frac{1}{2.78}\Delta\theta_F$$

For $\Delta\theta_F = -5°$, $\Delta\theta = -1.8°F$

For $K_c = 5$, $K = 8.9$

For $\Delta\theta_F = -5°$, $\Delta\theta = -0.51°F$

I-6. THE STABILITY PROBLEM

The previous example showed that the offset from load changes is inversely proportional to $1 + K$, where K is the overall gain, the product of the gain terms in the closed loop. If an overall gain of 20 to 50 could be used, the offset would generally be negligible. The trouble is that gains above a certain maximum lead to continuous cycling, just as with on-off control. Calculating the maximum gain and the optimum gain are major topics in future chapters. One example is presented here as an introduction to the

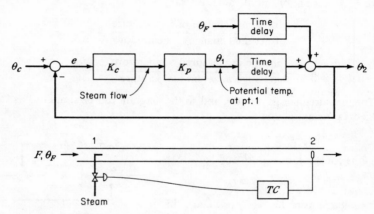

FIGURE E1-2a   Diagram for Example 1-2.

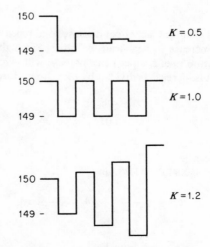

FIGURE E1-2b   Response curves for Example 1-2.

stability problem.   To permit a simple analysis, a process with a large time delay is used.

Example 1-2

A process stream is heated from 100 to 150°F by injecting steam into the flowing solution.   The temperature is measured 40 feet downstream from the injection point, and a proportional controller is used to adjust the flow of steam. Because the fluid velocity is 40 fpm, there is a time delay of 1 min from the injection point to the measurement point.   The other lags in the control loop are assumed negligible.   Show the response of the system to a change in feed temperature when the overall gain K is 0.5, 1.0, and 1.2.

Suppose that the feed temperature drops to 99°F.   One minute later, the measured temperature drops to 149°F.   With an overall gain of 0.5, a 1°F drop in process temperature causes the steam flow to increase enough to raise the temperature 0.5°F.   The effect of the increase in steam flow is felt 1 min later when the measured temperature jumps to 149.5°F.   This causes the steam valve to *close* by an amount that will change the temperature 0.25°F to 149.25°F.   Further changes occur each minute, and the temperature finally approaches 149.33°F.

With an overall gain of 1.0, a 1°F drop in temperature causes a steam-flow change equivalent to 1°F, and 1 min after the change in steam flow the error is zero.   However, when the error is reduced to zero, the steam flow is decreased to the initial value and the temperature drops to 149°F again.   Therefore the process oscillates between 149 and 150°F with a period of 2 min.

With an overall gain of 1.2, an error of $-1°F$ leads to a corrective action of 1.2°F, and 1 min after the correction is made, the error becomes $+0.2°F$.   A positive error makes the steam flow less than normal, and 1 min later the error is $-1°F - 1.2(0.2°F) = -1.24°F$.   The error becomes larger with each cycle, and the system is unstable.

The response of a pure time-delay process to a step change in load is a series of steps, as shown in Example 1-2.   For a more complicated process such as the jacketed kettle of Example 1-1, the response would be a damped sine wave if the controller gain were below the maximum value and an oscillation of increasing amplitude for gains above the maximum, as shown in Fig. 1-7.

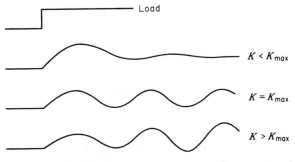

FIGURE 1-7   Effect of overall gain on response to load changes.

1-7. INTEGRAL AND DERIVATIVE ACTION

With integral control, the controller output is proportional to the integral of the error.

$$p = \frac{1}{T_i} \int e \, dt \tag{1-4}$$

where T_i = integral time.

There is no offset with integral control, since the output keeps changing as long as any error persists. However, the initial response to an error is slow, and proportional control is ordinarily used with integral control. The integral action corrects for the offset that usually occurs with proportional control only, and the effect is similar to manual adjustment or resetting of the set point after each load change. The terms "reset action" and "reset time" are widely used to characterize the integral action of a proportional-integral controller.

$$p = K_c \left(e + \frac{1}{T_R} \int e \, dt \right) \tag{1-5}$$

where T_R = reset time.

Derivative action is often added to proportional control to improve the response of slow systems. By increasing the output when the error is changing rapidly, derivative action anticipates the effect of large load changes and reduces the maximum error. The equation for an ideal three-mode controller is

$$P = K_c \left(e + \frac{1}{T_R} \int e \, dt + T_D \frac{de}{dt} \right) \tag{1-6}$$

where T_D = derivative time.

"Floating control" is a general term for controller action in which the rate of change of the controller output is determined by some function of the error. Integral control is proportional-speed floating control, since the rate of change is proportional to the error. With single-speed floating control, the output increases at a constant rate for a positive error and decreases at the same rate for a negative error.

PROBLEMS

1. The signal to a pneumatic proportional-integral controller is

$$e = 0.05 \sin \omega t$$

The controller gain is 3, and the reset time is 2 min.

a. Give the equation for the controller output in dimensionless terms and in pounds per square inch for $\omega = 0.1$ and $\omega = 1.0$ rad/min. The normal output pressure is 9 psi.

b. Sketch the outputs and the inputs for the two cases, and determine the phase angle.

2. A proportional-integral controller is tested by applying a step change in input equal to 10 or 20 per cent of the chart scale. For a gain of 2 and a reset time of 0.5 min, show the output expected for the two step changes. If it takes a few seconds to make the change in input, how would you determine the true gain and reset time from the output curve?

3. A proportional controller is used to control the flow of steam to a kettle which is continuously heating a thin slurry from 80 to 160°F. If the throttling range or proportional band is 25 per cent,† what offset results from a 10°F drop in inlet temperature?

$$\text{Flow rate} = 500 \text{ lb/hr}$$
$$c_p = 1 \text{ Btu/(lb)(°F)}$$
Normal valve position, half open
Normal controller pressure, 9 psi
Chart scale, 50–200°F

4. The level in an open tank is automatically controlled by adjusting a valve in the discharge line. The normal level is 10 ft, and the normal flow to and from the tank is 50 gpm. For a given head the flow through the valve is a linear function of the signal to the valve, and for a given opening the flow varies with the square root of the head. If the proportional controller changes the signal to the valve by 2 psi for a 1-ft change in level, what are the steady-state levels for input flows of 40 and 60 gpm? What would be the levels if the valve position were fixed?

5. The temperature in a laboratory bath is controlled by a thermometer set to have a differential gap of 1°F. When the heater is on, the bath temperature rises 3°F/min, and when the heater is off, the bath cools 5°F/min. Sketch the actual and the measured bath temperatures, assuming that the heater has no lag but that the thermometer temperature changes according to the equation

$$\frac{d\theta_m}{dt} = \frac{\theta - \theta_m}{0.5} \qquad \text{°F/min}$$

REFERENCES

1. Grebe, J. J., R. H. Boundy, and R. W. Cermak: The Control of Chemical Processes, *Trans. Am. Inst. Chem. Engrs.*, **29**:211 (1933).
2. Ivanoff, A.: Theoretical Foundations of the Automatic Regulation of Temperature, *J. Inst. Fuel*, **7**:117 (1934).
3. Callender, A., D. R. Hartree, and A. Porter: Time Lag in a Control System—I, *Phil. Trans. Roy. Soc. London*, ser. A, **235**:415 (1936).
4. Hartree, D. R., A. Porter, A. Callender, and A. B. Stevenson: Time Lag in a Control System—II, *Proc. Roy. Soc. (London)*, ser. A, **161**:460 (1937).

† Controller pressure goes from 3 to 15 psi and valve position from open to closed for a temperature interval equal to 25 per cent of full chart scale.

For small changes, a 1-psi change to the control valve changes the flow of steam by 5 per cent of the original value.

5. Minorsky, N.: Directional Stability of Automatically Steered Bodies, *J. Am. Soc. Naval Engrs.*, **34**:280 (1922).
6. Hazen, H. L.: Theory of Servomechanisms, *J. Franklin Inst.*, **218**:279 (1934).
7. Nyquist, H.: Regeneration Theory, *Bell System Tech. J.*, **11**:126 (1932).
8. Del Toro, Vincent, and S. R. Parker: "Principles of Control System Engineering," p. 252, McGraw-Hill Book Company, Inc., New York, 1960.
9. Campbell, D. P.: "Process Dynamics," p. 296, John Wiley & Sons, Inc., New York, 1958.

2 THE LAPLACE TRANSFORM

The Laplace transform is a mathematical tool which provides a systematic and relatively simple method of solving linear differential equations. Transforming a differential equation results in an algebraic equation, with the complex variable s replacing time as the independent variable. Solving this algebraic equation and performing the inverse transformation gives the solution of the original equation. Since extensive tables of transforms are available, equations for the transient response of control systems are nearly always obtained by using the Laplace-transform method. Another reason for the use of Laplace transforms is that the frequency response can be obtained directly from the system's transfer function, which is the ratio of the Laplace transform of the output to the Laplace transform of the input. The starting point for any quantitative study of a control system is to obtain the transfer functions for all parts of the block diagram.

The Laplace transform of a time function $f(t)$ is

$$F(s) \equiv \mathcal{L}[f(t)] \equiv \int_0^\infty f(t)e^{-st}\, dt \tag{2-1}$$

where $F(s)$ is a function of the complex variable s and $\mathcal{L}$ denotes the transformation process. Each term in the equation is transformed separately.

$$\mathcal{L}[f_1(t) + f_2(t)] = F_1(s) + F_2(s) \tag{2-2}$$

When a time function is multiplied by a constant, the transform is multiplied by the same constant.

$$\mathcal{L}[kf(t)] = kF(s) \tag{2-3}$$

The transform of a dependent variable x can be written as X or $X(s)$, where the use of capital letters and the addition of (s) are reminders that the equation has been transformed. However, the presence of s at any place in the equation shows that it has been transformed, and so the same symbol x can be used in both the original and the transformed equation.

A restriction on the use of Eq. (2-1) is that the function must be zero previous to zero time; that is $f(t) = 0$ for $t \leq 0$. The function $f(t) = A$ is interpreted as a step change from 0 to A at $t = 0$. The transform of a step change is

$$F(s) = \int_0^\infty Ae^{-st}\, dt = \frac{-A}{s}\,[e^{-st}]_0^\infty = \frac{A}{s} \tag{2-4}$$

Table 2-1. Laplace Transforms

$$f(t) \text{ for } t > 0$$
$$f(t) = 0 \text{ for } t \leq 0 \qquad\qquad F(s) = \int_0^\infty f(t)e^{-st}\, dt$$

	$f(t)$	$F(s)$
1.	A	$\dfrac{A}{s}$
2.	t	$\dfrac{1}{s^2}$
3.	t^2	$\dfrac{2}{s^3}$
4.	t^n	$\dfrac{n!}{s^{n+1}}$
5.	$e^{-\alpha t}$	$\dfrac{1}{s + \alpha}$
6.	$\dfrac{1}{T}e^{-t/T}$	$\dfrac{1}{Ts + 1}$
7.	$\sin \omega t$	$\dfrac{\omega}{s^2 + \omega^2}$
8.	$\cos \omega t$	$\dfrac{s}{s^2 + \omega^2}$
9.	$e^{-\alpha t}t^n$	$\dfrac{n!}{(s + \alpha)^{n+1}}$
10.	$e^{-\alpha t}\sin \omega t$	$\dfrac{\omega}{(s + \alpha)^2 + \omega^2}$
11.	$e^{-\alpha t}\cos \omega t$	$\dfrac{s + \alpha}{(s + \alpha)^2 + \omega^2}$
12.	$f'(t)$	$sF(s) - f_0$
13.	$f''(t)$	$s^2F(s) - sf_0 - f'_0$
14.	$\int f(t)$	$\dfrac{F(s)}{s} + \dfrac{\int f_0\, dt}{s}$
15.	$f(t - L)$	$e^{-Ls}F(s)$
16.	$\delta(t)$ (unit impulse)	1
17.	$1 - e^{-t/T}$	$\dfrac{1}{s(Ts + 1)}$

It is hardly ever necessary to perform the integration indicated by Eq. (2-1), since extensive tables of transforms are available. Table 2-1 is sufficient to transform all the equations in this text, and the most important transforms are also given in Eqs. (2-4) to (2-9).

The Laplace transform of a first derivative is

$$\mathcal{L}[f'(t)] = sF(s) - f_0 \qquad\qquad (2\text{-}5)$$

where f_0 is the initial value of $f(t)$ (more exactly, the value at $t = 0^+$).   For second or higher derivatives, the highest power of s in the transformed equation corresponds to the order of the derivative.

$$\mathcal{L}[f''(t)] = s^2 F(s) - s f_0 - f_0' \tag{2-6}$$

$$\mathcal{L}[f'''(t)] = s^3 F(s) - s^2 f_0 - s f_0' - f_0'' \tag{2-7}$$

In most process control problems, the initial value of the function and the initial values of its derivatives are zero, which makes the transforms much simpler.

The transform of an integral has s in the denominator,

$$\mathcal{L}[f^{-1}(t)] = \mathcal{L}[\smallint f(t)] = \frac{F(s)}{s} + \frac{\smallint f_0 \, dt}{s} \tag{2-8}$$

The transform for a function which is delayed L sec is expressed as

$$\mathcal{L}[f(t - L)] = e^{-Ls} F(s) \tag{2-9}$$

where $\qquad\qquad f(t - L) = 0 \qquad$ for $t < L$

A time delay or transportation lag in a control loop is represented by a block with the transfer function e^{-Ls}.

Example 2-1

Transform the following equation, and obtain the ratio of output to input, $Y(s)/X(s)$:

$$4 \frac{d^2 y}{dt^2} + 2 \frac{dy}{dt} + 3y = 5x$$

$$4[s^2 Y(s) - s y_0 - y_0'] + 2[s Y(s) - y_0] + 3 Y(s) = 5 X(s)$$

If the initial values are $y_0 = 0$, $y_0' = 0$,

$$(4s^2 + 2s + 3) Y(s) = 5 X(s)$$

The transfer function is the ratio of $Y(s)$ to $X(s)$,

$$\frac{Y(s)}{X(s)} = \frac{Y}{X} = \frac{5}{4s^2 + 2s + 3}$$

From a practical viewpoint, the Laplace transform can be treated as a tool without understanding the derivation or the exact meaning of the transformation, though it may help to think of s as equivalent in most cases to the operator D, which denotes differentiation.   The theory and application of Laplace transforms are discussed in the book by Churchill (1) and in some texts on automatic control (2, 3).

The general procedure in using Laplace transforms to solve a differential equation is to transform all terms of the equation, solve for the transformed variable, and perform the inverse transformation with the help of tables.

Example 2-2

Solve the following equation, using Laplace transforms:

$$25 \frac{d^2x}{dt^2} + x = 1 \qquad x_0 = 0, x'_0 = 0$$

Transforming each term and solving for X gives

$$25s^2X + X = \frac{1}{s}$$

$$X = \frac{1}{s(25s^2 + 1)}$$

$$x = \mathcal{L}^{-1}\left[\frac{1}{s(25s^2 + 1)}\right]$$

The table of transforms in the book by Nixon (4) permits writing the solution directly.

$$\mathcal{L}(1 - \cos \omega t) = \frac{1}{s(1 + s^2/\omega^2)}$$

$$x = 1 - \cos \tfrac{1}{5}t$$

The solution can also be obtained by using the shorter table of transforms in this book if the expression for X is separated into simpler terms.

Let

$$\frac{1}{s(25s^2 + 1)} = \frac{C_1}{s} + \frac{C_2}{(25s^2 + 1)}$$

By multiplying both sides by s and letting $s = 0$, the coefficient C_1 is shown to be 1.0. Substituting $C_1 = 1.0$ shows that C_2 must be $-25s$ for the equality in the preceding equation to be satisfied.

$$X = \mathcal{L}^{-1}\left(\frac{1}{s}\right) + \mathcal{L}^{-1}\left(\frac{-25s}{25s^2 + 1}\right)$$

$$x = 1 - \mathcal{L}^{-1}\left(\frac{s}{s^2 + \frac{1}{25}}\right) = 1 - \cos \tfrac{1}{5}t$$

The partial-fraction expansion can be used to simplify any rational proper fraction. The denominator is factored, and if there are n roots all of which are different, there are n terms in the expansion.

$$F(s) = \frac{A(s)}{B(s)} = \frac{A(s)}{K_0(s - a)(s - b)(s - c) \cdots (s - n)} \qquad (2\text{-}10)$$

$$\frac{A(s)}{B(s)} = \frac{K_1}{s - a} + \frac{K_2}{s - b} + \frac{K_3}{s - c} + \cdots + \frac{K_n}{s - n} \qquad (2\text{-}11)$$

Each constant is found by multiplying Eq. (2-11) by the corresponding root factor and setting s equal to the root. If several roots of the polynomial are equal, the expansion must include separate terms for the multiple roots with powers from 1 up to the number of equal roots and a special procedure is used to evaluate the constants (2, 3, 4).

$$F(s) = \frac{As}{K_0(s-a)^2(s-b)\cdots(s-n)} \tag{2-12}$$

$$F(s) = \frac{K_1}{(s-a)^2} + \frac{K_2}{s-a} + \frac{K_3}{s-b} + \cdots + \frac{K_{n+1}}{s-n} \tag{2-13}$$

Actually, the main problem in obtaining the inverse transformation usually lies in factoring the polynomial. Tables such as those worked out by Nixon (4) give the inverse transforms for most of the possible combinations of terms with up to 4 roots, and so a partial-fraction expansion is not necessary. For higher-order expressions, factoring is so difficult that approximations are usually made to simplify the equation or the equation is solved with an analog computer.

The initial and final values of a function may be obtained from the limits of the transformed equations. The initial value is found from the theorem

$$\lim_{t \to 0} f(t) = \lim_{s \to \infty} sF(s) \tag{2-14}$$

The final value is found by letting s approach zero,

$$\lim_{t \to \infty} f(t) = \lim_{s \to 0} sF(s) \tag{2-15}$$

Both theorems are useful in checking the transfer functions for systems whose initial and final values are known, and the final-value theorem is often used to obtain the steady-state error of a control system.

PROBLEMS

1. Write the Laplace transforms for the following equations, and solve for $X(s)$:

a. $\dfrac{d^2x}{dt^2} + 4\dfrac{dx}{dt} + 2x = 3t^3$, where $x_0 = 0$, $x_0' = 0$

b. $\dfrac{dx}{dt} + 2x = 5\sin 3t$, where $x_0 = 1$

c. $2\dfrac{d^3x}{dt^3} + 3\dfrac{dx}{dt} + x + 4 = 0$, where x_0'', x_0', $x_0 = 0$

d. $\dfrac{d^2x}{dt^2} + 2\dfrac{dx}{dt} + 6 = A\sin 2\omega t$, where $x_0' = 2$, $x_0 = 1$

2. Solve the following equations, using Laplace transforms, and check by differentiation:

a. $2\dfrac{d^2x}{dt^2} - 6\dfrac{dx}{dt} - 20x = 3$, where $x_0 = 0$, $x_0' = 0$

b. $5\dfrac{dx}{dt} + 2x = 1$, where $x_0 = 4$

c. $5\dfrac{dx}{dt} + x = te^{-t/2}$, where $x_0 = 0$

3. Use Laplace transforms and a partial-fraction expansion to solve these equations:

a. $2\dfrac{d^2x}{dt^2} + 4.5\dfrac{dx}{dt} + x = 1$, where $x_0, x_0' = 0$

b. $6\dfrac{dx}{dt} + x = 3t$, where $x_0 = 0$

4. Change the following expressions to forms that match those in the table of transforms in the book by Nixon (4):

a. $F(s) = \dfrac{2s^2 + 7.6s + 4.8}{2s^4 + 2s^3 + 2s^2 + s}$

b. $F(s) = \dfrac{5}{s(s^2 + 13s + 30)}$

c. $F(s) = \dfrac{5s + 2}{1 + 10s + 25s^2}$

5. Solve the following equations, using Laplace transforms, and show that the equations and all the boundary conditions are satisfied:

a. $3\dfrac{d^2x}{dt^2} + \dfrac{5\,dx}{dt} + 2x = 0$, where $x_0 = 0$, $x_0' = 0.5$

b. $\dfrac{d^2x}{dt^2} + 4x = 0$, where $x_0 = -8$, $x_0' = 2$

REFERENCES

1. Churchill, R. V.: "Modern Operational Mathematics in Engineering," chaps. 1, 2, McGraw-Hill Book Company, Inc., New York, 1944.
2. Del Toro, V., and S. R. Parker: "Principles of Control System Engineering," chap. 3, McGraw-Hill Book Company, Inc., New York, 1960.
3. Chestnut, H., and R. W. Mayer: "Servomechanisms and Regulating System Design," chap. 4, John Wiley & Sons, Inc., New York, 1951.
4. Nixon, F. E.: "Principles of Automatic Controls," chap. 3, app. 6, Prentice-Hall, Inc., Englewood Cliffs, N.J. 1953.

3 OPEN-LOOP RESPONSE OF SIMPLE SYSTEMS

3-1. FIRST-ORDER ELEMENTS

The first step in analyzing a control system is to determine the transfer functions for the various parts of the system. The transfer functions are obtained by applying either the general material or energy-balance equation or a force balance to each part of the system which has storage capacity for material or energy. For first-order elements the material or energy-balance equation is used.

$$\text{Inflow} - \text{outflow} = \text{rate of accumulation}$$
$$= \frac{d(\text{energy or mass})}{dt} \qquad (3\text{-}1)$$
$$= \frac{C\,d\theta}{dt}$$

where $C = \text{capacity} = \dfrac{d(\text{energy or mass})}{d\theta} = \dfrac{dQ}{d\theta}$

$\theta = $ dependent variable, such as temperature, level, or concentration

The thermal capacity of a system is the product of the mass and the specific heat and is usually given in Btu per degree Fahrenheit. Capacities for storage of material can be expressed in several ways, including cubic feet of liquid per foot of depth and moles A per mole A per cubic foot. These capacities are analogous to electrical capacitance, but note that they are defined as a *rate of change*, whereas electrical capacitance is usually defined as total charge divided by voltage. The electrical capacitance is generally independent of voltage, but thermal and mass capacities often depend on θ and cannot be calculated from Q/θ.

The inflow and outflow terms of the material-balance equation are expressed in terms of the dependent variable θ and one or more independent variables. The equation is then transformed and rearranged to show the Laplace transform of the dependent or output variable as a function of the transformed input variables. An element such as a tank may have several inputs, and a separate block on the block diagram is used for each input variable. A tank can also have several output streams, but there is only one output variable in Eq. (3-1) and in the corresponding transfer functions, namely, the process variable which appears in the accumulation term

21

(pressure, temperature, concentration, etc.). For example, consider level control in a storage tank that feeds several processes. The output variable is the level in the tank, and possible input variables are the flows to the tank, the pressure above the tank, and the positions of the valves in the exit lines.

Some systems have two or more types of storage capacity, and separate transfer functions are derived for each output variable, usually by assuming average constant values for all the other variables. For a tank where both the pressure and the temperature may change, the transfer functions for temperature change are obtained by using an average pressure, and an average temperature or density is used to derive equations for changes in pressure. This approach is satisfactory as long as the changes in variables are small, which is generally true for process control problems. The fact that we are dealing primarily with small changes also permits us to use linear approximations for exponential or other complex relationships, as will be shown in the sections on reactors and fluid-flow processes.

3-2. RESPONSE OF A THERMOMETER BULB

A thermometer bulb is a first-order system, one whose response can be described by a first-order linear differential equation. Consider a mercury thermometer suddenly immersed in water of temperature θ_1. Assume that the heat capacity of the glass is negligible and that the mercury is at a uniform temperature θ_2, as shown in Fig. 3-1.† The unsteady-state heat balance is

$$In - out = accumulation$$

$$UA(\theta_1 - \theta_2) - 0 = Mc_p \frac{d\theta_2}{dt} = C \frac{d\theta_2}{dt} \tag{3-2}$$

where M = mass of mercury
c_p = heat capacity of mercury
U = overall coefficient
A = heat-transfer area
t = time

The equation is rearranged so that all the terms with θ_2, the output variable, are on the left-hand side and the terms with θ_1, the input, are on the right-hand side.

$$\frac{Mc_p}{UA} \frac{d\theta_2}{dt} + \theta_2 = \theta_1$$

or

$$T \frac{d\theta_2}{dt} + \theta_2 = \theta_1 \tag{3-3}$$

---

† If the capacity of the glass is not negligible, the system is still first-order if the resistance of the glass is small compared with the film resistance, since the glass and the mercury would then be at the same temperature. If neither the resistance nor the capacity of the glass is negligible, the thermometer is treated as two interacting first-order elements, as shown in a later example.

The group Mc_p/UA has the units of time and is called the "time constant" of the system. The rate of response of the system, $d\theta_2/dt$, is inversely proportional to the time constant, and thus T is a measure of the time required for the system to adjust to a new input.   The time constant is the product of a resistance to heat flow, $1/UA$ or R, and a capacity for heat storage, Mc_p or C.   In electrical systems the time constant is a product of electrical resist-

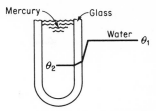

FIGURE 3-1   Temperature gradients for a mercury thermometer.

ance and electrical capacitance; in fluid-flow systems the time constant is the product of a fluid-flow resistance and a capacity to store fluid.   In many cases the system is recognized to be first-order, and the time constant is obtained directly from the resistance and capacity of the system without troubling to write out the material-balance equation.

To obtain the transfer function from Eq. (3-3), the initial value of θ_2 is taken as zero so that the initial-value term in the transformed equation can be dropped.   This just means that θ_2 now refers to deviations from the initial or normal value.   The Laplace transform of Eq. (3-3) is

$$Ts\theta_2 + \theta_2 = \theta_1 \tag{3-4}$$

In Eq. (3-4), the presence of s is a reminder that θ_2 and θ_1 are the transformed variables.   The equation is usually rearranged to show the ratio of the output to the input, which is the transfer function of the system.

$$\frac{\theta_2}{\theta_1} = \frac{1}{Ts + 1} = G \tag{3-5}$$

The response of the bulb to various input signals is obtained by substituting the transform of the particular input signal into Eq. (3-5) and taking the inverse transform of the resulting expression for θ_2.   Suppose that the input is suddenly increased from zero to B at $t = 0$.

For

$$\theta_1 = B \text{ (untransformed)}$$

$$\theta_1 = \frac{B}{s} = \theta_1(s) \text{ (transformed)} \tag{3-6}$$

$$\theta_2 = \frac{B}{s(Ts + 1)}$$

The actual temperature θ_2 is given by

$$\theta_2 = \mathcal{L}^{-1}\left[\frac{B}{s(Ts + 1)}\right] = B\mathcal{L}^{-1}\left[\frac{1}{s(Ts + 1)}\right] \tag{3-7}$$

By using the table of transforms, the solution of Eq. (3-7) is obtained directly.†

$$\theta_2 = B(1 - e^{-t/T}) \tag{3-8}$$

3-3. GENERAL RESPONSE TO STEP, RAMP, AND SINUSOIDAL INPUTS

STEP INPUT

The general solution for the response to a step input is given in terms of the fractional response y and is sketched in Fig. 3-2.

$$y = \frac{\theta_2}{B} = \frac{\theta_2}{\theta_{2 \text{ steady state}}} = 1 - e^{-t/T} \tag{3-9}$$

Equation (3-9) gives the response of any first-order process to a step input, and it is used so often that it should be memorized. When $t/T = 1$, the fractional response is $1 - e^{-1}$, or 0.632. Thus the time constant can be defined as the time for the system's output to reach 63.2 per cent of its final value after a step change in input. Experimental values of T are often taken as the time for 63 per cent response. Even if the system is only approximately first-order, the time for 63 per cent response can be considered an effective time constant for rough calculations. The terms "time constant,"

---

† Equation (3-7) could be handled by using partial fractions and a shorter table of basic transforms, though this takes more time.

$$\mathcal{L}^{-1}\left[\frac{1}{s(Ts+1)}\right] = \mathcal{L}^{-1}\left(\frac{1}{s} + \frac{-T}{Ts+1}\right) = 1 - \mathcal{L}^{-1}\left(\frac{1}{s+1/T}\right) = 1 - e^{-t/T}$$

Equations (3-3) can also be solved readily by several standard methods of handling differential equations. The advantage of using Laplace transforms and a complete table of inverse transforms will become more apparent as the equations become more complex.

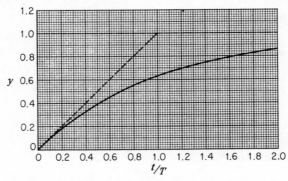

FIGURE 3-2   Fractional response of a first-order element to a step input.

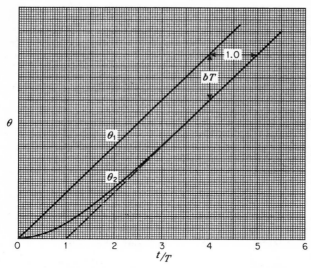

FIGURE 3-3   Response of a first-order element to a ramp input.

"first-order lag," "first-order element," and "exponential lag" all refer to portions of a system which have first-order behavior.

A more exact method of determining the time constant for true first-order systems is to use the slope of a semilog plot of the uncompleted response $1 - y$ or $B(1 - y)$ versus the time. A semilog plot also shows more clearly than an arithmetic plot any deviations from first-order behavior. A third way of determining the time constant is from the initial slope of the response curve, which is equal to the steady-state value over the time constant.

A quick way of checking what appear to be exponential-response curves is to compare the time needed to accomplish a certain fraction of the change remaining to be accomplished. For a first-order system, the time to go from 0 to 50 per cent is the same as the time to go from 50 to 75 per cent, and the time to go from 0 to 90 per cent is the same as the time to go from 90 to 99 per cent.

RAMP INPUT

The response of a first-order system to a ramp input is given next. For $\theta_1 = bt$ (untransformed) or $\theta_1 = b/s^2$ (transformed),

$$\theta_2 = \mathcal{L}^{-1}\left(\frac{b}{s^2}\frac{1}{Ts + 1}\right) = bT\left(e^{-t/T} + \frac{t}{T} - 1\right) \tag{3-10}$$

After the transient term $e^{-t/T}$ becomes negligible, the output is bT less than the input, or the output lags the input by a time equal to the time constant (see Fig. 3-3).

For $t \gg T$, $$\theta_2 = bt - bT = b(t - T) \tag{3-11}$$

SINUSOIDAL INPUT

Finally, consider the response of the first-order system to a sinusoidal input.

For $\theta_1 = A \sin \omega t$ (untransformed) or $\theta_1 = A/(s^2 + \omega^2)$ (transformed)

$$\theta_2 = \mathcal{L}^{-1}\left(\frac{A}{s^2 + \omega^2}\frac{1}{Ts + 1}\right) = \frac{A}{\omega}\mathcal{L}^{-1}\left[\frac{1}{(1 + s^2\omega^2)(Ts + 1)}\right] \quad (3\text{-}12)$$

The inverse transform is found from a complete table of transforms.

$$\theta_2 = \frac{A}{\omega}\left[\frac{T\omega^2}{1 + T^2\omega^2}e^{-t/T} + \frac{\omega \sin(\omega t - \phi)}{(1 + T^2\omega^2)^{\frac{1}{2}}}\right] \quad (3\text{-}13)$$

where $\phi = \tan^{-1} T\omega$.

Once the transient term has died out, the output is a sine wave with amplitude $A/(1 + T^2\omega^2)^{\frac{1}{2}}$ and a phase lag of $\phi°$, and this "steady-state" portion of the solution is called the "frequency response" of the system. Methods of plotting and using the frequency response form a large part of subsequent chapters. The complete solution of Eq. (3-12) is useful mainly in showing how many cycles are needed before the transient terms die out and the response becomes a true sine wave.

3-4. CONCENTRATION RESPONSE OF STIRRED TANKS

A well-mixed tank generally responds as a first-order system to changes in input concentration. Consider a tank used to damp concentration fluctuations in the feed to a reactor (Fig. 3-4). With no chemical reaction and a constant level, the material-balance equation is

$$Fc_1 - Fc_2 = V\frac{dc_2}{dt} \quad (3\text{-}14)$$

The initial value of the output variable is again taken as zero, so that c_2 represents the deviation from the normal value. The equation is transformed and rearranged to give the transfer function c_2/c_1. The time constant is

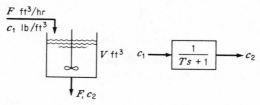

FIGURE 3-4   A blending tank.

V/F, the holdup time in the tank.

$$\frac{c_2}{c_1} = \frac{1}{Ts + 1} \tag{3-15}$$

where
$$T = \frac{V}{F}$$

The response of this system to step, ramp, and sinusoidal inputs is the same as that for the temperature bulb [Eqs. (3-9), (3-10), and (3-13)].

The assumption underlying Eq. (3-14) is that the concentration is uniform throughout the tank or that the input stream is instantaneously mixed with the liquid in the tank. Actually several seconds may be required to mix the contents of a tank, and the output concentration is not the same as the average concentration in the tank when both are changing. The effect of the mixing lag is discussed later in connection with pH control systems, for which the mixing lag is often appreciable when compared with the small holdup time. For the present examples the holdup time is assumed to be at least two orders of magnitude greater than the mixing lag, and the mixing lag is neglected.

Example 3-1

Gases A and B are fed continuously to a tank with a volume of 30 ft³. The normal tank conditions are 40 psia and 80°F, and the normal flow rates are $F_A = 40$ and $F_B = 10$ cfm measured at tank conditions. If the flow of B is suddenly increased to 12 cfm, when does the concentration of B reach 90 per cent of the new steady-state value?

The time constant for small changes in flow rate is the normal holdup time.

$$T = \frac{V}{F} = \frac{30}{40 + 10} = 0.6 \text{ min}$$

$$y = 0.9 = 1 - e^{-t/T}$$

$$\frac{t}{T} = 2.3$$

$$t = 1.4 \text{ min}$$

TIME CONSTANT WITH A FIRST-ORDER REACTION

If a stirred tank is used as a chemical reactor, the quantity reacting per unit time is an additional output term in the material balance for the reactant. For a first-order reaction, the reaction rate is proportional to the concentration of reactant in the tank, which is the same as the exit concentration for a well-mixed tank.

The quantity reacting is

$$r \text{ (lb/hr)} = k \text{ (hr}^{-1}) \, c_2 \text{ (lb/ft}^3) \, V \text{ (ft}^3) \tag{3-16}$$

The material balance for the reactant is

$$Fc_1 - Fc_2 - kc_2V = V\frac{dc_2}{dt} \tag{3-17}$$

Note that the reactor differs from the blending tank and the temperature bulb in that the output concentration does not equal the input concentration at steady state.   To solve for the effect of changes in input concentration, it is convenient to define both input and output variables as deviations from the normal values,

$$\Delta c_1 = c_1 - \bar{c}_1$$

$$\Delta c_2 = c_2 - \bar{c}_2$$

$$\frac{d\Delta c_2}{dt} = \frac{dc_2}{dt}$$

Substituting these definitions into Eq. (3-17) gives

$$F(\bar{c}_1 + \Delta c_1) - F(\bar{c}_2 + \Delta c_2) - k(\bar{c}_2 + \Delta c_2)V = V\frac{d\Delta c_2}{dt} \tag{3-18}$$

The normal quantity reacted is equal to the flow rate times the difference between the normal inlet and outlet concentrations,

$$F(\bar{c}_1 - \bar{c}_2) = k\bar{c}_2V$$

Therefore the normal values in Eq. (3-18) cancel, leaving

$$F\,\Delta c_1 - F\,\Delta c_2 - k\,\Delta c_2\,V = V\frac{d\Delta c_2}{dt} \tag{3-19}$$

Equation (3-19) is Eq. (3-17) with each c replaced by Δc.   Therefore the Δ is usually dropped, and the variables in Eq. (3-17) are understood to be deviations from the normal values.   With the initial values at zero, Eq. (3-17) is transformed and rearranged as follows:

$$Vsc_2 + (kV + F)c_2 = Fc_1$$

or

$$\frac{V}{kV + F}sc_2 + c_2 = \frac{Fc_1}{kV + F} \tag{3-20}$$

The transfer function for the reactor is written by using the time constant and the system gain.

$$\frac{c_1}{c_1} = \frac{K}{Ts + 1} \tag{3-21}$$

where

$$K = \frac{F}{kV + F}$$

$$T = \frac{V}{kV + F} = \frac{V/F}{kV/F + 1}$$

The effect of the chemical reaction is to make the time constant of the system less than the holdup time. Equation (3-21) shows that, as the rate constant becomes very large, the time constant approaches $1/k$. A fast reaction also means a low value of K, the steady-state gain, and this may help to explain the effect of k on the time constant. After a step change in c_1, the initial rate of change of c_2 is shown by Eq. (3-17) to be Fc_1/V, with or without a chemical reaction, since c_2 is initially zero. With a large k, the eventual change in c_2 is small, and so less time is required to accomplish a given fraction of the change.

Example 3-2

A first-order reaction is carried out in a stirred tank with a holdup time of 1.6 hr and a rate constant of 2 hr^{-1}. Show the effect of a sudden change in feed concentration from 0.50 to 0.48 mole/ft^3.

Initial conditions are

$$F(\bar{c}_1 - \bar{c}_2) = k\bar{c}_2 V$$

where $\bar{c}_1$ and $\bar{c}_2$ are the normal inlet and exit concentrations.

$$\frac{\bar{c}_2}{\bar{c}_1} = \frac{1}{1 + kV/F} = \frac{1}{1 + 3.2} = 0.238$$

$$\bar{c}_2 = 0.119$$

The transfer function is

$$\frac{c_2}{c_1} = \frac{K}{Ts + 1}$$

$$K = \frac{1}{1 + kV/F} = 0.238$$

$$T = \frac{V/F}{1 + kV/F} = \frac{1.6}{4.2} - 0.381 \text{ hr}$$

The transient response to a step input is $y = 1 - e^{-t/T}$. The steady-state change in concentration is $0.238(-0.02) = -0.00476$.

t/T	t, hr	y	c_2	$\bar{c}_2 + c_2$
0	0	0	0	0.119
1	0.38	0.63	-0.0030	0.116
2	0.76	0.86	-0.0041	0.1149
3	1.14	0.95	-0.0045	0.1145
∞	∞	1.0	-0.0048	0.1142

REACTIONS OTHER THAN FIRST-ORDER

If the reaction is zero-order, the time constant is again V/F, the holdup time, since changes in concentration do not change the reaction rate. For reactions which are second- or fractional-order, the exact transient response cannot be obtained by using transfer functions, since the differential equations are not linear. However, the response to small changes can be predicted quite well by using a linear approximation of the rate expression, as shown below:

For a second-order reaction

$$r = k_2 c_2{}^2 V \tag{3-22}$$

$$r \cong r_{av} + \left(\frac{dr}{dc}\right)_{av} (c_2 - c_{2,av}) \tag{3-23}$$

$$\left(\frac{dr}{dc}\right)_{av} = 2k_2 V c_{2,a} \tag{3-24}$$

$$r \cong r_{av} + 2k_2 V c_{2,av}(c_2 - c_{2,av}) \tag{3-25}$$

Letting c_2 denote changes in reactant concentration and $\bar{c}_2$ the normal or average value,

$$r \cong \bar{r} + (2k_2\bar{c}_2)Vc_2 \tag{3-26}$$

If the concentration is increased by 10 per cent, the linearized equation predicts a rate 20 per cent above the normal rate, compared with an exact value of 21 per cent above normal. This difference would be unimportant for most control calculations. Equation (3-26) can be used in a material balance similar to Eq. (3-17), or the transfer function can be written directly from Eq. (3-21) by substituting the term $2k_2\bar{c}_2$ (which has the units hr^{-1}) for the first-order rate constant k.

$$\frac{c_2}{c_1} = \frac{K}{Ts + 1}$$

where

$$K = \frac{F}{2k_2\bar{c}_2 V + F} \tag{3-27}$$

$$T = \frac{V/F}{2k_2\bar{c}_2 V/F + 1}$$

In Eqs. (3-27), c_1, c_2 are the transformed variables representing deviations from the normal input and output concentrations. The concept of linearization and the accuracy of the linearized equations are discussed further in the sections on pressure and level systems.

3-5. TEMPERATURE RESPONSE OF STIRRED TANKS

Consider a jacketed tank used to preheat a process stream (see Fig. 3-5). Assume that the capacity of the tank wall is negligible and that the temperature inside the jacket is uniform. With these assumptions and the

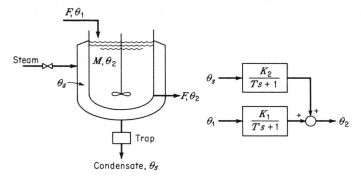

FIGURE 3-5   Diagrams for a jacketed kettle.

assumption of uniform-temperature fluid in the tank, the system is first-order.

$$F\rho c_p\theta_1 - F\rho c_p\theta_2 + UA(\theta_s - \theta_2) = Mc_p\frac{d\theta_2}{dt} \tag{3-28}$$

$$\frac{Mc_p}{F\rho c_p + UA}\frac{d\theta_2}{dt} + \theta_2 = \frac{F\rho c_p}{F\rho c_p + UA}\theta_1 + \frac{UA}{F\rho c_p + UA}\theta_s \tag{3-29}$$

$$T\frac{d\theta_2}{dt} + \theta_2 = K_1\theta_1 + K_2\theta_2 \tag{3-30}$$

$$(Ts + 1)\theta_2 = K_1\theta_1 + K_2\theta_s \tag{3-31}$$

The transfer functions for the tank are obtained by considering the initial values of θ_1, θ_2, and θ_s to be zero, which means that these variables stand for deviations from the average values. To show that the original heat balance is still correct even though three different base temperatures are used, Eq. (3-28) can be rewritten by using average temperatures (in, say, degrees Fahrenheit) and temperature deviations.

$$F\rho c_p(\bar\theta_1 + \Delta\theta_1) - F\rho c_p(\bar\theta_2 + \Delta\theta_2) + UA(\bar\theta_s + \Delta\theta_s - \bar\theta_2 - \Delta\theta_2)$$

$$= Mc_p\frac{d\theta_2}{dt} = Mc_p\frac{d\Delta\theta_2}{dt} \tag{3-32}$$

When $d\theta_2/dt = 0$, the temperatures are by definition the normal or average values.

$$F\rho c_p(\bar\theta_1 - \bar\theta_2) + UA(\bar\theta_s - \bar\theta_2) = 0 \tag{3-33}$$

Canceling the average values in Eq. (3-32) gives

$$F\rho c_p(\Delta\theta_1 - \Delta\theta_2) + UA(\Delta\theta_s - \Delta\theta_2) = Mc_p\frac{d\Delta\theta_2}{dt} \tag{3-34}$$

Equation (3-34) is exactly the same as Eq. (3-28), and the Δ's are therefore dropped for convenience. The variables in transformed equations are

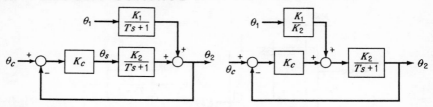

FIGURE 3-6 Equivalent block diagrams for proportional control of tank temperature.

nearly always deviations from the normal or initial values. A special symbol to denote deviations is used only when both the deviation and the absolute value enter the transformed equation.

If the response of the system to changes in feed temperature is desired, θ_s in Eq. (3-31) is set equal to zero and the standard solutions for step, ramp, or other changes in θ_1 are applied. For changes in steam temperature, the feed temperature is considered constant and the θ_1 term dropped. If the steam temperature is adjusted to control θ_2, changes in θ_1 would be load changes and the transfer function for θ_1 would be generally shown as a separate branch of the block diagram. However, since the time constant is the same for changes in either steam temperature or feed temperature, the load signal can be introduced ahead of the major block if multiplied by the gain ratio, as shown in Fig. 3-6.

RESPONSE TO CHANGES IN FLOW

The rate of flow to the tank might also be a variable, and if we consider only small changes in flow, the effect of these changes on the holdup time and the time constant can be neglected. Here we have to use separate symbols for the average flow $\bar{F}$ and the deviation ΔF, and the symbols $\bar{\theta}_1$ and $\bar{\theta}_2$ represent average values of the temperature on a common scale. It can be shown that the change in flow rate just adds one extra (input-output) term to Eq. (3-28).

$$\bar{F}\rho c_p(\theta_1 - \theta_2) + UA(\theta_s - \theta_2) + \Delta F\rho c_p(\bar{\theta}_1 - \bar{\theta}_2) = Mc_p\frac{d\theta_2}{dt} \qquad (3\text{-}35)$$

$$(Ts + 1)\theta_2 = K_1\theta_1 + K_2\theta_s + K_3\Delta F$$

where T, K_1, K_2 are the same as in Eqs. (3-29) and (3-30) and $K_3 = \dfrac{\rho c_p(\bar{\theta}_1 - \bar{\theta}_2)}{\bar{F}\rho c_p + UA}$.

The value of K_3 is negative, since an increase in flow rate decreases the output temperature.

For the stirred tank, the time constant for the response of the tank temperature is the same for changes in feed temperature, steam temperature, or flow rate, but the steady-state gains are different for each input variable. This is true for nearly all single-capacity systems;† the time constant is the

† See Sec. 14-7 for an exception.

same for all input variables (as long as the changes are small), but the gains are generally different and often not equal to unity. Recognizing that the time constant is the same for several inputs saves time in deriving transfer functions for control systems. Once the time constant is determined, it is necessary only to get the steady-state gain for different inputs, and these gains are often obtained from routine steady-state calculations rather than from the complete differential equation.

The stirred tank has a time constant for temperature changes of $Mc_p/(Fpc_p + UA)$ [Eq. (3-29)]. Note that the time constant approaches M/Fp or V/F, the holdup time, when the flow rate is high. At low flow rates, the time constant approaches Mc_p/UA, corresponding to that for a temperature bulb, which is a similar system with zero flow. It is important to remember that the time constant for temperature changes generally differs from that for concentration changes and from that for level changes, though these time constants are often the same order of magnitude as the holdup time.

3-6. LINEARIZATION OF LIQUID-LEVEL SYSTEMS

The flow of water through a valve or other constriction usually follows a square-root law.

$$F = b\sqrt{\Delta P} \qquad (3\text{-}36)$$

For the case of a tank discharging through a control valve to atmospheric pressure, the pressure drop in feet of fluid is the depth of liquid in the tank, h. (Any pressure drop in the discharge line can be added to that across the valve, since the frictional pressure drop for turbulent flow is also approximately proportional to the square of the flow rate.)

The exact equation for the system of Fig. 3-7 is

$$F_1 - b\sqrt{h} = A\left(\frac{dh}{dt}\right) \qquad (3\text{-}37)$$

The term b is a function of valve-stem position, which is held constant for the present analysis. Equation (3-37) can readily be solved to show the change in level for step changes in input flow or in valve position, but it is very difficult to solve the equation resulting from the combination of this equation with the linear equations for other parts of the control system. The outflow term is therefore replaced by an approximation which changes Eq. (3-37) to a linear first-order equation.

$$F_2 = F_{av} + \left(\frac{dF}{dh}\right)_{av}(h - h_{av}) \qquad (3\text{-}38)$$

Equation (3-38) is just the equation of the tangent to the flow curve at the average value of the flow. As long as deviations from the average are

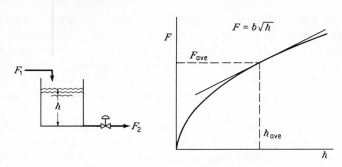

FIGURE 3-7   Linearization of a liquid-level system.

small, say, less than 20 per cent, the tangent is a good approximation of the curve.   Just drawing the tangent is often the easiest way of linearizing a process curve and shows at a glance the magnitude of the error involved for certain deviations from the average value.

Since we want the initial values of the variables to be zero, to simplify the transfer function, Eq. (3-38) is usually written with F and h understood to mean deviations from the average values.

$$F_2 = \frac{dF}{dh}\, h \tag{3-39}$$

$$A\frac{dh}{dt} + \frac{dF}{dh}\, h = F_1 \tag{3-40}$$

$$\frac{A}{dF/dh}\frac{dh}{dt} + h = \frac{F_1}{dF/dh} \tag{3-41}$$

The time constant of the tank is therefore $A(dh/dF)$, where A is the area and also the capacity in cubic feet of liquid per foot of height, and dh/dF is the resistance in feet of head per cubic foot per hour.   Although the resistance has the units of driving force divided by flow rate, it is *not* equal to the total driving force over the total flow rate.   In calculating time constants the resistance is always based on the slope of the process curve (Fig. 3-8a), which

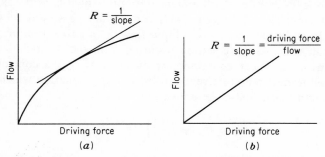

FIGURE 3-8   Dynamic resistance for (a) nonlinear and (b) linear systems.

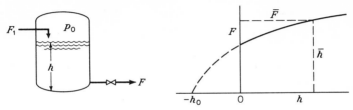

FIGURE 3-9   Flow curve for a pressurized tank.

differs from the usual definition of electrical resistance or resistance to heat transfer. The two definitions are the same only if the flow is directly proportional to the driving force, as in Fig. 3-8b.

The general definition of resistance is the rate of change of driving force with flux,

$$R = \frac{d(\text{driving force})}{d(\text{flow or flux})} = \frac{dh}{dF} \tag{3-42}$$

If the equation for the flux is known, it is convenient to express the resistance in terms of the average flux and average driving force. When the flow follows a square-root law [Eq. 3-36],

$$\frac{dF}{dh} = \frac{b}{2\sqrt{h}} = \frac{\bar{F}}{\sqrt{\bar{h}}2\sqrt{h}} \tag{3-43}$$

$$\left(\frac{dh}{dF}\right)_{\text{av}} = \frac{2\bar{h}}{\bar{F}} \tag{3-44}$$

$$T = \frac{2A\bar{h}}{\bar{F}} = \frac{2V}{\bar{F}} \tag{3-45}$$

Thus for an open tank discharging to the air the time constant is twice the holdup time. If the tank is under a constant pressure as in Fig. 3-9, the time constant is even greater. If h_0, the pressure head above the tank, is equal to the tank depth, the time constant is four times the holdup time, as shown by Eq. (3-46).

$$T = A\frac{dh}{dF} = 2A\frac{\bar{h} + h_0}{\bar{F}} \tag{3-46}$$

The following example compares the exact transient response for an open tank with that calculated from the linearized equation.

Example 3-3

A tank has an area of 1.0 ft^2, a normal depth of 4 ft, and a normal discharge rate of 20 ft^3/hr. How does the depth change with time if the flow to the tank is suddenly increased to 25 ft^3/hr?

Exact Solution

Outflow $F = 10\sqrt{h}$ since $\bar{F} = 20$, $\bar{h} = 4$

$$1.0\frac{dh}{dt} = 25 - 10\sqrt{h} \qquad h = 4 \text{ at } t = 0, h = 6.25 \text{ at } t = \infty$$

Let

$$H = \sqrt{h} \qquad dH = \frac{dh}{2\sqrt{h}} = \frac{dh}{2H}$$

$$25 - 10H = \frac{2H\,dH}{dt}$$

$$\frac{H\,dH}{12.5 - 5H} = dt$$

$$\tfrac{1}{25}[12.5 - 5H - 12.5\ln(12.5 - 5H)] = t + c$$

Substituting $H = 2$ at $t = 0$ gives $c = -0.358$. The solution is plotted in Fig. 3-10.

Linearized Solution

The time constant can be based on the initial value of dF/dh, the final value, or an average.

The initial slope is

$$\frac{dF}{dh} = \frac{10}{2\sqrt{4}} = 2.5 \text{ ft}^3/(\text{hr})(\text{ft})$$

$$\text{Estimated } h_{\text{final}} = 4 + \frac{25 - 20}{2.5} = 6.0 \text{ ft}$$

$$T = A\frac{dh}{dF} = \frac{1}{2.5} = 0.4 \text{ hr}$$

Using the general solution $y = 1 - e^{-t/T}$,

$$y' = \frac{h - 4}{6 - 4} = 1 - e^{-t/0.4}$$

Based on the average value of the slope (a chord slope),

$$T = A\frac{\Delta h}{\Delta F} = 1\frac{6.25 - 4.0}{25 - 20} = 0.45 \text{ hr}$$

$$y = \frac{h - 4}{6.25 - 4} = 1 - e^{-t/0.45}$$

The curves in Fig. 3-10 show that there is very little difference in the exact and linearized solutions even though the change in h is over 50 per cent.

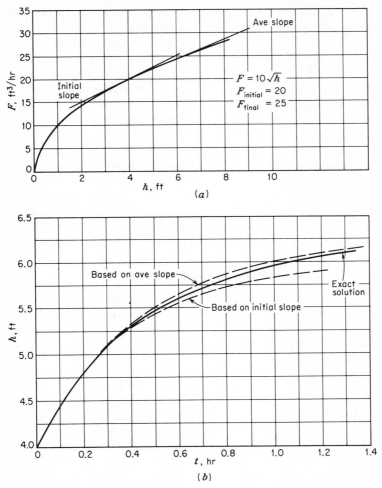

FIGURE 3-10   Response of a liquid-level process, Example 3-3.   (a) Flow curve;   (b) response to a step input in flow.

Use of the average slope gives slightly better results than use of the initial slope.   The time constant based on the initial slope could be used with the correct steady-state value of h to improve the linearized solution.

3-7. RESPONSE OF PRESSURE SYSTEMS

The problem of pressure control is similar to that of level control in that the flow equations must generally be linearized to obtain the system transfer function.   The fact that both the inflow and the outflow usually depend on the tank pressure and that some tanks have multiple inlets and outlets makes

F_1 scfm → V, P_1 psia → ⋈ → F_2 scfm P_2 psia

FIGURE 3-11   A pressure system with one resistance.

pressure systems more complex than level systems. Only rarely is the time constant a simple multiple of the hold-up time. The possibilities of sonic flow and of temperature change during throttling are added complications.

A pressure system somewhat like the level system treated in Eqs. (3-36) to (3-45) is a tank operating slightly above atmospheric pressure, with a feed rate fixed by upstream conditions only (Fig. 3-11). (A previous process or a constant-output compressor might fix the flow to the tank.)

The capacity of the tank is the change in the amount of gas stored per unit change in pressure and is usually expressed in cubic feet of gas (at normal tank temperature and atmospheric pressure) per pound per square inch. The resistance is dP_1/dF_2, the reciprocal of the slope of the flow curve. If the tank temperature is constant,† the capacity is the volume divided by atmospheric pressure and the time constant is

$$T = \frac{V}{14.7} \frac{\text{scf}}{\text{psi}} \frac{dP_1}{dF_2} \frac{\text{psi}}{\text{scfm}} \tag{3-47}$$

For turbulent flow through the valve, the flow varies with the square root of the pressure drop, and the resistance to use is twice the pressure drop divided by the flow. Also, the numerator and denominator of Eq. (3-47) are multiplied by the absolute pressure P_1 to show how the time constant depends on the holdup time.

$$T = \frac{VP_1}{14.7} \frac{2\,\Delta P}{F_2 P_1} = \frac{2\,\Delta P}{P_1} T_H \tag{3-48}$$

(The tank holds $VP_1/14.7$ scf, and so $VP_1/14.7F_2$ is the holdup time.)

When the pressure drop is small, the time constant is much less than the holdup time because most of the gas in the tank would still be there when P_1 dropped to the level of P_2; the capacity for storage comes only from pressures above atmospheric. A comparable liquid-level system would be a tank with 10 ft of head discharging to a tank on the same level with 8 ft of head.

When the upstream pressure is greater than about twice the downstream pressure, sonic flow exists in the valve and the flow is directly proportional to the upstream pressure, as shown in Fig. 3-12. The resistance is the ratio of absolute upstream pressure to flow rate (not the ratio $\Delta P/F$). For sonic flow $R = P_1/F_2$

$$T = \frac{V}{14.7} \frac{P_1}{F_2} = T_H \tag{3-49}$$

† The temperature in a large tank drops as the pressure decreases because the gas expands almost adiabatically. This makes the capacity somewhat less than $V/14.7$. The bellows and other small tanks of a pneumatic controller operate almost isothermally except at very high frequencies.

The time constant is equal to the holdup time for this special case where the input is independent of tank pressure and the output directly proportional to tank pressure. When gas is discharged at sonic velocity, the expansion is usually adiabatic and the flow rate for given upstream conditions is readily obtained from the handbooks. For sonic flow in very small outlets, the flow is between the value for adiabatic flow and the somewhat higher value for isothermal flow, and the exact value is hard to predict.

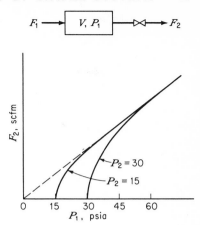

FIGURE 3-12   Flow curves for high-pressure drops.

The response of tanks with only one opening is important in the design of pneumatic controllers. Such dead-end systems are comparable with thermometer bulbs in that flow through the resistance may be in either direction. If the Reynolds number is so small that the pressure drop is proportional to the flow, the resistance is constant and the analysis straightforward. In the more general case, the resistance or slope of the flow curve changes with flow rate, and since the flow rate eventually goes to zero after a step change in input pressure, a time constant calculated from any one slope has no particular significance. Using a time constant based on the average slope gives a fair approximation of the transient response to step inputs, as shown in the following example. When such an element is part of a control loop, two possible time constants should be calculated based on the initial slopes for large and small pressure changes. The similar problem of "plus-minus" flows in air lines is discussed in Ref. 1.

Example 3-4

Compare the exact and the linearized solutions for the transient response of the system in Fig. 3-13 to a small step change in P_1.

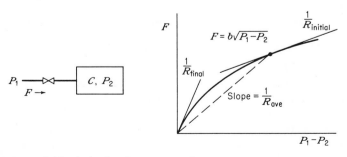

FIGURE 3-13   A dead-end pressure system.

Exact Solution

$$C \frac{dP_2}{dt} = b\sqrt{P_1 - P_2} \qquad P_2 = 0 \text{ at } t = 0$$

$$\frac{dP_2}{\sqrt{P_1 - P_2}} = \frac{b}{C} dt$$

$$-2\sqrt{P_1 - P_2} = \frac{bt}{C} - 2\sqrt{P_1}$$

Linearized Solution

$$R = \frac{P_1}{b\sqrt{P_1}} = \frac{\sqrt{P_1}}{b}$$

$$C \frac{dP_2}{dt} = \frac{P_1 - P_2}{R}$$

$$T = RC = \frac{C\sqrt{P_1}}{b}$$

$$y = \frac{P_2}{P_1} = 1 - e^{-t/T}$$

	Linearized solution	Exact solution
t/T	P_2/P_1	P_2/P_1
0.2	0.18	0.19
0.5	0.40	0.44
1	0.63	0.75
2	0.86	1.0
3	0.95	

The response curves are fairly close up to 50 per cent recovery, but the linearized equation predicts too slow a response at high recoveries. The "exact" solution predicts that $P_2 = P_1$ at a finite time; actually the flow through the resistance would be laminar at very low pressure drops, and a first-order equation would result. With a first-order system, infinite time is required for complete recovery.

TANKS WITH TWO OR MORE RESISTANCES

When the flows into and out of the tank are both influenced by the tank pressure, both flow resistances affect the time constant. In the following derivation, P_1, P_2, and P_3 refer to changes in the pressures upstream, in the tank, and downstream. [If the student has any doubt about the use of three different base pressures, he should set up the complete equation with all pressures in pounds per square inch absolute and show how the average values cancel to give Eq. (3-50).]

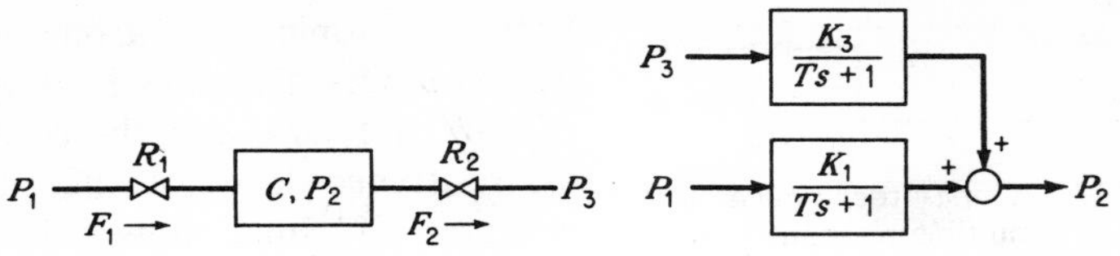

FIGURE 3-14   A pressure system with two resistances.

For the system of Fig. 3-14,

$$\text{Accumulation} = \text{extra input} - \text{extra output}$$

$$C\frac{dP_2}{dt} = \frac{P_1 - P_2}{R_1} - \frac{P_2 - P_3}{R_2} \tag{3-50}$$

where $R_1 = \dfrac{d(P_1 - P_2)}{dF_1}$

$R_2 = \dfrac{d(P_2 - P_3)}{dF_2}$

Collecting terms,

$$\frac{C\, dP_2}{dt} + P_2\left(\frac{1}{R_1} + \frac{1}{R_2}\right) = \frac{P_1}{R_1} + \frac{P_3}{R_2} \tag{3-51}$$

or

$$T\frac{dP_2}{dt} + P_2 = K_1 P_1 + K_3 P_3 \tag{3-52}$$

where $T = C\dfrac{R_1 R_2}{R_1 + R_2}$

$K_1 = \dfrac{R_2}{R_1 + R_2}$

$K_3 = \dfrac{R_1}{R_1 + R_2}$

The block diagram for the process shows how the output pressure depends on both P_1 and P_3.   Either of these could be the controlled variable or the load variable, or both could be load variables if the controller acts to change R_1 or R_2.

If there are several inlets and outlets the system is still first-order.   For the system of Fig. 3-15, it follows from Eq. (3-52) that

$$T\frac{dP_3}{dt} + P_3 = K_1 P_1 + K_2 P_2 + K_4 P_4 + K_5 P_5 \tag{3-53}$$

where $T = \dfrac{C}{\Sigma(1/R)}$

$K_1 = \dfrac{1}{R_1 \Sigma(1/R)}$, etc.

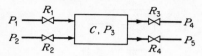

FIGURE 3-15   A first-order system with multiple resistances.

If there is sonic flow in one of the inlet resistances, that flow is independent of tank pressure, and the corresponding resistance does not affect the time constant. If sonic flow exists downstream, say, in R_4, the resistance does influence the time constant but the gain for the corresponding pressure term, K_5, is zero.

RESPONSE TO CHANGES IN VALVE POSITION

If the valve in the discharge line is used to control the tank pressure, the transfer function relating pressure and valve-stem position is obtained by linearizing the valve characteristics as shown in Fig. 3-16.

$$C \frac{dP_2}{dt} = \frac{P_1 - P_2}{R_1} - \frac{P_2 - P_3}{R_2} + x \left(\frac{\partial F}{\partial x} \right)_{\Delta P} \qquad (3-54)$$

where x = fractional change in valve position.

The resistances R_1 and R_2 are evaluated at the average valve positions, and the rate of change of flow with position is evaluated at the average pressure drop.   Since the last term in Eq. (3-54) does not include P_2, the time constant for slight changes in valve position is the same as for changes in P_1 or P_3. The process gain, or change in pressure per unit change in valve position, depends on the characteristics of both valves.

$$\frac{P_2}{x} = \frac{K_p}{Ts + 1} \qquad (3-55)$$

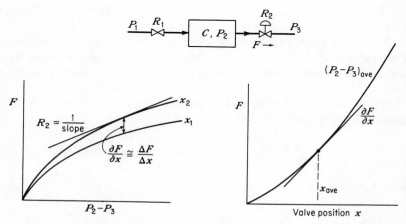

FIGURE 3-16   Linearization of valve characteristics.

where $K_p = \dfrac{\partial F/\partial x}{\Sigma(1/R)}$

$$T = \frac{C}{\Sigma(1/R)}$$

3-8. PURE-CAPACITY ELEMENTS

If the flows of liquid to and from a tank are independent of the liquid level, the tank is a pure-capacity or integrating element. The differential equation for the system of Fig. 3-17a is

$$A \frac{dh}{dt} = F_1 - F_2 \tag{3-56}$$

or
$$Ash = F_1 - F_2$$

The response of the system to a step change in either of the input variables is a ramp change in level.

$$h = \frac{F_1}{A} t \quad \text{or} \quad h = -\frac{F_2}{A} t \tag{3-57}$$

The system shows no self-regulation, since after a slight change in input the level changes at a constant rate and never reaches a new steady-state value until the tank overflows or runs dry.

Pure-capacity elements can also arise in thermal systems. If a batch of material is heated by electric-resistance heaters and the heat losses are small, the batch acts as a pure capacity. The resistance to heat transfer does not affect the rate of heat transfer since the heater temperature rises when the power is increased. A similar situation can occur when the flow of steam to a heating coil is fixed by a flow controller or by a sonic restriction. The coil temperature adjusts until all the entering steam condenses; and a step increase in steam rate leads to a ramp change in batch temperature. There is a small time constant associated with the mass of the heating coil which slightly delays the start of the ramp response.

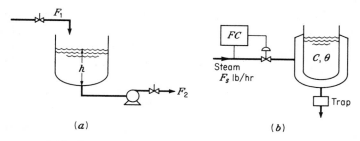

FIGURE 3-17   Pure-capacity systems. (a) Tank with inflow and out-flow independent of head; (b) tank with heat flux independent of tank temperature.

3-9. SECOND-ORDER SYSTEMS; THE MANOMETER

If the mechanical parts of a system or the fluid in a system are subject to accelerations, the dynamic behavior is described by second-order equations. A mass suspended from a spring is the classic example of a second-order system, but a manometer is chosen as an example of more interest to chemical engineers.   The basic equation is a force balance, and to simplify matters, the frictional pressure drop is assumed proportional to the velocity, and all the fluid is assumed to accelerate uniformly.

For the manometer shown in Fig. 3-18

$$\frac{AL\rho}{g_c} \frac{d^2h}{dt^2} = A\left(P - 2h\rho \frac{g}{g_c}\right) - RA \frac{dh}{dt} \tag{3-58}$$

where A = cross-sectional area

ρ = liquid density (density of gas above fluid is negligible)

P = applied pressure

R = frictional resistance

With laminar flow, the resistance is given by the Hagen-Poiseuille equation.

$$\frac{\Delta P}{L} = \frac{32V\mu}{D^2 g_c} \quad \text{or} \quad R = \frac{32L\mu}{D^2 g_c} \tag{3-59}$$

Substituting into Eq. (3-58) and rearranging gives

$$\frac{L}{2g} \frac{d^2h}{dt^2} + \frac{16L\mu}{\rho g D^2} \frac{dh}{dt} + h = \frac{Pg_c}{2\rho g} = h_i \tag{3-60}$$

Equation (3-60) may be written in a standard form.

$$\frac{1}{\omega_n{}^2} \frac{d^2h}{dt^2} + \frac{2\zeta}{\omega_n} \frac{dh}{dt} + h = h_i \tag{3-61}$$

where ω_n = natural frequency, rad/sec

ζ = damping coefficient

The significance of ω_n and ζ becomes apparent after considering the solution to Eq. (3-61) for a step change in input pressure.   With a damping coefficient less than 1, the output overshoots the final value and oscillates before coming to equilibrium.   The system is said to be "underdamped."

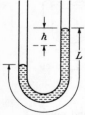

For $\zeta < 1.0$, $\quad \dfrac{h}{h_i} = 1 + \dfrac{1}{\sqrt{1 - \zeta^2}} e^{-\zeta \omega t}$

$$\times \sin\left(\omega_n \sqrt{1 - \zeta^2}\, t - \phi\right) \tag{3-62}$$

FIGURE 3-18   A   U-tube   where $\phi = \tan^{-1} \dfrac{\sqrt{1 - \zeta^2}}{-\zeta}$

manometer.

With a damping coefficient of zero, the response is an undamped sine wave of frequency ω_n and amplitude $2h_i$.

With a damping coefficient of 1.0, the system is critically damped and comes to equilibrium without overshooting.

For $\zeta = 1.0$, $\qquad\qquad \dfrac{h}{h_i} = 1 - (1 + \omega_n t)e^{-\omega_n t}$ $\qquad\qquad$ (3-63)

If the damping coefficient is greater than 1.0, the system is overdamped and comes to equilibrium slowly. When $\zeta > 1$, the quadratic term can be factored and the solution is given in terms of the two roots.

For $\zeta > 1.0$, $\qquad \dfrac{s^2}{\omega^2} + \dfrac{2\zeta}{\omega} s + 1 = (T_a s + 1)(T_b s + 1)$

$$\frac{h}{h_i} = 1 + \frac{1}{T_b - T_a}(T_a e^{-t/T_a} - T_b e^{-t/T_b})$$

(3-64)

Experimental values of ω_n and ζ can easily be obtained from underdamped response curves of the type shown in Fig. 3-19. The damping coefficient is found either from the decay ratio, which is the ratio of successive peak heights, or from the maximum overshoot.

$$\text{Decay ratio} = e^{-2\pi\zeta/\sqrt{1-\zeta^2}}$$

(3-65)

$$\frac{\text{Max overshoot}}{\text{Final valve}} = e^{-\pi\zeta/\sqrt{1-\zeta^2}} = \sqrt{\text{decay ratio}}$$

(3-66)

The frequency of the damped oscillations is only slightly lower than the natural frequency for values of ζ between 0 and about 0.5.

$$\text{Frequency of damped oscillation} = \omega_n \sqrt{1 - \zeta^2}$$

(3-67)

The response of a second-order system to sinusoidal inputs also shows the significance of ω_n and ζ. For the underdamped case the amplitude ratio is greater than 1 for frequencies near the natural frequency, and the peak amplitude ratio increases with decreasing damping coefficient. Equations and graphs for the amplitude ratio and phase lag are presented in the chapter on frequency response.

Typical U-tube manometers filled with water or mercury would be underdamped, as shown by Table 3-1. In practice the damping coefficients would usually be somewhat higher than those given in the table because of turbulence at high velocities and the energy lost when the flow reverses direction. A more complete theory and some supporting data are given by Biery (2). To avoid oscillations, the manometers used with mercury-type

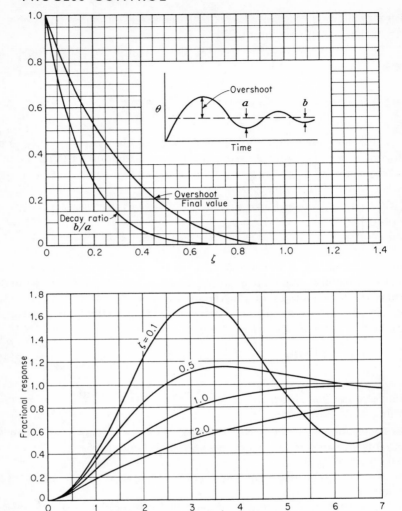

FIGURE 3-19   Step response of a second-order system.

flowmeters have an adjustable constriction at the base of the float chamber which greatly increases the friction loss and which usually makes the damping coefficient greater than 1.0. However, this extra damping often makes such flowmeters the slowest elements in flow control loops.

There are relatively few examples where second-order *components* have a significant effect on the dynamics of process control loops, though, of course, they are the major consideration in the control of moving machinery. Pneumatic controllers and transmitters have moving parts, but the natural frequencies are generally so much higher than the critical frequency of the

Table 3-1. Theoretical Damping Coefficients and Natural Frequencies for Manometers

D, cm	L, cm	ρ, g/cm³	μ, cp	$\omega_n = \sqrt{\dfrac{2g}{L}}$, rad/sec	$\zeta = \dfrac{8L\mu}{D^2\rho g}\sqrt{\dfrac{2g}{L}}$
0.2	200	13.5	1.6	3.13	0.15 (mercury)
0.5	200	13.5	1.6	3.13	0.024 (mercury)
0.5	200	1.0	1.0	3.13	0.15 (water)

process that the controller dynamics can be neglected. For some fast processes such as flow control with short pneumatic transmission lines, the critical frequency of the process may be close to the natural frequency of the instruments or the transmission line, and damping must be added to get good control. Frequent use of second-order equations is made in characterizing closed-loop control systems. Even though a control system is really described by a third- or higher-order equation, the shape of the transient response can generally be characterized satisfactorily by two parameters, a frequency and a damping coefficient.

3-10. RESPONSE OF NONINTERACTING FIRST-ORDER ELEMENTS IN SERIES

Consider two first-order systems that are connected so that the output variable of the first is the input variable of the second. If the response of the first system does not depend on conditions in the second system, the systems are noninteracting and the two transfer functions are just multiplied to get the overall response of the system. A heated tank and an immersed thermometer (Fig. 3-20) can be considered noninteracting systems, since the flow of heat to the thermometer is a negligible term in the heat balance for the tank.

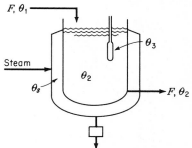

$$\frac{\theta_2}{\theta_s} = \frac{K_1}{T_1 s + 1} \qquad (3\text{-}68)$$

where T_1 = tank time constant,

$$\frac{\theta_3}{\theta_2} = \frac{1}{T_2 s + 1} \qquad (3\text{-}69)$$

where T_2 = bulb time constant. Combining the two transfer functions,

$$\frac{\theta_3}{\theta_s} = \frac{\theta_2}{\theta_s}\frac{\theta_3}{\theta_2} = \frac{K_1}{(T_1 s + 1)(T_2 s + 1)} \qquad (3\text{-}70)$$

FIGURE 3-20 A tank-thermometer system with negligible interaction.

(The system transfer function can also be obtained by combining the differential equations for tank and bulb to give a second-order equation and then taking the Laplace transform.)

Equation (3-70) shows that the overall response is that of an overdamped second-order system, and the solution for a step change in θ_s is given by Eqs. (3-64). The initial slope of the response curve is zero, since the slope depends on the difference $\theta_2 - \theta_3$ and both these temperatures are zero at the start. After the response is about 20 per cent complete, the curve is not very different from that for a first-order system, as shown by the typical curves in Fig. 3-21. The response is the same whether the larger time constant is first or last in the process arrangement.

It is often necessary to estimate the time constants of a system from the transient response. If the system is thought to have two major time constants, the sum $T_1 + T_2$ can be estimated from the time for 73 per cent recovery, since all the curves of Fig. 3-21 intersect at about $t/(T_1 + T_2) = 1.3$ and $y = 0.73$. The ratio of the time constants can then be obtained from the value of y at $t/T_1 + T_2 = 0.5$, the point where the curves of Fig. 3-21 are farthest apart. If the fractional recovery at this point is less than 0.26, the system contains three or more time constants or an unaccounted-for transportation lag. Another method that has been used to estimate the two or three largest time constants involves linear extrapolation on semilog plots, but this method is inaccurate when the time constants are the same order of magnitude.

If there are several equal time constants in series, the output is almost zero for a considerable time after a step change in the input, as shown by

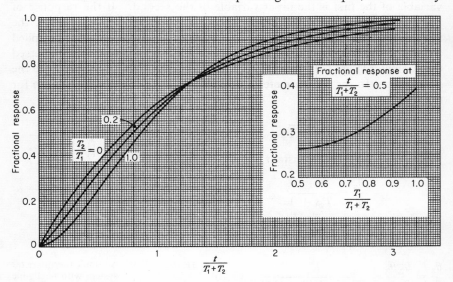

FIGURE 3-21   Step response of two noninteracting time constants in series.

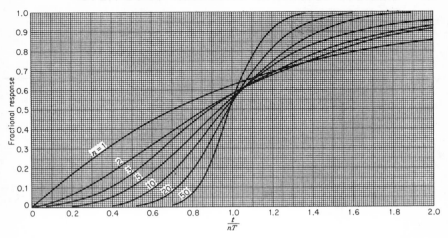

FIGURE 3-22   Step response for n equal noninteracting time constants in series.

the curves of Fig. 3-22.   With typical experimental data, it is difficult to distinguish the very slow initial response from the flat response of a process with pure time delay.   Fortunately, using an effective time delay and one or two effective time constants to characterize such systems gives about the same results as using several equal time constants, as far as closed-loop performance is concerned.   Based on the time for 5 per cent response, the effective time delay is about $0.5nT$ for $n = 10$ and approaches nT as n becomes very large.   Whether one or two effective time constants are used to fit the response above $y = 0.05$ depends on the accuracy required.

Other examples of noninteracting systems include a series of stirred reactors for which temperature or concentration is the pertinent variable. Pressure tanks in series are noninteracting only if there is sonic flow from one tank to the next.   A distillation tower is the best example of a series of liquid-flow elements which are noninteracting.   The holdup of each plate changes with flow rate, but the flow rate is not influenced by the holdup on the plate below.   In analog computers, combinations of electrical resistances and capacitances are made noninteracting by using an amplifier between each stage.   The amplifier draws so little current that it isolates each RC element from the next element in line.   The arrangement of thermal, flow, and electrical elements for interacting and noninteracting systems is summarized in Fig. 3-23.

3-11.   RESPONSE OF INTERACTING ELEMENTS IN SERIES

The majority of thermal systems containing two or more capacities are interacting, since the heat fluxes depend on temperature differences and not

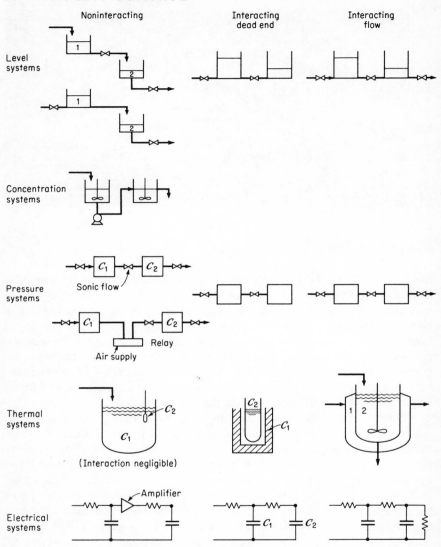

FIGURE 3-23   Interacting and noninteracting systems.

just on the temperature of one part of the system. The simplest type of interaction is that typified by a thermometer bulb and well. This is called a "dead-end system," since the equation for the bulb has no output term. In the following derivation the only resistances are assumed to be those in the fluid outside the well and in the fluid between the well and the bulb (see Fig. 3-24), and the capacity of these thin fluid layers is neglected. The heat balance for the bulb is the same as before:

$$M_2 c_{p_2} \frac{d\theta_2}{dt} = h_2 A_2 (\theta_1 - \theta_2) \tag{3-71}$$

$$R_2 C_2 \frac{d\theta_2}{dt} + \theta_2 = \theta_1 \tag{3-72}$$

or $$(R_2 C_2 s + 1)\theta_2 = \theta_1$$

The heat balance for the well has both inflow and outflow terms.

$$(M_1 c_{p_1}) \frac{d\theta_1}{dt} = h_1 A_1 (\theta_i - \theta_1) - h_2 A_2 (\theta_1 - \theta_2) \tag{3-73}$$

$$C_1 \frac{d\theta_1}{dt} = \frac{\theta_i - \theta_1}{R_1} - \frac{\theta_1 - \theta_2}{R_2} \tag{3-74}$$

or $$C_1 s \theta_1 + \theta_1 \left(\frac{1}{R_1} + \frac{1}{R_2} \right) = \frac{\theta_i}{R_1} + \frac{\theta_2}{R_2}$$

To relate θ_2 and θ_i, θ_1 must be eliminated by combining Eqs. (3-72) and (3-74). The use of transforms makes this quite simple.

$$(R_2 C_2 s + 1)\theta_2 = \frac{\theta_i / R_1 + \theta_2 / R_2}{C_1 s + 1/R_1 + 1/R_2} \tag{3-75}$$

Bringing the factor with s to the left side of the equation and multiplying both sides by R_1 gives

$$\left(R_1 C_1 s + 1 + \frac{R_1}{R_2} \right)(R_2 C_2 s + 1)\theta_2 - \frac{R_1}{R_2} \theta_2 = \theta_i \tag{3-76}$$

or $$[R_1 C_1 R_2 C_2 s^2 + (R_1 C_1 + R_2 C_2 + R_1 C_2)s + 1]\theta_2 = \theta_i$$

or $$[T_1 T_2 s^2 + (T_1 + T_2 + R_1 C_2)s + 1)]\theta_2 = \theta_i$$

In using the last version of Eq. (3-76), remember that T_2 is the time constant of the bulb while in the well, which is not the same as the time constant of the bare bulb. The time constant T_1 is the value for an empty well, and $R_1 C_2$ has the dimensions of time but no physical significance.

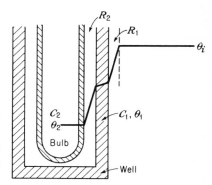

FIGURE 3-24   Temperature gradients for a bulb in a well.

Equation (3-76) can be factored to give two effective time constants, since, as shown below, the damping coefficient is always greater than 1.†

$$\left(\frac{1}{\omega}\right)^2 = T_1T_2 \qquad \frac{2\zeta}{\omega} = T_1 + T_2 + R_1C_2 \qquad (3\text{-}77)$$

$$\zeta = \frac{T_1 + T_2}{2\sqrt{T_1T_2}} + \frac{R_1C_2}{2\sqrt{T_1T_2}} \geq 1 \qquad (3\text{-}78)$$

Therefore Eq. (3-76) can be written

$$(T_a s + 1)(T_b s + 1)\theta_2 = \theta_i \qquad (3\text{-}79)$$

The quadratic formula can be used to get T_a and T_b, or they can be obtained fairly quickly by trial, since they are real positive numbers.

Example 3-5

Obtain the effective time constants for a dead-end interacting system with the following resistances and capacities:

$$R_1 = 1 \qquad C_1 = 1 \qquad T_1 = 1$$
$$R_2 = 2 \qquad C_2 = 2 \qquad T_2 = 4$$
$$R_1C_2 = 2$$

From Eq. (3.76),

$$\frac{\theta_2}{\theta_i} = \frac{1}{4s^2 + 7s + 1} = \frac{1}{(6.4s + 1)(0.62s + 1)}$$

Because of the interaction, the ratio of the effective time constants, T_b/T_a, is always greater than the ratio T_2/T_1. This makes the transient response more like that of a single first-order element. A greater spread of the time constants also increases the stability of a closed-loop system [see Eq. (4-41)], and if the interaction is ignored, the calculated controller settings are therefore conservative. The interaction can be neglected if the product R_1C_2 is much smaller than T_1 plus T_2, as would be true for a thermometer bulb in a large tank ($R_1 \ll R_2$ and $C_2 \ll C_1$).

In some cases, there are flows in and out of all elements of an interacting system. The example of pressure tanks in series is used to develop the

† The geometric mean is less than or equal to the arithmetic mean; so

$$\frac{T_1 + T_2}{2\sqrt{T_1T_2}} \geq 1$$

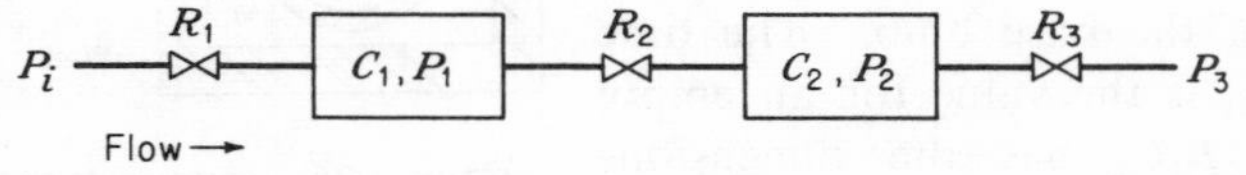

FIGURE 3-25   Interacting pressure systems.

equations.   For the system of Fig. 3-25,

$$C_1 \frac{dP_1}{dt} = \frac{P_i - P_1}{R_1} - \frac{P_1 - P_2}{R_2}$$

$$\left(C_1 s + \frac{1}{R_1} + \frac{1}{R_2}\right)P_1 = \frac{P_i}{R_1} + \frac{P_2}{R_2} \qquad (3\text{-}80)$$

$$C_2 \frac{dP_2}{dt} = \frac{P_1 - P_2}{R_2} - \frac{P_2 - P_3}{R_3}$$

$$\left(C_2 s + \frac{1}{R_2} + \frac{1}{R_3}\right)P_2 = \frac{P_1}{R_2} + \frac{P_3}{R_3} \qquad (3\text{-}81)$$

Combining Eqs. (3-80) and (3-81) to eliminate P_1,

$$P_2 \left[\frac{R_1 C_1 R_2 C_2 R_3 s^2}{R_1 + R_2 + R_3} + \frac{(R_1 C_1 R_3 + R_1 C_1 R_2 + R_2 C_2 R_3 + R_1 C_2 R_3)s}{R_1 + R_2 + R_3} + 1\right]$$

$$= P_i \frac{R_3}{R_1 + R_2 + R_3} + \frac{P_3(R_1 + R_2)}{R_1 + R_2 + R_3}\left(\frac{R_1 C_1 R_2 s}{R_1 + R_2} + 1\right) \quad (3\text{-}82)$$

In this case the products $R_1 C_1$ and $R_2 C_2$ are not the time constants of the tanks taken separately: with flow through one tank, the time constant is $C_1/(1/R_1 + 1/R_2)$, as derived in Eqs. (3-50) and (3-51).   However, the quadratic term in Eq. (3-82) still has a damping coefficient greater than 1 and can be factored to give effective time constants T_a and T_b.

Example 3-6

Obtain the effective time constants for an interacting flow system which has two equal capacities and three equal resistances.

$$C_1 = C_2 = 2$$
$$R_1 = R_2 = R_3 = 1$$

From Eq. (3-82), $\qquad P_2\left(\frac{4s^2}{3} + \frac{8s}{3} + 1\right) = \frac{P_i}{3} + \tfrac{2}{3}P_3(s + 1)$

For changes in P_i, $\qquad \dfrac{P_2}{P_i} = \dfrac{1}{3(2s + 1)(\tfrac{2}{3}s + 1)}$

For changes in P_3, $\qquad \dfrac{P_2}{P_3} = \dfrac{2(s + 1)}{3(2s + 1)(\tfrac{2}{3}s + 1)}$

With one tank, the time constant would be $2/(1 + 1) = 1$, which is between the values of 2 and $\tfrac{2}{3}$ for the interacting system.

A stirred tank with a water jacket is another example of a system with interacting elements.   If the jacket fluid is assumed to be well mixed and if the wall capacity is neglected, the equations are similar to Eqs. (3-80) and (3-81) and the overall transfer functions of Eq. (3-82) can be used.

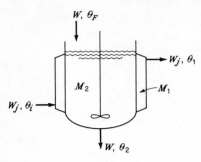

FIGURE 3-26 A jacketed tank, an interacting thermal system.

For the jacket of the kettle in Fig. 3-26,

$$M_1 c_{p_1} \frac{d\theta_1}{dt} = W_j c_{p_1}(\theta_i - \theta_1)$$
$$- UA(\theta_1 - \theta_2) \quad (3\text{-}83)$$

or $$C_1 \frac{d\theta_1}{dt} = \frac{\theta_i - \theta_1}{R_1} - \frac{\theta_1 - \theta_2}{R_2}$$

For the tank,

$$M_2 c_{p_2} \frac{d\theta_2}{dt} = UA(\theta_1 - \theta_2)$$
$$+ W c_{p_2}(\theta_F - \theta_2) \quad (3\text{-}84)$$

$$C_2 \frac{d\theta_2}{dt} = \frac{\theta_1 - \theta_2}{R_2} - \frac{\theta_2 - \theta_F}{R_3}$$

The transfer function for changes in cooling water temperature is

$$\frac{\theta_2}{\theta_i} = \frac{K_1}{(T_a s + 1)(T_b s + 1)} \quad (3\text{-}85)$$

where T_a and T_b are factors of the quadratic term in Eq. (3-82) and

$$K_1 = \frac{R_3}{R_1 + R_2 + R_3}$$

3-12. RESPONSE OF A DISTRIBUTED SYSTEM

If several identical interacting elements are arranged in series, the response is practically the same as that of a distributed system, which is one where resistance and capacity are associated with each incremental length of the system.   A pneumatic transmission line is a distributed flow system, and every pipe and tank wall has distributed thermal capacity and resistance. The response of an isolated distributed element is readily obtained from published solutions for transient conduction or diffusion in slabs or other shapes (3, 4).   The response of a system that has both distributed and lumped elements is difficult to obtain, and the distributed element is often replaced by a series of interacting RC stages to simplify the analysis.

The step response of an isolated distributed system is shown in Fig. 3-27. The fractional response is based on the temperature at the midpoint of a thick slab (or the temperature at the outer edge of an insulated wall).   The response is a function of the dimensionless group $\alpha t/(r_m)^2$, which is equivalent to the group t/RC, where R is the total resistance and C the total capacity of the slab.

$$\frac{\alpha t}{(r_m)^2} = \frac{k}{\rho c_p} \frac{t}{(r_m)^2} = \frac{t}{(\rho c_p r_m)(r_m/k)} = \frac{t}{RC} \quad (3\text{-}86)$$

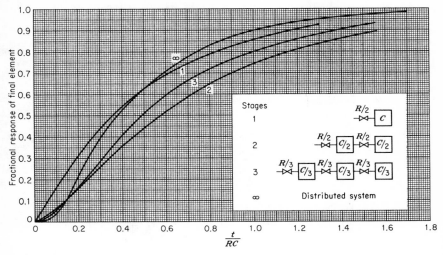

FIGURE 3-27   Step response of distributed and interacting systems.

The response is about 63 per cent complete at $t = RC/2$, or the effective time constant for a first-order approximation is $RC/2$. This result is rationalized by saying that the average conduction distance is only half the thickness; so the total capacity is placed at the middle of the section in a one-capacity model. The slow initial response of a distributed system is better fitted by using models with two or more capacities, as shown in Fig. 3-28. The curves for two- and three-capacity models fall below the one-capacity curve at large times, because the capacities are placed at the end of the sections. If the response at large times is more important than the initial response, the capacities could be located in the middle of the sections, as shown in Fig. 3-28d.

If the distributed element is the shell of a heat exchanger or the wall of an insulated tank, it acts as a side capacity. The total flow of heat to the

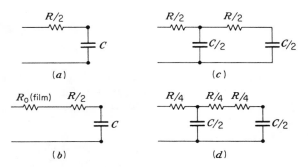

FIGURE 3-28   Models of distributed systems.

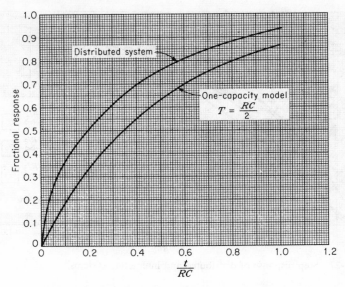

3-29   Average response of a distributed system to a step input.

wall rather than the temperature on the far side of the wall is the important parameter, since the wall temperature is not the input to any other part of the process. As shown in Fig. 3-29, the one-capacity model gives too low a value for the heat absorbed. However, if there is an appreciable film or contact resistance at the wall, a one-capacity model is usually satisfactory.

For studies with an electric analog, a distributed element can be replaced by several RC stages to give results of the required accuracy. If the resistance or capacity of the distributed element is relatively small compared with that of other parts of the system, 1 or 2 stages are enough. If the distributed element is a reactor, heat exchanger, or packed column, 5 to 10 stages may be needed for an accurate simulation.

PROBLEMS

1. A cylindrical tank with an area of 20 ft² and a height of 10 ft has a normal liquid depth of 6 ft. The flow to the tank is 10 ± 1 cfm, and the liquid is discharged through a control valve to a process at atmospheric pressure.

Calculate the time constant if the tank is

a. Open to the atmosphere.

b. Closed, with a constant pressure of 10 psig above the liquid.

c. Closed, with a fixed amount of air sealed above the liquid. The initial pressure is 10 psig.

2. A solution is heated continuously in a 100-gal tank by injecting live steam into the tank. The signal from a thermometer in the outlet line is sent to a proportional controller which regulates the flow of steam through a control valve. The flow to the tank is 25 gpm, and the solution is heated from 70 to 150°F.

Set up a complete block diagram for the process, and fill in as many of the transfer functions as you can. Show all possible load variables in the diagram.

3a. An open cylindrical tank 4 ft in diameter is filled with water to a depth of 3 ft. The water flows out through a 2-in. control valve at an average rate of 30 gpm. If the input is increased from 30 to 40 gpm, what is the new depth in the tank and about how long will it take to accomplish 95 per cent of the change in level?

b. Calculate the transfer function relating tank level to valve position, and compare with the transfer function relating level and input flow.

4. Aqueous streams A and B flow continuously into a stirred tank and react rapidly, releasing 500 Btu/lb of solution A. The reaction goes substantially to completion, and the heat release is therefore independent of small changes in temperature. Water is circulated through a cooling coil at such a rate that the temperature rise of the cooling water is very small.

F_a = 1,000 lb/hr feed to kettle (solution A)
F_b = 1,000 lb/hr feed to kettle (solution B)
c_p = 1 for both streams
W = 5,000 lb holdup in kettle
A = 30 ft² area of cooling coil
U = 600 Btu/(hr)(ft²)(°F)
θ_a = 80°F = inlet temperature of streams A and B
θ_1 = 70°F = cooling-water temperature

a. What is the output temperature?
b. What is the output temperature 5 min after the cooling-water temperature suddenly drops to 60°F?

5. A 20-ft³ tank is supplied with 200 scfm of air by a compressor. The tank is kept at 50 psia by a pressure controller and valve bleeding air to the atmosphere. The tank supplies about 80 scfm to process 2 operating at about 40 psia and 100 scfm to process 3 operating at 45 psia. The drop in pressure from the tank to these processes occurs in the pipelines and in manually adjusted valves. The high pressure drop across the bleed valve means that sonic velocity exists in this valve, and the mass flow rate is proportional to the upstream pressure.

a. Write the complete differential equation for this system, considering the possible load changes to be changes in the pressure of both processes 2 and 3. What is the effective time constant of the system for changes in P_3, the pressure in process 3?

b. If the controller gain is 10, calculate the overall open-loop gain of this system. The full chart width corresponds to a pressure range of 50 psia (15 to 65 psia). An equal-percentage control valve is used, with the characteristic that a 1 per cent change in valve position increases the flow 5 per cent.

6. The following data were obtained by making a step input in steam flow to a process and recording the output temperature with a filled-bulb thermometer.

What are the effective time constants for the process?

Time, min	Output, °F
0	50
2	51
4	54
6	58
8	62
10	65
12	68
14	71
16	73
20	76
30	79

7. A U-tube manometer is made of 0.5-cm tubing and has a mercury column 100 cm long.   Each leg of the manometer is connected by 500 cm of the same sized tubing to a pressure tap in a water pipe.   Assuming laminar flow for both water and mercury, how does the presence of water above the mercury affect the critical frequency and the damping coefficient?

8. Consider a batch reactor 6 ft in diameter and 10 ft high.   The temperature is controlled by circulating water through the jacket.   The overall coefficient is 120 Btu/(hr)(ft²)(°F), and the holdup time in the jacket is 2 min.   Neglecting the wall capacity and the change in heat generated with temperature, calculate the effective time constants of the system.   How much error would be introduced if the interaction were neglected?

REFERENCES

1. Buckley, P. S., and J. M. Mozley: "Instrumentation Systems Handbook of Automation, Computations, and Control," vol. 3, chap. 7, John Wiley & Sons, Inc., New York, 1961.
2. Biery, J. C.: Numerical and Experimental Study of Damped Oscillating Manometers: I, Newtonian Fluids, *A.I.Ch.E.J.*, **9**:606 (1963).
3. McAdams, W. H.: "Heat Transmission," 3d ed., pp. 36–38, McGraw-Hill Book Company, New York, 1954.
4. Crank, John: "The Mathematics of Diffusion," pp. 46, 55, Oxford University Press, London, 1956.

4 TRANSIENT RESPONSE OF CONTROL SYSTEMS

4-1. SERVO-OPERATION AND REGULATOR OPERATION

A closed-loop control system consists of a process and a controller that automatically adjusts one of the inputs to the process in response to a signal fed back from the process output. The performance of the system can be judged by the transient response of the output to specific changes in input. The change in input may be a change in set point or a change in any one of several load variables. If the purpose of the control system is to make the process follow changes in set point as closely as possible, the operation is called "servo-operation." The term "regulator operation" is used when the main problem is to keep the output almost constant in spite of changes in load. The designer must be aware of the purpose of the control system, since the system that gives optimum servo-operation will generally not be the best for regulator operation. A large capacity or inertia makes systems sluggish for servo-operation, but it often helps to minimize the error in regulator operation.

Servo-operation is typical of electrical and mechanical systems and has received the most attention in texts on automatic control, but regulator operation is more important for process control studies. A typical continuous chemical process operates with a constant set point for hours or days at a time, and the changes in set point are generally minor adjustments. The changes in load variables such as uncontrolled flows, temperatures, and pressures are much more frequent and cause larger errors than the changes in set point. Some batch processes do require a continuous change in temperature or other variable, which suggests that the control system be designed for servo-operation. However, if the set point changes quite slowly and steadily, the errors from load changes may be as large as the error caused by the changing set point and the response for both regulator and servo-operation should be considered. For some of the simpler cases considered in this chapter, equations for both types of operation are presented, but for most of the examples the emphasis is on regulator operation.

4-2. TYPES OF INPUTS

The term "transient response" could mean the response of a control system to any type of input, but it usually refers to a step change in the set point or a load variable.   A step change is used partly for convenience; the solutions for this input are easier to obtain than for any other type of disturbance. The step change is also the most severe type of disturbance, and the response to the step change shows the maximum error that could occur for a given eventual change in load.   If several control systems or controller settings are being compared, the system with the best response to a step change in a load variable will generally have the best response to random fluctuations of that variable.   As far as stability is concerned, it does not matter which input is changed or what type of change is made, as long as the system is linear—a closed-loop system unstable to one input is unstable to all inputs.

If very close control is needed, the response to ramp, pulse, sinusoidal, or other types of disturbances could also be calculated to help evaluate the control system. ' With servo-operation the probable types of input changes are usually known, but for regulator operation the nature of the load fluctuations is hardly ever specified when the control system is being designed. Without such information, the evaluation is usually based on the step response or, in some cases, on both the step response and the frequency response.

4-3. GENERAL EQUATIONS FOR THE TRANSIENT RESPONSE

The general equations for the response of closed-loop systems are obtained by algebraic manipulation of the signals and transfer functions of the block diagram.   Figure 4-1 shows a single-loop system with one load variable and no measurement lag.   For this system,

$$\theta = eG + \theta_L G_L \qquad e = \theta_c - \theta$$
$$= \theta_c G - \theta G + \theta_L G_L$$
$$= \theta_c \frac{G}{1+G} + \theta_L \frac{G_L}{1+G} \tag{4-1}$$

The variables θ_c, θ_L, and θ are changes from the normal values of the set point, load variable, and process output, and they are all initially zero. The response to changes in set point is obtained by setting $\theta_L = 0$, replacing θ_c by the Laplace transform of θ_c, which is $1/s$ for a unit step change, and using the inverse transformation to get θ as a function of time.   For changes in load, θ_c is zero, and the transform for the specific load change is used. In general, there might be several load variables, but only one would be

considered at a time. The measurement lag is neglected, to simplify Fig.
4-1; its effect is considered in Sec. 4-11.
In the following cases the lag in the
controller is also neglected, and the
transfer functions for an ideal controller are used.

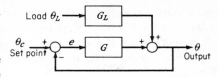

FIGURE 4-1   Closed-loop control system.

4-4. PROPORTIONAL CONTROL OF A SINGLE-CAPACITY PROCESS

Consider the case where a first-order process and a proportional controller
are the only elements in the control system (Fig. 4-2). The time constant
for load changes is the same as for changes in the manipulated variable, as
was proved in Chap. 3. The valve gain is combined with the process gain
K_p, and the valve dynamics are neglected.

For a change in set point

$$\frac{\theta}{\theta_c} = \frac{G}{1 + G} = \frac{K}{Ts + 1} \frac{1}{(Ts + 1 + K)/(Ts + 1)} \tag{4-2}$$

where $K = K_c K_p$

If $\theta_c = A$, $\mathcal{L}(\theta_c) = A/s$,

$$\theta = \frac{A}{s} \frac{K}{Ts + 1 + K} \tag{4-3}$$

To put Eq. (4-3) in a familiar form, the numerator and denominator are
divided by $K + 1$.

$$\theta = \frac{AK}{1 + K} \frac{1}{s\left(\dfrac{T}{1 + K}s + 1\right)} \tag{4-4}$$

or

$$\theta = \frac{AK}{K + 1} \mathcal{L}^{-1}\left[\frac{1}{s(T's + 1)}\right] \tag{4-5}$$

where $$T' = \frac{T}{K + 1}$$

In Eq. (4-4) θ is a transformed variable, and in Eq. (4-5) it is the actual
output. From a table of transforms,
or perhaps, in this simple case, from
memory, the solution of Eq. (4-5) is

$$\theta = \frac{AK}{K + 1}(1 - e^{-t/T'}) \tag{4-6}$$

The solution has the same form as
the solution for a step input to an

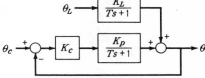

FIGURE 4-2   Proportional control of a
single-capacity process.

open-loop first-order system, but the effective time constant is lower than T by a factor $1 + K$. Note that the speed of response depends on the overall gain K, which is the product of the controller gain, the valve gain, and the process gain. The fractional response is sketched for several gains in Fig. 4-3.

The response to changes in load is obtained from the second term of Eq. (4-1).

$$\frac{\theta}{\theta_L} = \frac{G_L}{1 + G} = \frac{K_L}{Ts + 1 + K} \tag{4-7}$$

where $K = K_c K_p$, as before. For $\theta_L = A$, $\mathcal{L}(\theta_L) = A/s$,

$$\theta = \frac{A}{s}\frac{K_L}{Ts + 1 + K} \tag{4-8}$$

Again the factor $1 + K$ is used to reduce the equation to a familiar form.

$$\theta = \frac{AK_L}{1 + K}\frac{1}{s\left(\dfrac{T}{1 + K}s + 1\right)} \tag{4-9}$$

$$\theta = \frac{AK_L}{1 + K}(1 - e^{-t/T'}) \tag{4-10}$$

The effective time constant $T' = T/(1 + K)$ is the same for load changes as for changes in set point.

The initial slope of the response curve for load changes is independent of the overall gain, as shown in Fig. 4-4. The initial slope is constant because the controller does not act until the load change has started to take effect.

$$\frac{d\theta}{dt} = \frac{AK_L}{1 + K}\left(\frac{1 + K}{T}e^{-t/T'}\right) \tag{4-11}$$

$$\left(\frac{d\theta}{dt}\right)_0 = \frac{AK_L}{T} \tag{4-12}$$

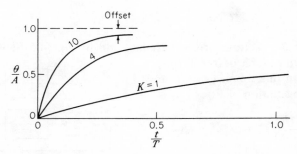

FIGURE 4-3   Response of a single-capacity process to a step change in set point.

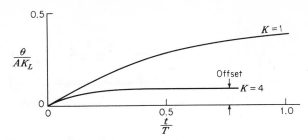

FIGURE 4-4   Response of a single-capacity process to a step change in load.

Equations (4-6) and (4-10) and Figs. 4-3 and 4-4 show that for both servo-operation and regulator operation increased controller gain decreases the steady-state error and also decreases the time required to reach a new equilibrium after a change in input.   The controller gain to use for such a process would seem to be the maximum possible value for the controller that is available, since the equations predict no oscillation as the gain approaches infinity.   Of course, a real process would have other lags or delays in the control valve, in the instrument itself, and in the connecting tubing, all of which would affect the response at very high gains and lead to unstable oscillation if too high a gain were used.   The maximum gain or the optimum gain cannot be calculated unless these other lags are specified.   However, if the largest time constant is two or three orders of magnitude greater than any of the other lags, the maximum gain is so high that there is usually no point in calculating it.   A very narrow-band proportional controller could be used, and perhaps even on-off control would be satisfactory.

For some systems it is not desirable to have the output respond very rapidly to changes in input, and a low controller gain is chosen.   The function of many level and pressure control systems is to damp fluctuations in the flow out of a vessel.   If a gradual change in outflow is desired following a sudden change in flow into the vessel, the controlled variable (level or pressure) must change gradually over a wide range to utilize the damping capacity of the vessel.   The method of calculating the best controller gain for such a system is discussed in the chapter on level control.

4-5. PROPORTIONAL CONTROL OF A TWO-CAPACITY PROCESS

LOAD CHANGE BEFORE THE LAST ELEMENT

For the system shown in Fig. 4-5, the response to a unit change in load will be obtained.   The process could be a jacketed kettle where the temperature is controlled by adjusting the flow of coolant to the jacket.   The load variable might be the flow rate or temperature of the process stream

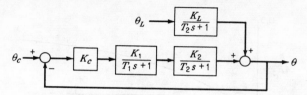

FIGURE 4-5   Proportional control of a two-capacity process
(intermediate load change).

fed to the kettle.   The time constant of the kettle, T_2, is the same for load
changes and changes in jacket temperature, but the gains K_L and K_2 are
different.

$$\frac{\theta}{\theta_L} = \frac{G_L}{1+G} = \frac{K_L}{T_2s+1} \frac{(T_1s+1)(T_2s+1)}{(T_1s+1)(T_2s+1)+K} \qquad (4\text{-}13)$$

where $K = K_c K_1 K_2$.

For $\theta_L = 1$, $\theta = \dfrac{1}{s} \dfrac{K_L(T_1s+1)}{T_1T_2s^2+(T_1+T_2)s+(1+K)} \qquad (4\text{-}14)$

The term $K+1$ is used as a factor to make the denominator of Eq. (4-14)
fit the standard form for second-order equations.

$$\theta = \frac{K_L}{K+1} \frac{T_1s+1}{s\left(\dfrac{T_1T_2}{K+1}s^2 + \dfrac{T_1+T_2}{K+1}s + 1\right)} \qquad (4\text{-}15)$$

$$\theta = \frac{K_L}{K+1} \frac{T_1s+1}{s\left(\dfrac{s^2}{\omega^2} + \dfrac{2\zeta}{\omega}s + 1\right)} \qquad (4\text{-}16)$$

where $\omega = \sqrt{\dfrac{K+1}{T_1T_2}}$

$\zeta = \dfrac{T_1+T_2}{2\sqrt{T_1T_2}} \dfrac{1}{\sqrt{K+1}} = \dfrac{1+R}{2\sqrt{R}} \dfrac{1}{\sqrt{K+1}}$

$R = \dfrac{T_2}{T_1}$

The inverse transform of this equation is obtained from the tables;
the form of the solution depends on the value of the damping coefficient.

For $\zeta < 1$,

$$\theta = \frac{K_L}{K+1}\left[1 + \frac{(1-2\zeta\omega T_1 + \omega^2 T_1^2)^{1/2}}{(1-\zeta^2)^{1/2}} e^{-\zeta\omega t} \sin(\omega\sqrt{1-\zeta^2}\, t + \phi)\right]$$

$$(4\text{-}17)$$

where $\phi = \tan^{-1}\dfrac{\omega T_1\sqrt{1-\zeta^2}}{1-\zeta\omega T_1} - \tan^{-1}\dfrac{\sqrt{1-\zeta^2}}{-\zeta}$

For $\zeta = 1$, the quadratic term becomes $(s/\omega + 1)^2$ or $(T_a s + 1)^2$, and the solution is

For $\zeta = 1$, $\theta = \dfrac{K_L}{K+1} \left\{ 1 + \left[\dfrac{T_1 - T_a}{(T_a)^2} t - 1 \right] e^{-t/T_a} \right\}$ (4-18)

where $T_a = \dfrac{1}{\omega} = \dfrac{\sqrt{T_1 T_2}}{K+1} = \dfrac{2 T_1 T_2}{T_1 + T_2}$

For $\zeta > 1$, the quadratic term is factored to $(T_a s + 1)(T_b s + 1)$, and the solution is

For $\zeta > 1$, $\theta = \dfrac{K_L}{K+1} \left(1 + \dfrac{T_a - T_1}{T_b - T_a} e^{-t/T_a} - \dfrac{T_b - T_1}{T_b - T_a} e^{-t/T_b} \right)$ (4-19)

If the time constants T_1 and T_2 are equal, the response is always oscillatory, since $\zeta < 1$.

For $T_1 = T_2$, $\zeta = \dfrac{1}{\sqrt{K+1}}$ (4-20)

For other ratios of T_2 to T_1, the response may be underdamped or overdamped, depending on the value of K. Increasing the controller gain decreases the damping coefficient and also decreases the steady-state offset. Figure 4-6 shows typical results for the underdamped and critically damped cases, for an example where $T_2/T_1 = 0.25$.

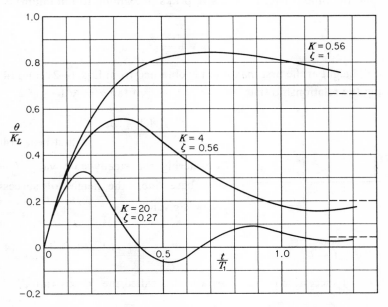

FIGURE 4-6   Response to step change in load for system of Fig. 4-5 when $T_2 = 0.25 T_1$.

In plotting the curves for the underdamped cases, it is often sufficient to locate the first few maxima and minima. To obtain the peak heights, Eq. (4-17) is shortened and differentiated.

$$\theta = A(1 + Be^{-\zeta\omega t}\sin\alpha) \qquad (4\text{-}21)$$

$$\frac{d\theta}{dt} = AB(-\zeta\omega)e^{-\zeta\omega t}\sin\alpha + ABe^{-\zeta\omega t}(\cos\alpha)\omega\sqrt{1-\zeta^2} \qquad (4\text{-}22)$$

when $d\theta/dt$ equals zero.

$$\zeta\sin\alpha = \sqrt{1-\zeta^2}\cos\alpha \qquad (4\text{-}23)$$

or
$$\tan\alpha = \frac{\sqrt{1-\zeta^2}}{\zeta} = \tan(\omega\sqrt{1-\zeta^2}\,t + \phi)$$

The first maximum occurs when

$$t = \frac{\tan^{-1}\dfrac{\sqrt{1-\zeta^2}}{\zeta} - \phi}{\omega\sqrt{1-\zeta^2}} \qquad \text{(angles in radians)} \qquad (4\text{-}24)$$

The first minimum occurs when α is π rad or $180°$ greater than for the first peak, and the time between successive peaks (maximum to minimum) is

$$\Delta t = \frac{\pi}{\omega\sqrt{1-\zeta^2}} \qquad (4\text{-}25)$$

The peak height at the first maximum is obtained from Eqs. (4-21) and (4-23) by using the relationship that $\sin\alpha = \sqrt{1-\zeta^2}$ if $\tan\alpha = \sqrt{1-\zeta^2}/\zeta$,

$$\theta = A(1 + Be^{-\zeta\omega t}\sqrt{1-\zeta^2})$$
$$\text{at first peak} \quad (4\text{-}26)$$

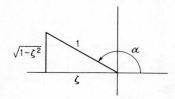

Considering the steady-state value $\theta = A$ as a base line, the height of successive peaks is given by

$$\frac{\theta_{max} - A}{A - \theta_{min}} = \frac{e^{-\zeta\omega t_{max}}}{e^{-\zeta\omega t_{min}}} = e^{-\zeta\omega\,\Delta t} = e^{-\zeta\pi/\sqrt{1-\zeta^2}} \qquad (4\text{-}27)$$

The ratio of successive positive deviations is called the "decay ratio,"

$$\text{Decay ratio} = e^{-2\pi\zeta/\sqrt{1-\zeta^2}} \qquad (4\text{-}28)$$

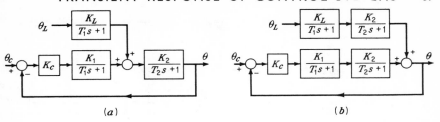

FIGURE 4-7 Proportional control of a two-capacity process (load change before first time constant).

The equations for the decay ratio and the frequency of the damped oscillation are the same as those given in Chap. 3 for the open-loop response of an underdamped second-order system. The main difference between the two cases is that the peak value is less than twice the final value for the open-loop system, whereas a much higher ratio is possible for the closed-loop system of Fig. 4-5. The peak overshoot of the closed-loop system depends on the ratio T_2/T_1 as well as on the damping coefficient. If T_2 is relatively small, the peak overshoot approaches $\theta_L K_L$, since the corrective action taken by the controller is delayed by the large time constant T_1. If the second time constant were much larger than the first, the peak error would not exceed twice the final offset. If the load change occurs at the start of the process, as shown in Fig. 4-7, the peak error is always less than twice the final offset.

LOAD CHANGE BEFORE FIRST ELEMENT

Figure 4-7 shows two equivalent block diagrams for the case where the load change occurs at the first element of a two-capacity process. For the jacketed-kettle example, such a load change might be a change in the cooling-water temperature, if the controller adjusts the flow rate of the coolant. The transient response is obtained by using Fig. 4-7b and Eq. (4-1).

$$\frac{\theta}{\theta_L} = \frac{G_L}{1+G} \quad \text{where } G_L = \frac{K_L K_2}{(T_1 s + 1)(T_2 s + 1)} \tag{4-29}$$

$$\frac{\theta}{\theta_L} = \frac{K_L K_2}{(T_1 s + 1)(T_2 s + 1) + K_c K_1 K_2} \tag{4-30}$$

$$\frac{\theta}{\theta_L} = \frac{K_L K_2}{K + 1} \left(\frac{1}{\dfrac{T_1 T_2}{K + 1} s^2 + \dfrac{T_1 + T_2}{K + 1} s + 1} \right) \tag{4-31}$$

where $\qquad K = K_c K_1 K_2$

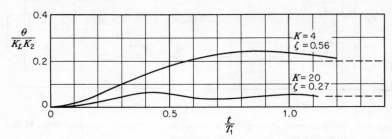

FIGURE 4-8   Response to step change in load for system of Fig. 4-7 when $T_2 = 0.25T_1$.

Equation (4-31) has the same form as the general equations for an open-loop second-order process, and, except for the constant term, the solutions are given in Eqs. (3-62) to (3-64). Only the solution for the underdamped case and a step change in input is repeated here.

$$\theta = \frac{K_L K_2}{K+1}\left[1 + \frac{e^{-\zeta\omega t}}{\sqrt{1-\zeta^2}}\sin(\sqrt{1-\zeta^2}\,\omega t - \phi)\right] \qquad (4\text{-}32)$$

where $\qquad \phi = \dfrac{\tan^{-1}\sqrt{1-\zeta^2}}{-\zeta}$

$$\text{Time at max deviation} = t_{max} = \frac{\pi}{\omega\sqrt{1-\zeta^2}} \qquad (4\text{-}33)$$

$$\frac{\text{Max deviation}}{\text{Final deviation}} = 1 + e^{-\zeta\pi/\sqrt{1-\zeta^2}} \qquad (4\text{-}34)$$

The damping coefficient, the frequency, and the decay ratio are the same as for the system of Fig. 4-5, since they depend only on the elements in the closed loop. Typical curves are shown in Fig. 4-8.

Increasing the controller gain for a system with two time constants decreases the offset and the peak error. Increasing the gain also decreases the integral of the absolute error (based on the steady-state value), since the increased frequency and the lower peak error more than offset the effect of a greater number of oscillations. Thus the best controller gain based on the step response might seem to be the highest possible value. However, the response to periodic disturbances is poor if the gain is very high (see Chap. 7). Furthermore, real systems have other small lags or time delays which lead to instability at very high gains. In practice the controller gain is therefore chosen to give a reasonable damping coefficient, say, 0.2 to 0.3, if only the two largest time constants of the system are known.

4-6. PROPORTIONAL CONTROL OF A THREE-CAPACITY PROCESS

If a process has three first-order elements in series and a load change occurs after the first element (see Fig. 4-9), the transfer function is

$$\frac{\theta}{\theta_L} = \frac{G_L}{1 + G} = \frac{K_L K_3}{(T_2 s + 1)(T_3 s + 1)}$$

$$\times \frac{(T_1 s + 1)(T_2 s + 1)(T_3 s + 1)}{T_1 T_2 T_3 s^3 + (T_1 T_2 + T_1 T_3 + T_2 T_3)s^2 + (T_1 + T_2 + T_3)s + 1 + K}$$

$$(4\text{-}35)$$

where $K = K_c K_1 K_2 K_3$.

To solve this equation, the cubic term must be factored into a linear term plus a quadratic term (the underdamped case) or into three linear terms (the overdamped case). When a unit step change is made in θ_L, the solution for the underdamped case is obtained from the following transform:

$$\mathcal{L}\left[\frac{as + 1}{s(Ts + 1)(s^2/\omega^2 + 2\zeta s/\omega + 1)}\right]$$

$$= 1 + \frac{1}{\sqrt{1 - \zeta^2}}\left(\frac{1 - 2a\zeta\omega + a^2\omega^2}{1 - 2T\zeta\omega + T^2\omega^2}\right)^{\!\!1/2} e^{-\zeta\omega t} \sin\left(\omega\sqrt{1 - \zeta^2}\, t + \phi\right)$$

$$+ \frac{T\omega^2(a - T)e^{-t/T}}{1 - 2T\zeta\omega + T^2\omega^2} \quad (4\text{-}36)$$

where $\phi = \tan^{-1}\dfrac{a\omega\sqrt{1 - \zeta^2}}{1 - a\zeta\omega} - \tan^{-1}\dfrac{T\omega\sqrt{1 - \zeta^2}}{1 - T\zeta\omega} - \tan^{-1}\dfrac{\sqrt{1 - \zeta^2}}{-\zeta}$

The step response in general terms is

$$\theta = \frac{K_L K_3}{1 + K}\left(1 + Be^{-\zeta\omega t}\sin\alpha + Ce^{-t/T}\right) \quad (4\text{-}37)$$

Figure 4-10 shows typical response curves for the case where the three time constants are equal. As the gain is increased, the peak height and

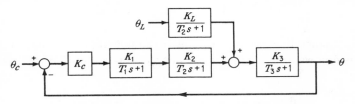

FIGURE 4-9 Proportional control of a three-capacity process.

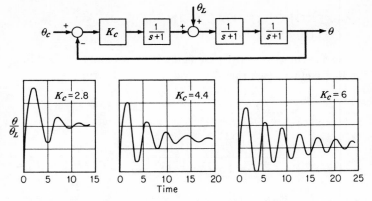

FIGURE 4-10   Transient response of a three-capacity system.

final offset are reduced, but the oscillations take longer to damp out.   At a gain of 8, $\zeta = 0$, and undamped oscillations result from the change in load; with a slightly higher gain ζ is negative, and the oscillations become larger with each cycle.   The system is inherently unstable for gains greater than 8, and any slight change in load or set point leads to continuous cycling limited only by the maximum and minimum controller output.

4-7. STABILITY LIMIT

A closed-loop system with three or more first-order elements in series is unstable if the overall gain exceeds a certain value.   A physical explanation for instability is given in the chapter on frequency response.   In this chapter the mathematical reason for instability is given, and rules for simple systems that would be stable in the open loop are developed.   The more general stability criteria of Routh and of Nyquist are presented in the Appendix.

The transfer functions for the closed-loop response to changes in set point or load were shown to be

$$\frac{\theta}{\theta_c} = \frac{G}{1 + G} \qquad \frac{\theta}{\theta_L} = \frac{G_L}{1 + G} \tag{4-1}$$

For any change in θ_c or θ_L, the equation for θ can be written in general terms as follows, where n and m are integers and A and B are constants:

$$\theta = \frac{A_m s^m + A_{m-1} s^{m-1} + \cdots + A_1 s + A_0}{B_n s^n + B_{n-1} s^{n-1} + \cdots + B_1 s + B_0} \tag{4-38}$$

The denominator can be factored to show the roots of the polynomial, which may be real, imaginary, or complex numbers or zero.

$$\theta = \frac{A_m s^m + A_{m-1} s^{m-1} + \cdots + A_1 s + A_0}{(s - r_1)(s - r_2) \cdots (s - r_n)} \tag{4-39}$$

The inverse transformation yields terms containing $e^{r_1 t}, e^{r_2 t}, \ldots, e^{r_n t}$. If r_1 is a real negative number, the term containing $e^{r_1 t}$ decays to zero. If r_1 and r_2 are a pair of complex numbers, the corresponding terms are $e^{(x \pm jy)t}$, which represent a sine wave that also decays to zero if the real part x is negative. If any of the roots is positive or has a positive real part, the term containing that root eventually becomes very large and the system is unstable.

A control system is stable if the denominator of the closed-loop transfer function has no positive roots or no roots with positive real parts. The denominator of the transfer function set equal to zero is the characteristic equation. The maximum gain can be found by factoring the characteristic equation for various gains or by determining the conditions that just make one of the roots positive.

For proportional control of a process with three time constants, the denominator of the closed-loop transfer function is

$$K + (T_1 s + 1)(T_2 s + 1)(T_3 s + 1)$$

which can be factored into a first-order term and a quadratic term. When the gain exceeds the critical value, the damping coefficient in the quadratic term becomes negative or the two complex roots acquire positive real parts. The maximum gain is found by setting $\zeta = 0$ in the following equation:

$$K + (T_1 s + 1)(T_2 s + 1)(T_3 s + 1) = (Ts + 1)\left(\frac{s^2}{\omega^2} + \frac{2\zeta}{\omega} s + 1\right)(K + 1) \tag{4-40}$$

By equating coefficients and using the ratios of time constants, the maximum gain and critical frequency are shown to be

$$1 + K_{max} = (1 + R_2 + R_3)\left(1 + \frac{1}{R_2} + \frac{1}{R_3}\right) \tag{4-41}$$

$$\omega_c = \frac{1}{T_1} \sqrt{\frac{1 + K_{max}}{R_2 + R_3 + R_2 R_3}} \tag{4-42}$$

where $R_2 = \dfrac{T_2}{T_1}$

$R_3 = \dfrac{T_3}{T_1}$

The stability of the system depends on the overall gain and the ratios of the time constants. The maximum gain becomes very large if any one of the three time constants is either much larger or much smaller than the others.

4-8. INTEGRAL CONTROL

With integral control the output of the controller is proportional to the integral of the error.

$$p = \frac{1}{T_i} \int e \, dt \qquad (4\text{-}43)$$

$$\frac{P}{E} = \frac{1}{T_i s} \qquad (4\text{-}44)$$

where T_i = integral time.

There is no offset with integral control, since the controller output keeps changing until the error is reduced to zero. Integral control is also called "proportional-speed floating control," since the rate of change of controller output is proportional to the error. When integral control is combined with proportional control, the term "reset action" is often used, since adding integral action is equivalent to manually resetting the set point after each change in load. Integral action is usually used along with other control modes, and only the response of a simple process is shown here for the case of pure integral control.

The following equations give the response to a step change in load for a single-capacity process with integral control. The equations can be compared with Eqs. (4-7) to (4-10) for proportional control of the same process.

$$\frac{\theta}{\theta_L} = \frac{G_L}{1 + G} = \frac{K_L}{Ts + 1} \frac{T_i s (Ts + 1)}{T_i T s^2 + T_i s + K} \qquad (4\text{-}45)$$

$$\frac{\theta}{\theta_L} = \frac{K_L T_i}{K} \frac{s}{(T_i T/K)s^2 + (T_i/K)s + 1} \qquad (4\text{-}46)$$

For a unit step change in θ_L, $\theta_L = 1/s$.

$$\theta = \frac{K_L T_i}{K} \mathcal{L}^{-1} \left[\frac{1}{(T_i T/K)s^2 + (T_i/K)s + 1} \right] \qquad (4\text{-}47)$$

The shape of the response curve depends on the damping coefficient in the quadratic term. The solution for the underdamped case is

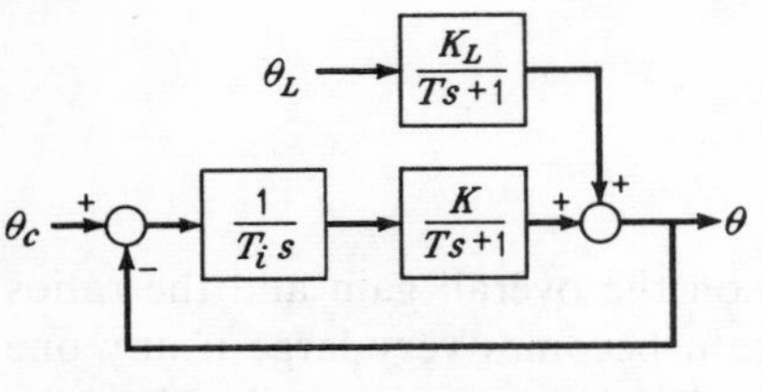

$$\theta = \frac{K_L T_i}{K} \left(\frac{\omega}{\sqrt{1 - \zeta^2}} e^{-\zeta \omega t} \right.$$

$$\left. \times \sin \omega \sqrt{1 - \zeta^2}\, t \right) \qquad (4\text{-}48)$$

where $\omega = \sqrt{K/T_i T}$

$$\zeta = \frac{1}{2} \sqrt{\frac{T_i}{TK}}$$

FIGURE 4-11   Integral control of a single-capacity process.

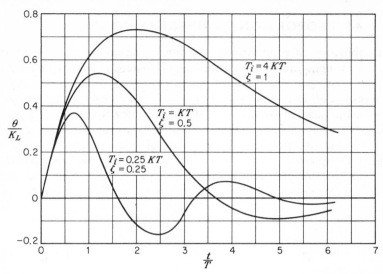

FIGURE 4-12   Step response of a single-capacity process with integral control.

With critical damping, $\zeta = 1$ or $T_i = 4KT$.

$$\theta = K_L \frac{t}{T} e^{-t/2T} \qquad \text{for } \zeta = 1 \tag{4-49}$$

The effect of the damping coefficient on the response is shown in Fig. 4-12. The curves are similar to those for proportional control of a two-capacity process (Fig. 4-6), except that there is no offset with integral control. Decreasing the reset time lowers the damping coefficient and makes the oscillations more pronounced. In spite of this, the integral of the absolute error decreases with decreasing reset time, since the lower peak height and higher frequency more than compensate for the increased number of oscillations. Based on the criterion of a minimum integrated error, the lowest possible reset time would theoretically be best. However, systems with two or more time constants would show unstable oscillations at very low reset times. Rather than trying to allow exactly for the small additional lags that are always present in a real system, the integral time could be chosen to give a reasonable decay ratio based on the major time constant. For a decay ratio of 0.25, the damping coefficient is 0.22 [Eq. (4-28)], and the reset time is calculated from

$$T_i = 0.19TK \tag{4-50}$$

As mentioned in the discussion of proportional control, the purpose of level or pressure control systems may be to minimize the rate of change of flow out of a tank. Integral control adjusted for critical damping or overdamping might be satisfactory for these applications.

4-9. PROPORTIONAL-INTEGRAL CONTROL

The combination of proportional and integral control is very widely used, since a two-mode controller is only slightly more expensive than a proportional controller and the reset action completely or almost completely eliminates the offset following load changes. (The offset would be reduced to zero with most electronic controllers or a few pneumatic controllers which have ideal integral action. With most pneumatic controllers, an offset of about one per cent of the potential value remains after a load change, as explained in Sec. 6-4.) In most electronic and pneumatic controllers, the integral action varies with the controller gain as well as with the adjustable integral time, and the characteristic equation is

Change in output = proportional action + integral action

$$p \quad = \quad K_c e \quad + \quad \frac{K_c}{T_R} \int e \, dt \qquad (4\text{-}51)$$

or

$$\frac{P}{E} = K_c \left(1 + \frac{1}{T_R s} \right) = \frac{K_c(T_R s + 1)}{T_R s}$$

The reset time T_R can be defined as the time required for the output due to integral action to just equal the output due to proportional action following a step change in error. This definition is the basis for an easy method of calibrating the controller. With the feedback loop opened, a constant error is imposed on the system, and the time for the initial corrective action to be doubled is measured. In some older controllers, the reset or integral rate is expressed in min^{-1}, the reciprocal of the reset time. A controller could be designed to have the integral action independent of the gain, but there is no obvious advantage to such a combination.

SINGLE-CAPACITY PROCESS

The equation for proportional-integral control of a single-capacity process is given below. Although most such processes do not need reset action because a high gain is used, this is the only case for which direct

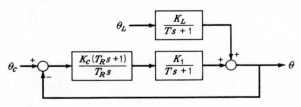

FIGURE 4-13 Proportional-integral control of a single-capacity process.

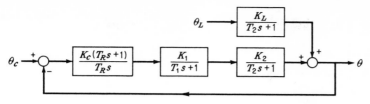

FIGURE 4-14   Proportional-integral control of a two-capacity process.

solutions can be written without factoring cubic or higher-order terms.

$$\frac{\theta}{\theta_L} = \frac{G_L}{1 + G} = \frac{K_L}{Ts + 1} \frac{T_R s(Ts + 1)}{(T_R s)(Ts + 1) + K(T_R s + 1)}$$

$$= \frac{K_L T_R}{K} \frac{s}{\dfrac{T_R T}{K} s^2 + T_R \dfrac{K + 1}{K} s + 1} \qquad K = K_c K_1 \qquad (4\text{-}52)$$

The response is almost the same as with integral control alone [Eq. (4-46)], but the damping coefficient is increased by a factor $K + 1$.

$$\omega = \sqrt{\frac{K}{T_R T}}$$

$$\zeta = \frac{K + 1}{2} \sqrt{\frac{T_R}{TK}} \qquad (4\text{-}53)$$

Equation (4-53) shows that the damping coefficient goes through a minimum with increasing values of the gain, which is rather unusual.   In most systems, increasing the gain leads to lower damping coefficients and eventually to unstable oscillations.

TWO-CAPACITY PROCESS

The equations for proportional-integral control of a two-capacity process are of the same complexity as those for porportional control of a three-capacity process; a third-order term must be factored to secure the transient response.

$$\frac{\theta}{\theta_L} = \frac{K_L}{T_2 s + 1} \frac{T_R s(T_1 s + 1)(T_2 s + 1)}{K(T_R s + 1) + T_R s(T_1 s + 1)(T_2 s + 1)}$$

$$= \frac{K_L}{K} \frac{T_R s(T_1 s + 1)}{\dfrac{T_R T_1 T_2}{K} s^3 + \dfrac{T_R(T_1 + T_2)}{K} s^2 + \dfrac{K + 1}{K} T_R s + 1} \qquad (4\text{-}54)$$

Generally, the controller settings are chosen to give an underdamped response, and for a step change in θ_L the appropriate transform is

$$\mathcal{L}^{-1}\left[\frac{as + 1}{(Ts + 1)\left(\dfrac{s^2}{\omega^2} + \dfrac{2\zeta}{\omega} s + 1\right)}\right]$$

$$= \frac{\omega}{\sqrt{1 - \zeta^2}}\left(\frac{1 - 2a\zeta\omega + a^2\omega^2}{1 - 2T\zeta\omega + T^2\omega^2}\right)^{\!\!\frac{1}{2}} e^{-\zeta\omega t}\sin\left(\omega\sqrt{1 - \zeta^2}\,t + \phi\right)$$

$$+ \frac{(T - a)\omega^2}{1 - 2T\zeta\omega + T^2\omega^2}e^{-t/T} \quad (4\text{-}55)$$

where
$$\phi = \tan^{-1}\frac{a\omega\sqrt{1 - \zeta^2}}{1 - a\zeta\omega} - \tan^{-1}\frac{T\omega\sqrt{1 - \zeta^2}}{1 - T\zeta\omega}$$

The limiting controller settings for stable operation can be found by setting the damping coefficient equal to zero or by applying the Routh criterion (Appendix 2).

$$T_{R,\min} = \frac{K}{K + 1}\frac{T_1 T_2}{T_1 + T_2} \quad (4\text{-}56)$$

Since the overall gain would usually be greater than 1.0, the minimum reset time is slightly less than the smaller of the two time constants. Equation (4-56) gives the limiting reset time rather than the maximum gain, since, for some cases, there is no limit to the gain that can be used. If the reset time is greater than the group $T_1 T_2/(T_1 + T_2)$, the control system will never show unstable oscillations.

The reason why a limiting gain exists in some cases and not in others is best explained by using the frequency-response methods presented in later chapters. A brief summary of the analysis is given here. A system can be unstable at high gains if the phase lag can exceed 180°. The phase lag for a single time constant is zero at low frequencies and approaches 90° at high frequencies; so at least three time constants must be included in the control loop to demonstrate instability with proportional control. With pure integral control, the controller introduces 90° additional lag, and thus systems with only two time constants will be unstable for high integral rates. For a system with two time constants and proportional-integral control, the phase lag of the controller changes from 90 to 0° with increasing frequency, and a phase lag of 180° is possible only for certain values of the integral time.

With three capacities and a proportional-integral controller in the control loop, the equations for the transient response become too complex to present here. The inverse transforms are tabulated, but, at least for

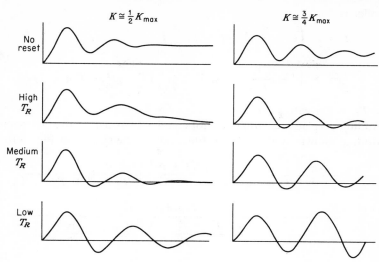

FIGURE 4-15   Effect of gain and reset time for a process with three time constants.

process control problems, it is rarely worth the effort to factor the fourth-order equation. If the exact solution is needed for several controller settings, the response is usually obtained with an analog computer. If only approximate solutions are needed, the transient response can be predicted from the closed-loop frequency response, as shown later. Another approach is to neglect the smallest time constant for transient calculations, once reasonable values of the controller settings have been determined from frequency-response analysis for the complete process. The approximate transient response should not be used to get optimum controller settings, since a two-capacity process may not show unstable oscillations at high gains.

The general effects of changing the gain and changing the reset time for a process with three time constants are shown in Fig. 4-15, where K_{max} is the maximum gain for just proportional action. With reset action added, the system becomes unstable at a lower gain than with only proportional control, and the stability decreases with decreasing reset time. However, a moderate reset rate is needed to return the system to the control point in a reasonable time. Methods of determining the optimum settings are discussed in Chap. 9.

4-10. DERIVATIVE ACTION

When derivative control is used, it is usually added to proportional control or to proportional-integral control. The derivative action varies directly with the gain setting, and the transfer function for an ideal three-mode controller is

$$p = K_c \left(e + \frac{1}{T_R} \int e \, dt + T_D \frac{de}{dt} \right)$$

$$\frac{P}{E} = K_c \left(1 + \frac{1}{T_R s} + T_D s \right) = \frac{K_c}{T_R s} (T_D T_R s^2 + T_R s + 1)$$

(4-57)

where T_D = derivative time.

Adding derivative action does not change the order of the characteristic equation, as shown by the transfer function for the system of Fig. 4-16.

$$\frac{\theta}{\theta_L} = \frac{K_L}{T_2 s + 1} \frac{T_R s(T_1 s + 1)(T_2 s + 1)}{T_R s(T_1 s + 1)(T_2 s + 1) + K(T_D T_R s^2 + T_R s + 1)}$$

or

$$\frac{\theta}{\theta_L} = \frac{K_L}{K} \frac{T_R s(T_1 s + 1)}{\dfrac{T_R T_1 T_2}{K} s^3 + \dfrac{T_R(T_1 + T_2 + KT_D)}{K} s^2 + \dfrac{K + 1}{K} T_R s + 1}$$

(4-58)

This equation is like Eq. (4-54), and the corresponding transform, Eq. (4-55), is applicable. Derivative action increases the stability of the system and permits either a higher gain or a lower reset time to be used. The minimum reset time for stable operation is

$$T_{R,\min} = \frac{K}{K + 1} \frac{T_1 T_2}{T_1 + T_2 + KT_D}$$

(4-59)

[compare with Eq. (4-56)]. The critical frequency of the system can be calculated from the maximum gain and the values of the two time constants.

$$\omega_c = \sqrt{\frac{K_{\max} + 1}{T_1 T_2}}$$

(4-60)

Although Eq. (4-60) does not include T_R or T_D, increasing either the reset time or the derivative time increases the maximum gain and thus the critical frequency. If the derivative time is large enough, the system would theoretically be stable for all gains.

Derivative action also increases the stability of systems with three or more capacities, and with ideal control a three-capacity system could be

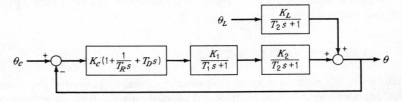

FIGURE 4-16   Proportional-integral-derivative control of a two-capacity process.

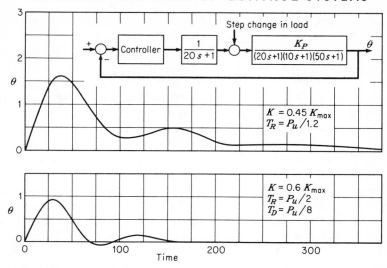

FIGURE 4-17   Effect of derivative action on the response to a step in load.

stabilized for all gains by using a high reset time and a derivative time greater than the smallest time constant.   Although no controller has ideal behavior, adding derivative action to a system containing mainly first-order elements does greatly improve the controllability.   Typical response curves for a four-capacity process are shown in Fig. 4-17.   The curves were obtained with a pneumatic analog and a Taylor Transcope controller using the recommended settings given in Chap. 9.   Note that both the peak error after an upset and the period of oscillation are reduced.   Derivative action makes the error integral, $\int |e| \, dt$, about 3.7-fold lower.

There is relatively little benefit in using derivative action if the system has a large dead time; whether or not derivative action is worthwhile for these cases is best determined using frequency-response analysis.   Derivative action is usually omitted for very fast systems such as flow control or pressure control, since these systems usually have noisy signals and the noise is amplified by derivative action (see Chap. 13).

THE START-UP PROBLEM

Derivative action can help to reduce overpeaking on start-up of batch processes.   Consider a reactor where the temperature must be increased rapidly from 20 to 80°C and then held at 80°C to complete the reaction. Assume the best settings for constant-temperature operation are a bandwidth of 6°C and a reset time of 5 min.   If the run is started with these settings, the large initial error causes the controller to saturate; i.e., the output is constant at the maximum value, and the control valve is wide open.   If only proportional action were used, the controller would stay saturated until the temperature reached the lower limit of the control band, or about 77°C for this example. However, in a proportional-integral controller, the cumulative effect of the error integral, $K_c/T_R \int e\, dt$, is to prolong the saturation period, and the control valve does not close appreciably until the set-point temperature is reached or exceeded. Because of the lags in the system, the maximum temperature reached may be several degrees above the set point.

To avoid overpeaking, the valve must start to close at a temperature a few degrees below the set point; the proper value depends on the system lags and on the rate of change of temperature near the set point. This anticipatory action can be obtained with a cascade-type controller, in which the signal is sent first to a proportional-derivative stage and then to an integral stage. The signal to the integral stage and the controller output start to change one derivative time before the time when the uncontrolled temperature would reach the set point (1). Unfortunately, in most commercial pneumatic controllers (such as the one in Fig. 6-9) the integral and derivative sections are in parallel, and the controller stays saturated until the set point is reached; thus derivative action comes into play too late to have much effect on the start-up.

4-11. EFFECT OF MEASUREMENT LAG

In the previous cases, the output of the process was fed back to the controller without any intermediate lag or time delay. In some cases the measurement lag is one of the significant time constants, and the system diagram in Fig. 4-18 is used.

The measurement lag may be a first-order system such as a thermometer bulb, $G_m = 1/(T_m s + 1)$, a time delay, $G_m = e^{-Ls}$, or a combination of elements. The general equation for the output is

$$\theta = \theta_L G_L + (\theta_c - \theta_m)G \tag{4-61}$$

Since $\theta_m = \theta G_m$,

$$\theta = \theta_L G_L + \theta_c G - \theta G G_m$$

For load changes

$$\frac{\theta}{\theta_L} = \frac{G_L}{1 + G G_m} \tag{4-62}$$

For set-point changes

$$\frac{\theta}{\theta_c} = \frac{G}{1 + G G_m} \tag{4-63}$$

Note that θ is the actual value of the output variable, not the value that would be recorded or indicated by the controller. The equation for the measured response is

$$\frac{\theta_m}{\theta_L} = \frac{G_L G_m}{1 + G G_m} \tag{4-64}$$

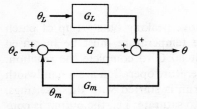

FIGURE 4-18   Process with a measurement lag.

Example 4-1

To take a simple example, consider proportional control of a single-capacity process with a first-order measurement lag.

$$G = \frac{K}{Ts + 1} \qquad G_L = \frac{K_L}{Ts + 1} \qquad G_m = \frac{1}{T_m s + 1}$$

From Eq. (4-62)

$$\frac{\theta}{\theta_L} = \frac{K_L}{Ts + 1} \frac{(Ts + 1)(T_m s + 1)}{TT_m s^2 + (T + T_m)s + K + 1}$$

$$= \frac{K_L}{K + 1} \frac{T_m s + 1}{\dfrac{TT_m s^2}{K + 1} + \dfrac{(T + T_m)s}{K + 1} + 1} \tag{4-65}$$

Equation (4-65) is like Eq. (4-15), which gives the transfer function for load changes occurring between the elements of a two-capacity process. The solution for the underdamped case [Eq. (4-17)] and the typical curves of Fig. 4-6 show that the peak overshoot following a load change may be several times the final value if T_m is large compared with T. [T_m is equivalent to T_1 in Eq. (4-16) and Fig. 4-6.] The *measured* overshoot is less than twice the final value, since the equation for θ_m/θ_L is of the same form as that for a change in set point. When the measurement lag and the process lag are equal and the controller gain is adjusted to give a decay ratio of 0.25, the corresponding equations for a unit change in load are

$$\theta = 0.046\left[1.0 + 4.65e^{-t/T}\sin\left(261°\frac{t}{T} - 12°\right)\right]$$

$$\theta_m = 0.046\left[1.0 + 1.02e^{-t/T}\sin\left(261°\frac{t}{T} - 102°\right)\right]$$

These equations are plotted in Fig. 4-19.

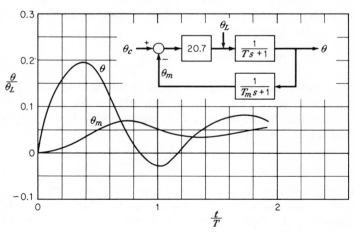

FIGURE 4-19   Effect of measurement lag on the transient response (Example 4-1 with $T = T_m$).

The frequency of the damped oscillations and the decay ratio are the same for both θ and θ_m, but, as shown by Example 4-1, the actual overshoot may be several times the measured overshoot. When the measurement lag is the same order of magnitude as the other lags in a control system, the measured output is not a reliable measure of the control-system performance. In some cases apparent improvements in controllability result from increasing the measuring lag, but the actual error is usually as great as or greater than before.

4-12. EFFECT OF TIME DELAY

In many process control systems, there is a time delay between elements of the control loop. A time delay exists if the temperature of a stream leaving a heat exchanger is measured a few feet downstream from the end of the exchanger. Analytical instruments for reactors or distillation columns are usually placed in a separate sample loop, and there is a time delay before changes in system concentrations are transmitted to the instrument cell. With plug-flow reactors, the time delay is equal to the residence time in the reactor. The terms dead time and transportation lag also refer to a pure time delay. For systems with distributed resistance and capacity or systems with many time constants in series, the initial response to a step change is sometimes imperceptible; while these systems have no true time delay, the response can be approximated by using an effective time delay plus one or more time constants.

A time delay in the control loop is represented by a block with the transfer function e^{-Ls}. Frequency-response analysis shows that a time delay has a destabilizing effect on the control systems, more so than adding a time constant of the same magnitude. The transient response of the system is not easily obtained, because the number of roots is infinite, as shown by expansion of the exponential term.

$$e^{-Ls} = 1 - Ls + \frac{(Ls)^2}{2!} - \frac{(Ls)^3}{3!} + \cdots \qquad (4\text{-}66)$$

The approximate transient response could be obtained by using a few terms from the above series, but other approximations for the time delay are easier to handle. The following Pade approximant (2) is simple to use, and often accurate enough.

$$e^{-Ls} \cong \frac{1 - Ls/2}{1 + Ls/2} \qquad (4\text{-}67)$$

Another method of allowing for time delay is given by Cohen and Coon (3). They calculated response curves for various controller modes after neglecting the higher harmonics, whose amplitudes were only 4 to 8 per cent of the fundamental. If accurate response curves for several conditions are needed, an analog computer with a magnetic-tape delay should be used; if only one solution is needed, a stepwise calculation of the type shown in Example 4-2 may be simpler.

Example 4-2

A process has a time delay of 1 min, followed by a first-order lag of 1 min. The measurement, controller, and valve lags are negligible. For a gain of 1.2, about half the maximum value, calculate the response to a step change in load occurring after the time delay.

Approximate Solution

Let

$$e^{-Ls} = \frac{1 - 0.5s}{1 + 0.5s}$$

$$\frac{\theta}{\theta_L} = \frac{G_L}{1 + G} = \frac{K_L}{K + 1} \frac{(s + 1)(1 + 0.5s)}{(s + 1)(1 + 0.5s) + 1.2(1 - 0.5s)}$$

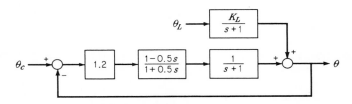

FIGURE E4-2*a*   Diagram for Example 4-2.

For $\theta_L = \dfrac{1}{s}$, $\theta = \dfrac{K_L}{2.2} \mathcal{L}^{-1}\left[\dfrac{1}{s}\dfrac{1 + 0.5s}{(0.227s^2 + 0.409s + 1)}\right]$

From Eq. (4-17), $\theta = \dfrac{K_L}{2.2}[1 + Ae^{-\zeta \omega t}\sin(\omega\sqrt{1 - \zeta^2}\,t + \phi)]$

where $\zeta = 0.43$
$\omega = 2.10 \text{ min}^{-1}$
$\phi = \tan^{-1}1.74 - \tan^{-1}(-2.23)$
$= -120° + 65° = -55°$
$A = 1.21$
$\theta = \dfrac{K_L}{2.2}[1 + 1.21e^{-0.86t}\sin(110°t - 55°)]$

The solution is plotted in Fig. 4-20.

Stepwise Solution

For the first minute, the input to the time constant is fixed, and the response is exponential.

$$y = 1 - e^{-t/T} = 1 - e^{-t}$$

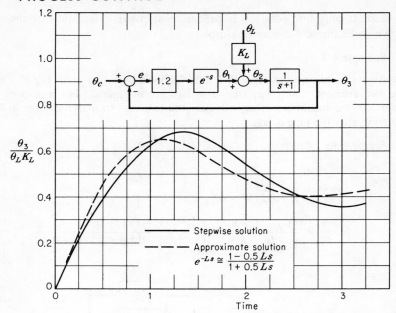

FIGURE 4-20   Response of a process with dead time to a step change in load (Example 4-2).

After 1 min the equation used is

$$T\frac{d\theta_3}{dt} = \theta_2 - \theta_3$$

$$\Delta\theta_3 = \frac{\Delta t}{T}(\theta_2 - \theta_3)_{av}$$

where $\theta_2 = 1 - 1.2(\theta_3$ at $t - 1$ min$)$.

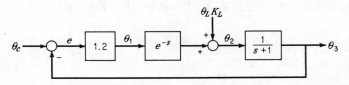

FIGURE E4-2b   Diagram for Example 4-2.

Fairly large time increments can be used if the calculated change in θ_3 is based on the average value of $\theta_2 - \theta_3$ over the increment, and the solution takes only an hour or two.   A portion of the solution is shown in Table 4-1.

The approximate solution comes fairly close to the more exact stepwise solution even though the step response of the Pade approximant is far from that of a true time delay.   If the time delay were the last element in the system, the predicted response to a step-load change would have been more in error.

Table 4-1

Time	θ_1	θ_2	θ_3	$\theta_2 - \theta_3$	$(\theta_2 - \theta_3)_{av}$	$\Delta\theta_3$	θ_3
0	0	1	0				
0.2	0	1	0.181				
0.4	0	1	0.329				
0.6	0	1	0.449				
0.8	0	1	0.550				
1.0	0	1	0.632	0.368			
1.2	−0.217	0.783	∼0.68	0.103	0.235	0.047	0.679
1.4	−0.395	0.605	∼0.685	−0.08	0.012	0.002	0.681
1.6	−0.538	0.462	∼0.65	−0.188	−0.134	−0.027	0.654
1.8	−0.660	0.340	∼0.60	−0.26	−0.224	−0.045	0.609
			∼0.61	−0.27	−0.229	−0.046	0.608
2.0	−0.757	0.243	∼0.55	−0.307	−0.287	−0.057	0.551

PROBLEMS

1. The level in a storage tank is controlled by throttling the flow of fluid to the tank. Fluid is pumped from the tank by a constant-displacement pump, and the flow does not depend on the level in the tank.

a. Show that the system will oscillate continuously if an integral controller is used.

b. Would the system be unstable if a proportional-reset controller were used?

2. The major elements of a process control system are given below. The time constants are in minutes. Calculate the transient response for a step change in load if the controller gain is half the maximum value.

Valve:
$$G_v = \frac{4}{0.1s + 1}$$

Process:
$$G_P = \frac{1.5}{(2s + 1)(10s + 1)}$$

Transfer function for load changes:
$$G_L = \frac{3}{10s + 1}$$

Controller:
$$G_c = K_c$$

3. A process has two major time constants and is controlled with a proportional controller.

$$G_1 = \frac{0.5}{2s + 1} \qquad G_2 = \frac{5}{1s + 1} \qquad \text{(times in minutes)}$$

a. What is the recommended controller gain for a damping coefficient of 0.3?

b. Sketch the response to a step change in set point and to a step change in load made before the first element. Give numerical values for the frequency, peak overshoot, decay ratio, and steady-state value.

c. Repeat (*b*) for load change made between the first and second element. (Assume $K_L = 1$ for both load changes.)

4. A closed-loop system has time constants of 1 min and 10 min and a proportional controller. Obtain the response to a ramp change in set point at a controller gain that gives a damping coefficient of 0.3. Would a higher or lower gain be advantageous?

5. Compare the step response of a true time delay with that of the Pade approximant given in Eq. (4-67).

REFERENCES

1. Caldwell, W. I., G. A. Coon, and L. M. Zoss: "Frequency Response for Process Control," p. 227, McGraw-Hill Book Company, New York, 1959.
2. Truxal, J. G.: "Automatic Feedback Control System Synthesis," p. 550, McGraw-Hill Book Company, New York, 1955.
3. Cohen, G. H., and G. A. Coon: Theoretical Consideration of Retarded Control, *Trans. ASME,* **75**:827 (1953).

5 FREQUENCY-RESPONSE ANALYSIS

Frequency response means the response of a system or a single element to sinusoidal inputs covering a wide range of frequencies. The advantage of using the frequency response to analyze or design control systems is that the system response is obtained easily from the response of the individual elements, no matter how many elements are included. By contrast, calculations of the transient response are quite tedious with only three components in the system and are too difficult to be worthwhile for four or more components. Even if an analog computer is used to get the exact transient response, it is often worthwhile to calculate the frequency response as a check and as a quick means of focusing attention on the most important cases.

The basic results obtained from a frequency-response analysis are values of the maximum gain for stable operation and the critical frequency of the system. From these two numbers, reasonable values of the three controller parameters can be predicted (slightly better predictions of the optimum settings are possible by considering in addition the slopes of the frequency-response curves at the critical frequency). The critical frequency is also important as a measure of the speed of response, since the frequency of the damped oscillations at reasonable controller gains is usually 0.7 to 0.9 times the critical frequency. In many cases, the optimum settings and the speed of response are all that is needed to compare proposed systems or to judge the merit of a proposed change in an existing system. In general, any change that either doubles the permissible gain or doubles the critical frequency can be considered a twofold improvement in controllability, since the error integral following a load change varies almost inversely with the product of the maximum gain and the critical frequency [see Eq. (5-26)].

The maximum gain and the critical frequency are obtained from the *open-loop* frequency response, which is the response of the system with the control loop broken at some point. Usually the loop is broken after the controller, as shown in Fig. 5.1, so that the input is a sinusoidal signal to the control valve, and the output is either the signal from the controller or the process output, which is the input to the controller. For the examples in this chapter, only ideal proportional control is considered, and the controller output bears a constant relationship to the controller input; the

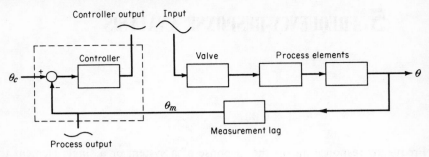

FIGURE 5-1   Open-loop frequency response.

amplitude is greater by the gain factor K_c, and the output lags the input by 180°. Since the controller gain is often undetermined at the start of a problem, it is more convenient to use the process output for the frequency-response plots.

5-1. FREQUENCY RESPONSE OF A FIRST-ORDER SYSTEM

When a sinusoidal input is applied to a linear system, the output has steady-state and transient terms [Eq. (3-13)].   After a few cycles the transient dies out, and the output is a sine wave of the same frequency.   This eventual response can be characterized by the amplitude ratio and the phase angle. Typical curves for a first-order process are shown in Fig. 5-2.   At low frequency, the output is almost equal to the input, and there is only a slight phase lag.   At high frequencies the fluctuations in the input are severely damped because of the capacity in the system, and the output lags the input by almost 90°.

The equations for the response of a first-order system were given in

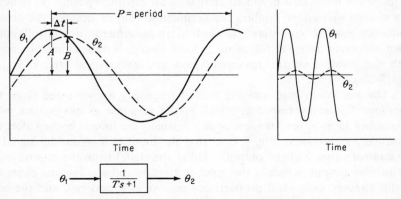

FIGURE 5-2   Response of a first-order system to sinusoidal inputs.

Chap. 3 [Eq. (3-13)].   The steady-state portion of the solution for $\theta_1 = A \sin \omega t$ is

$$\theta_2 = \frac{A \sin (\omega t + \phi)}{(1 + \omega^2 T^2)^{\frac{1}{2}}} \qquad (5\text{-}1)$$

where $\phi = -\tan^{-1} \omega T$.

The amplitude ratio, B/A in Fig. 5-2, decreases indefinitely with increasing frequency.

$$\text{A.R.} = \frac{1}{(1 + \omega^2 T^2)^{\frac{1}{2}}} \qquad (5\text{-}2)$$

The phase angle is always negative for a first-order system, and the negative angle is called a "phase lag."   The phase lag is $360°(\Delta t/P)$ in Fig. 5-2 and approaches a limit of $90°$ at high frequencies.

$$\text{Phase lag} = -\phi = \tan^{-1} \omega T$$
$$\lim_{\omega \to \infty} \phi = -90° \qquad (5\text{-}3)$$

The reason for a limit to the phase lag is evident from Fig. 5-2 and the differential equation for a first-order system [Eq. (3-3)].   Since $d\theta_2/dt$ is positive when θ_1 exceeds θ_2, the curves for θ_2 and θ_1 intersect at a maximum or minimum in θ_2.   As the amplitude of θ_2 becomes small, the intersection approaches the midpoint of the θ_1 wave and the phase lag approaches one-fourth of a cycle.

5-2. BODE DIAGRAMS

A convenient method of presenting the response data at various frequencies is to use a log-log plot for the amplitude ratios, accompanied by a semilog plot for the phase angles.   Such plots are called "Bode diagrams," after H. W. Bode, who did basic work on the theory of feedback amplifiers. By using ωT as a parameter, a general plot for first-order systems is obtained (Fig. 5-3).   Since the amplitude ratio approaches 1.0 at low frequencies and $1/\omega T$ at high frequencies, the straight-line portions of the response if extended would intersect at $\omega T = 1.0$.   The frequency corresponding to $\omega T = 1$ is called the "corner frequency," and the amplitude ratio is 0.707 at this point.   The phase lag is $45°$ at the corner frequency, and the phase curve is symmetrical about this point.

The system shown in Fig. 5-2 has a gain of 1, which means that the output equals the input as the frequency approaches zero.   If the system gain K is greater than 1, the output amplitude is greater than the input amplitude at low frequencies and a more comprehensive definition of amplitude ratio is needed.   The amplitude ratio is defined as the ratio of output amplitude to input amplitude at a given frequency, divided by the ratio of the amplitudes at zero frequency.   This is equivalent to dividing the measured ratio B/A by the gain K, which makes the amplitude ratio

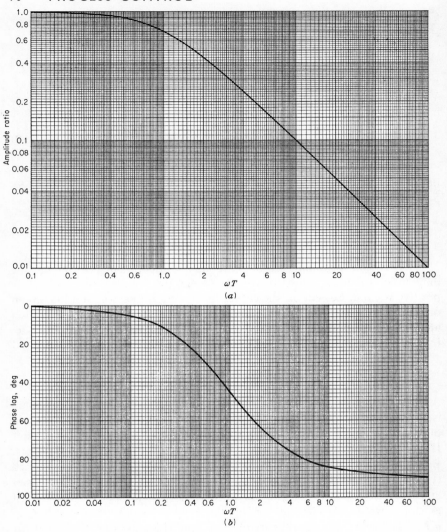

FIGURE 5-3   Bode plot for first-order system.   (a) Amplitude ratio versus ωT;   (b) phase lag versus ωT.

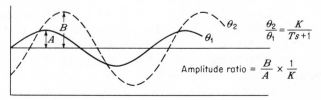

$$\frac{\theta_2}{\theta_1} = \frac{K}{Ts + 1}$$

$$\text{Amplitude ratio} = \frac{B}{A} \times \frac{1}{K}$$

FIGURE 5-4   Open-loop response of a process with a gain greater than 1.

dimensionless and makes Eq. (5-2) and the curves of Fig. 5-3 applicable to any first-order process.

In some texts, the amplitude ratio is defined just as B/A, or output over input, which leads to amplitude ratios which may not be dimensionless (degrees Fahrenheit per pounds per square inch, pounds per square inch per gallons per minute, etc.). There is nothing unsound about this procedure, but the use of different units for each process element makes it somewhat harder to plot the overall response of the system.

5-3. REVIEW OF COMPLEX NUMBERS

The frequency response of second-order systems and other process elements can be obtained by solving the differential equations for a sinusoidal input, but a much simpler procedure is to use complex numbers to obtain the amplitude ratio and phase lag directly from the transfer function. Before proceeding, the main features of complex numbers are reviewed briefly.

A complex number can be written as the sum of a real number and an imaginary number.

$$\mathbf{z} = x + jy \tag{5-4}$$

where $j = \sqrt{-1}$.

The complex number represents a vector which has a component $+x$ on the real axis and a component $+y$ on the imaginary axis (Fig. 5-5). The magnitude of the vector is $\sqrt{x^2 + y^2}$, and the angle is $\tan^{-1}(y/x)$. The number may be written in polar form or in exponential form,

$$\mathbf{z} = |z| \underline{/\phi} = \sqrt{x^2 + y^2} \tan^{-1}\frac{y}{x} \tag{5-5}$$

$$\mathbf{z} = |z| e^{j\phi} \tag{5-6}$$

To add complex numbers, the real parts and imaginary parts are combined separately after putting the numbers in the form of Eq. (5-4).

$$\mathbf{z}_1 + \mathbf{z}_2 = (x_1 + jy_1) + (x_2 + jy_2) = (x_1 + x_2) + j(y_1 + y_2) \tag{5-7}$$

To get the product of two complex numbers, the magnitudes are multiplied and the phase angles added, as shown by multiplying the numbers in exponential form.

$$\mathbf{z}_1\mathbf{z}_2 = |z_1| e^{j\phi_1} |z_2| e^{j\phi_2} = |z_1| |z_2| e^{j(\phi_1 + \phi_2)} \tag{5-8}$$

The conjugate of a complex number is obtained by changing the sign of the imaginary part. Conjugate numbers are used in clearing complex fractions.

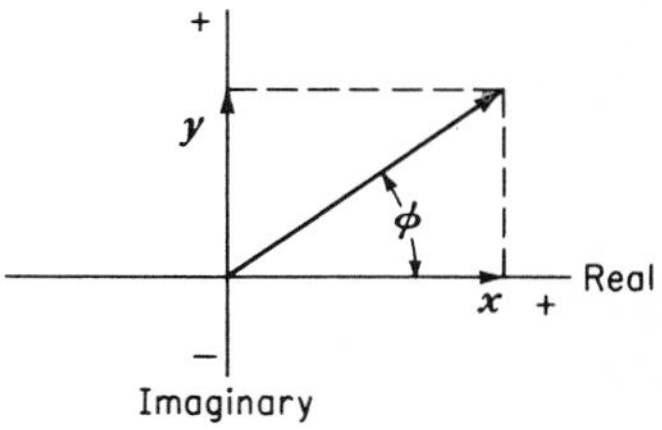

FIGURE 5-5 Plot of a complex number.

$$\frac{1}{x + jy}\frac{x - jy}{x - jy} = \frac{x - jy}{x^2 + y^2} \tag{5-9}$$

5-4. USING COMPLEX NUMBERS TO GET FREQUENCY RESPONSE

The frequency response of an element or an entire system can be obtained directly from the transfer function without performing the inverse transformation or otherwise integrating the corresponding differential equation. If the transform variable s is replaced by $j\omega$, the resulting complex number shows the amplitude and phase angle corresponding to a sinusoidal input with a frequency of radians per unit time. The derivation of this substitution is given in Refs. 1 and 2 and in many other texts on servomechanisms or mathematics. No proof is given here except to show that this method gives the correct result for a first-order process.

For $\theta_2/\theta_1 = K/(Ts + 1)$

$$\frac{\theta_2}{\theta_1} = \frac{K}{1 + j\omega T}\frac{1 - j\omega T}{1 - j\omega T} = \frac{K}{1 + \omega^2 T^2} - \frac{jK\omega T}{1 + \omega^2 T^2} \tag{5-10}$$

The vector representation of Eq. (5-10) is given in Fig. 5-6. The vector can be expressed in polar form to show the amplitude and phase lag, which are the same as the values given by Eqs. (5-2) and (5-3).

$$\frac{\theta_2}{\theta_1} = \frac{K}{(1 + \omega^2 T^2)^{\frac{1}{2}}} \underline{/-\tan^{-1}\omega T} \tag{5-11}$$

Transfer functions for some systems have terms such as $Ts + 1$ in the numerator. This represents a vector with a positive phase angle, or the term is said to contribute phase lead to the system.

$$Ts + 1 = 1 + j\omega T = \sqrt{1 + \omega^2 T^2}\ \underline{/\tan^{-1}\omega T} \tag{5-12}$$

The amplitude ratio approaches ωT at high frequencies, and the phase lead approaches $90°$.

The term $1/s$ in a transfer function is a vector with a length $1/\omega$ and a phase lag of $90°$.

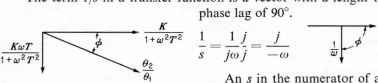

$$\frac{1}{s} = \frac{1}{j\omega}\frac{j}{j} = \frac{j}{-\omega} \tag{5-13}$$

FIGURE 5-6   Vector representation of a first-order system.

An s in the numerator of a transfer function is a vector with length ω and a phase lead of $90°$.

5-5. FREQUENCY RESPONSE OF A SECOND-ORDER ELEMENT

The frequency response for a second-order element is found by substituting $j\omega$ for s in the general equation for underdamped systems.

$$\frac{\theta_2}{\theta_1} = \frac{1}{s^2/\omega_n{}^2 + 2\zeta s/\omega_n + 1} = \frac{1}{-\omega^2/\omega_n{}^2 + 2\zeta j\omega/\omega_n + 1} \qquad (5\text{-}14)$$

Multiplying numerator and denominator by $1 - \omega^2/\omega_n{}^2 - 2\zeta j\omega/\omega_n$ gives

$$\frac{\theta_2}{\theta_1} = \frac{1 - (\omega/\omega_n)^2}{[1 - (\omega/\omega_n)^2]^2 + 4\zeta^2(\omega/\omega_n)^2} - j\frac{2\zeta\omega/\omega_n}{[1 - (\omega/\omega_n)^2]^2 + 4\zeta^2(\omega/\omega_n)^2} \qquad (5\text{-}15)$$

or

$$\frac{\theta_2}{\theta_1} = \frac{1}{\{[1 - (\omega/\omega_n)^2]^2 + 4\zeta^2(\omega/\omega_n)^2\}^{1/2}} \bigg/\!\!-\tan^{-1}\frac{2\zeta\omega/\omega_n}{1 - (\omega/\omega_n)^2} \qquad (5\text{-}16)$$

The amplitude ratios and phase angles are plotted for several values of ζ in Figs. 5-7 and 5-8. At $\omega = \omega_n$, the amplitude ratio is $1/2\zeta$, and the phase

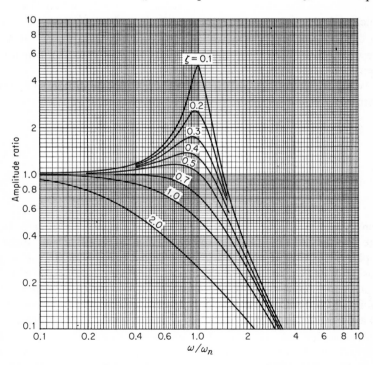

FIGURE 5-7   Bode plot for second-order systems, amplitude ratio versus ω/ω_n.

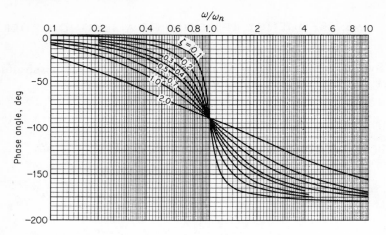

FIGURE 5-8   Bode plot for second-order systems, phase angle versus ω/ω_n.

angle is $-90°$. At very high frequencies the amplitude ratio approaches $1/(\omega/\omega_n)^2$, and the phase angle approaches $-180°$. When the damping coefficient is less than 0.707, the amplitude ratio reaches a maximum of $1/(2\zeta\sqrt{1-\zeta^2})$ at a frequency $\omega = \omega_n\sqrt{1-2\zeta^2}$. This maximum is called a "resonant peak," and the height of the peak is an important factor in the design of systems with second-order elements.

When the damping coefficient is greater than 1, the frequency response is obtained by combining the first-order curves for the two factors.

$$\text{If } \zeta > 1 \qquad \frac{1}{s^2/\omega_n^2 + 2\zeta s/\omega_n + 1} = \frac{1}{(T_a s + 1)(T_b s + 1)} \qquad (5\text{-}17)$$

5-6. RESPONSE OF ELEMENTS IN SERIES

When there are several elements in a system, the overall transfer function is the product of the individual transfer functions. The amplitude ratios and phase lags for the system are obtained by multiplying all the amplitude ratios and adding all the phase lags. The overall response is therefore independent of the order of the elements in the system. This method of combining the responses can be justified by Eq. (5-8) or by common sense; with a sinusoidal input, the output of each element is a sine wave that forms the input to the next element, and the phase lags are therefore cumulative. By definition, the overall amplitude ratio approaches 1 at low frequencies. The overall output/input ratio (in, say, degrees Fahrenheit per pounds per square inch) can be obtained by multiplying the amplitude ratio by the product of the gain terms.

$$\text{A.R.}_{\text{overall}} = (\text{A.R.})_1(\text{A.R.})_2(\text{A.R.})_3 \tag{5-18}$$

$$\phi_{\text{overall}} = \phi_1 + \phi_2 + \phi_3 \tag{5-19}$$

$$\frac{\text{Output}}{\text{Input}} = K_1(\text{A.R.})_1 K_2(\text{A.R.})_2 K_3(\text{A.R.})_3$$

$$= K_1 K_2 K_3 (\text{A.R.})_{\text{overall}}$$

$$= K(\text{A.R.})_{\text{overall}} \tag{5-20}$$

On a Bode diagram, the amplitude ratios can be combined by adding the distances from the line A.R. $= 1$, since this is equivalent to adding the logarithms. The frequency response for a system with three first-order elements is sketched in Fig. 5-9. The amplitude graph has a limiting slope of -3, and the overall phase lag approaches $270°$ at high frequencies.

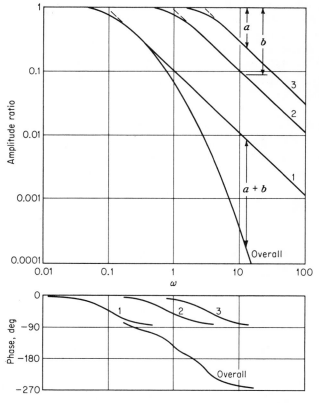

FIGURE 5-9   Open-loop frequency response for a system with three time constants.

5-7. THE STABILITY CRITERION

The maximum gain for a feedback control system is found from the over-all frequency response for the open-loop system. The maximum overall gain is the reciprocal of the amplitude ratio at the frequency where the phase lag is 180°. The maximum controller gain is calculated from the maximum overall gain and the process gains.†

$$K_{max} = \frac{1}{\text{A.R. at } 180° \text{ lag}} \tag{5-21}$$

$$K = K_c(K_1 K_2 K_3 \cdots K_n)$$

$$K_{c,\,max} = \frac{K_{max}}{K_1 K_2 K_3 \cdots K_n} \tag{5-22}$$

The frequency corresponding to 180° phase lag is the resonant frequency of the system, and fluctuations at or near this frequency are usually amplified by the control system, just as they would be for an underdamped second-order system. Consider what happens to the system of Fig. 5-10 if a small fluctuation at the critical frequency enters the system as a load disturbance. The output θ is 180° behind the signal θ_L; that is, θ is a minimum when θ_L is a maximum. Although there is no delay in the controller action, the controller output is 180° behind process output θ, since the controller output increases when θ decreases. Therefore the controller output $K_c(-\theta)$ is in phase with the load θ_L, and the combined input may be much greater than the original load. The eventual value of the output amplitude is the total input times the amplitude ratio and the process gains. For the system of Fig. 5-10,

$$\theta = (K_c\theta + B)K_p(\text{A.R.})_{\text{overall}} \tag{5-23}$$

or

$$\theta = \frac{BK_p(\text{A.R.})_{\text{overall}}}{1 - K_c K_p(\text{A.R.})_{\text{overall}}} \tag{5-24}$$

† The constants K, K_1, K_c are really the zero-frequency gains or steady-state gains. The term "gain" may refer either to those constants or to the output/input ratio at a given frequency.

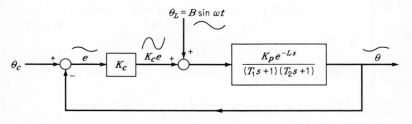

FIGURE 5-10   Response of a control system at the critical frequency.

As the product $K_c K_p (\text{A.R.})_{\text{overall}}$ (the overall gain of the system) approaches 1, a slight disturbance at the critical frequency results in very large fluctuations of the process output.   If the overall gain exceeds 1, an infinitesimal disturbance is magnified on each cycle through the control system and the controller signal eventually oscillates between zero and maximum output. It does not matter where the disturbance is introduced, and the disturbance need not be a true sine wave at the critical frequency.   Random fluctuations in load will always have a component of this frequency, and amplification of this component leads to unstable operation if too high a gain is used.

Some open-loop systems are unstable by themselves (the transfer function has positive roots) but can be stabilized by the proper control action.   The Bode plot and the stability criterion of Eq. (5-21) do not apply to such cases.   The stability must be checked by using the Routh criterion or the Nyquist method, which are described in Appendixes 1 and 2.   Examples of inherently unstable chemical reactors are discussed in Chap. 15, and other examples of systems that are open-loop unstable or conditionally stable are presented in Refs. 3 and 4.

Example 5-1

Calculate the maximum controller gain for a pressure control system which has two noninteracting first-order elements and a control valve, which also behaves approximately as a first-order system.   The time constant for the valve and transmission line is 2 sec, and a 1 per cent change in valve-stem position changes the flow 1.5 per cent, based on the average value of the flow.   The first tank has a time constant of 10 sec, and a 1 per cent increase in the controlled flow increases the pressure by 1.2 psi.   The second tank has a time constant of 5 sec, and the pressure increases by 0.8 psi following an increase of 1 psi in the first tank.   A pressure

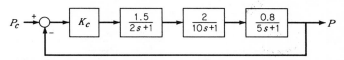

transmitter with negligible lag and an input range of 60 psi sends a 3- to 15-psi signal to the proportional controller.

Solution

$$G_1 = \frac{K_v}{T_v s + 1} = \frac{1.5}{2s + 1}$$

$$G_2 = \frac{K_2}{T_1 s + 1} = \frac{\dfrac{1.2 \text{ psi}/60 \text{ psi}}{0.01}}{10s + 1} = \frac{2}{10s + 1}$$

$$K_2 = \frac{2\% \text{ change in pressure}}{1\% \text{ change in flow}}$$

$$G_3 = \frac{K_3}{T_3 s + 1} = \frac{0.8 \text{ psi}/1 \text{ psi}}{5s + 1} = \frac{0.8}{5s + 1}$$

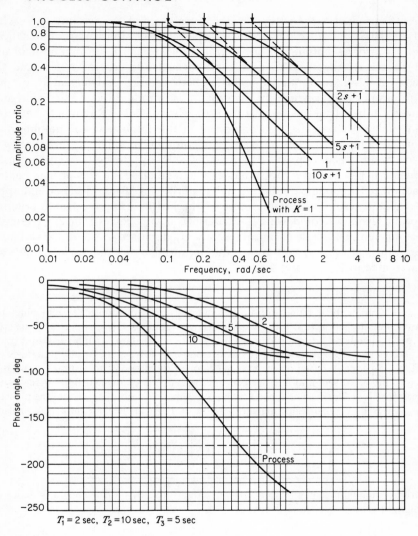

$T_1 = 2\,\text{sec},\ T_2 = 10\,\text{sec},\ T_3 = 5\,\text{sec}$

FIGURE 5-11   Bode diagram for Example 5-1.

The phase and amplitude-ratio curves for the three time constants are plotted in Fig. 5-11 by using the corner frequencies of 0.1, 0.2, and 0.5 rad/sec and the general curves of Fig. 5-3.   The overall phase lag is 180° at a critical frequency of 0.41 rad/sec, and the amplitude ratio at this frequency is 0.085.

$$K_{\max} = \frac{1}{0.085} = 12$$

$$K_{c,\,\max} = \frac{12}{(1.5)(2)(0.8)} = 5$$

The frequency response of the controller was not included in Fig. 5-11, because the controller was assumed to have no phase lag except for the inherent 180° lag, which is taken into account in Eq. (5-22).   Including the controller gain and the process gains in the Bode diagram would just shift the amplitude curve up by a factor K_cK_p, and this curve would then be the overall gain of the system.   For overall gains of less than 1 at $= 0.41$ rad/sec, the process would be stable.

A different method of plotting the frequency response is to plot the gain of each element, i.e., the ratio of output to input, rather than the amplitude ratio.   The individual curves would then start at the values of K_1, K_2, etc., rather than at 1.0.   Figure 5-12 shows the gain curves for Example 5-1 on this basis.   The combined curve has the same shape as the overall curve in Fig. 5-11, but it starts at 2.4 instead of 1.0.   The maximum controller gain is the reciprocal of the process-gain curve at the critical frequency, or $K_{c, \text{max}}$ is $1/0.2 = 5$, which agrees with the previous calculation. The main advantage of using the amplitude ratios as shown in Fig. 5-11 is that the curves all start at the same level and do not intersect, which makes them easier to plot.

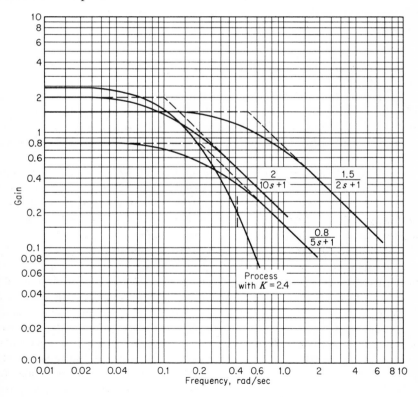

FIGURE 5-12   Alternative Bode diagram for Example 5-1.

5-8. GAIN AND PHASE MARGINS

There are two methods of using the open-loop frequency response to predict good controller settings. The controller gain can be set at a certain fraction of the maximum value, usually about 0.5, which corresponds to an overall gain of 0.5 at the critical frequency (180° phase lag). The difference between 1.0 and the overall gain at the critical frequency is a measure of stability and is called the "gain margin" by some authors. Others define the gain margin as the actual value of the overall gain at 180° phase lag, and many servomechanism texts use a third definition, the ratio of the maximum gain to the actual gain, or in decibels $20 \log (K_{\max}/K)$. The other method of providing a safety margin is to make the overall gain 1.0 at a frequency where the phase lag is appreciably less than 180°. If the gain is 1.0 at 150° phase lag, the phase margin is 30° and the corresponding frequency is the gain crossover frequency. One advantage of using phase margins is that the frequency of the damped oscillations for the closed-loop system is almost the same as the gain crossover frequency. Further comparisons of these two and other methods are given in the chapter on optimum controller settings.

5-9. COMPARING CONTROL SYSTEMS

In the introduction to this chapter, the product of the maximum gain and the critical frequency was suggested as an index of controllability for comparing different systems. The theory underlying this index, some of its limitations, and some examples of its use are presented in the following paragraphs.

A control system designed for regulator operation must minimize the effect of load changes on the process output. Assuming that the economic loss caused by deviations from the control point is proportional to the error and that positive and negative errors are equally costly, the time integral of the absolute error, $\int |e| \, dt$, for a given load change is a logical measure of the performance of the control system. For nearly all processes for which control calculations are justified, integral action is used to eliminate any offset (usually with proportional action), and so a minimum offset would not be a realistic criterion. At the optimum controller settings, the response to a load change is similar to that of an underdamped second-order system; the frequency is 10 to 30 per cent less than the critical frequency, and the decay ratio is usually about $\frac{1}{4}$. The decay ratio is the height of successive positive peaks, which is almost the same as the ratio of the areas under successive peaks. The error integral is therefore roughly proportional to the height of the first peak times the reciprocal of the critical frequency for systems adjusted to give the same decay ratio.

The height of the first peak following a unit load change depends mainly on the load gain K_L, the overall gain K, and the location of the load disturbance. For load changes at the start of the process, the response of a system with proportional control is similar to that of a second-order system as shown in Fig. 3-19. For a gain that gives a decay ratio of $\frac{1}{4}$, the damping coefficient is 0.22, and the peak error is about 1.5 times the steady-state error. Since the offset or steady-state error for load changes is $K_L/(1 + K)$, the peak error is approximately $1.5K_L/(1 + K)$. Adding integral action of course eliminates the offset, but the peak error is about the same as before (see Fig. 5-13) since the integral action has little effect by the time the first peak is reached. Therefore the estimate of peak error is based on the offset that would occur if only proportional control were used.

For load changes at the start of the system

$$\text{Peak error} \cong 1.5 \frac{K_L}{1 + K} \tag{5-25}$$

In trying to improve the performance of a control system, the gain for load changes K_L will probably not be changed. Changes can be made in the number or magnitude of the intermediate time constants, which will change the critical frequency and the recommended gain. The ratio of the error integrals can be estimated from

$$\frac{\int |e|\, dt\ \text{I}}{\int |e|\, dt\ \text{II}} \cong \frac{(1 + K)\omega_c\ \text{II}}{(1 + K)\omega_c\ \text{I}} \tag{5-26}$$

For processes which have no dead time or distributed elements, the maximum overall gain is generally greater than 10, and the optimum gain is

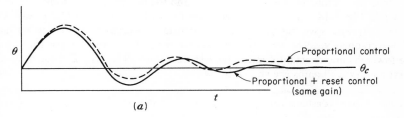

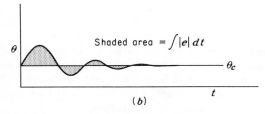

FIGURE 5-13 Comparing control systems using the error integral. (a) Typical process response for step change in load; (b) similar process with both K and ω_c twice as large.

about half the maximum value; so the product of the critical frequency and the maximum gain can be used as a relative measure of the error integral.

If $K_{max} > 10$,

$$\frac{\int |e|\, dt \text{ I}}{\int |e|\, dt \text{ II}} \cong \frac{K_{max}\omega_c \text{ II}}{K_{max}\omega_c \text{ I}} \tag{5-27}$$

Equation (5-26) is more accurate than Eq. (5-27) and could be made even more accurate by using the gain crossover frequencies instead of the critical frequencies.† However, these equations are intended for quick order-of-magnitude calculations and not for accurate comparisons. If the error ratio must be determined to two significant figures, more exact methods of predicting the transient response must be used. The method presented here should not be used if the measurement lag is one of the major lags in either of the systems being compared, since the actual error would be greater than the measured error by different amounts. The method overestimates the effect of controller gain if the major load disturbances occur toward the end of the series of process elements. A load change at the last element leads to a peak error which may be several times the term $K_L/(1 + K)$ (compare Figs. 4-6 and 4-8). A method of predicting the transient response for these cases is given in Chap. 7.

Example 5-2

A process has three first-order elements with time constants of 10, 5, and 2 min. Would lowering the largest time constant to 5 min have as much effect on controllability as lowering the second largest time constant to 2.5 min?

Solution

The Bode plots for the original system and the two proposed modifications are shown in Fig. 5-14. The critical frequency for both cases B and C is greater than for case A, since lowering any of the time constants decreases the total phase lag at a given frequency. The amplitude curves for B and C are above those for A, since the lowered time constant contributes less damping. The critical frequencies and maximum overall gains are shown below.

Case	Time constants, min	ω_c, rad/min	(A.R.)$_c$	K_{max}	$\omega_c K_{max}$
A	10, 5, 2	0.41	0.085	12	4.9
B	10, 2.5, 2	0.53	0.075	14	7.4
C	5, 5, 2	0.49	0.10	10	4.9

Modification B would increase the critical frequency by 30 per cent and increase the maximum gain about 20 per cent. The error integral for an upset would be about two-thirds as great as before. Modification C would increase the critical frequency by 20 per cent, but the lower maximum gain would cancel this advantage, and the net effect would be almost no change in controllability.

† The gain crossover frequency is close to the frequency of oscillation of the closed loop.

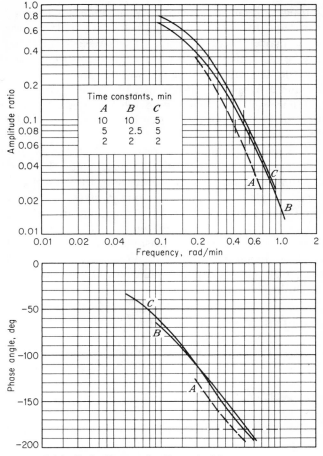

FIGURE 5-14   Bode diagram for Example 5-2.

The above example illustrates a general rule that reducing the second largest time constant does more to improve a control system than reducing the largest time constant.   The largest time constant often contributes almost 90° phase lag at the critical frequency; lowering this time constant by a factor of 2 has little effect on the critical frequency but a large effect on the amplitude ratio.   If the largest time constant T_1 is much greater than the others, decreasing T_1 actually leads to poorer control, since the lower permissible gain more than offsets the slight increase in frequency.   Lowering the second largest time constant T_2 by the same factor means a greater shift in critical frequency, since the phase curve for T_2 is steeper than that for T_1 at the critical frequency.

The above reasoning suggests that *increasing* the largest time constant would often improve controllability.   However, increasing the largest time constant is usually the most expensive way of improving the system, since this would perhaps involve doubling the size of the tank, reactor, or heater that forms the major process element. The second largest time constant, which might be associated with the valve, the measuring element, or the jacket of a kettle, can often be changed without great expense. The valve lag can be reduced by use of a positioner, the lag of a thermowell reduced by using a thinner well or a higher fluid velocity past the well, and the lag in a jacket or cooling coil can be reduced by using smaller holdup or higher fluid velocity.

If a system had only three time constants, decreasing the smallest one would have a large effect on the critical frequency and the maximum gain. As the third time constant T_3 approached zero, the critical frequency and maximum gain would both approach infinity, since the two largest time constants would contribute 180° lag only at infinite frequency. However, since a real process has other small time constants or time delays, the advantage of reducing or eliminating one of the small time constants is not very great. To evaluate the effect of such changes, the gain and frequency corresponding to 30° phase margin should be used if only three time constants are included in the calculations.

5-10. EFFECT OF TIME DELAY

The frequency response of a true time delay or transportation lag is an amplitude ratio of 1.0 and a phase lag equal to the product of the delay and the frequency.

For a time delay of L sec,

$$\begin{aligned} \text{A.R.} &= 1.0 \\ \phi_L &= \omega L \text{ rad} \\ &= 57.3\omega L° \\ &= 360 f L° \end{aligned} \tag{5-28}$$

Since the phase lag increases without limit, any process with a time delay has a finite critical frequency and a finite maximum gain. The controllability can always be improved by decreasing the time delay in the system, since the total phase lag is decreased and the overall amplitude ratio is unchanged. This increases the critical frequency and usually increases the maximum gain, since A.R. for most systems decreases with increasing ω. If there is more than one delay in the control loop, they can be lumped together to calculate the overall response of the system.

Example 5-3

A control system has time constants of 2 min and 0.2 min and a time delay of 0.5 min. How much better would control be if the time delay were halved?

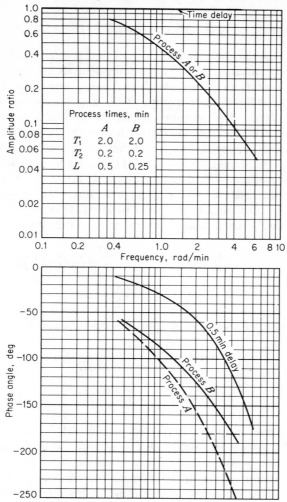

FIGURE 5-15   Bode diagram for Example 5-3.

Solution

The Bode diagrams for both systems are shown in Fig. 5-15. The pertinent results are:

System	T_1, min	T_2, min	L, min	ω_c, rad/min	K_{max}	$K_{max}\omega_c$
A	2	0.2	0.5	2.5	5.5	14
B	2	0.2	0.25	4.1	10.5	43

A twofold decrease in the time delay means about a threefold increase in the controllability. Note that this improvement is much greater than that calculated for halving one of the time constants in Example 5-2.

5-11. RESPONSE OF DISTRIBUTED SYSTEMS

A distributed system, such as a thick wall that is a barrier to heat transfer or a pneumatic transmission line, is equivalent to an infinite number of interacting elements in series. With increasing frequency, the phase lag increases without limit, and the slope of the amplitude-ratio curve becomes progressively steeper.[†] The simplest case to consider is the one where the distributed element does not interact with previous or succeeding elements, which means that the input is fixed and there is no gradient at the far end of the element. This condition is fulfilled if there is a heavy layer of insulation on the far side of the wall or if the volume at the end of the transmission line is small relative to the volume of gas in the line. The frequency response for this case is taken from Farrington's solution (5), with the terms R and C used for the total resistance and total capacity of the system.

$$\text{A.R.} = \frac{1}{[\cosh^2 (\omega RC/2)^{1/2} - \sin^2 (\omega RC/2)^{1/2}]^{1/2}} \tag{5-29}$$

$$\phi = -\tan^{-1} \left[\tan\left(\omega \frac{RC}{2} \right)^{1/2} \tanh \omega \frac{RC}{2} \right] \tag{5-30}$$

At high frequencies, these values approach the following limits:

$$\lim_{\omega \to \infty} \text{A.R.} = \frac{1}{\cosh (\omega RC/2)^{1/2}} \tag{5-31}$$

$$\lim_{\omega \to \infty} \phi = - \left(\frac{\omega RC}{2} \right)^{1/2} \tag{5-32}$$

The frequency response is plotted in Fig. 5-16. At low frequencies, the distributed system has almost the same response as the lumped system ($T = RC/2$), and a distributed element could be replaced by a single time constant for analog studies if the element contributed less than 50° phase lag. Using a time delay and a time constant fits the phase curve and the amplitude curve fairly well over the frequency range of interest.

If heat or fluid flows in series through a distributed element and an upstream or downstream element, the elements interact and there is no simple equation or graphical solution for the response. The theory for electrical transmission lines can be applied if an exact solution is needed, or a lumped approximation of the distributed element can be used. The response of pneumatic transmission lines with capacity at the end of the line is discussed in Chap. 10. In most examples of heat transfer from one fluid to another through a wall, the wall resistance is small compared with

[†] Tubular heat exchangers, packed-bed reactors, and packed absorbers are distributed systems of a more complex type. The response of exchangers is discussed in Chap. 11.

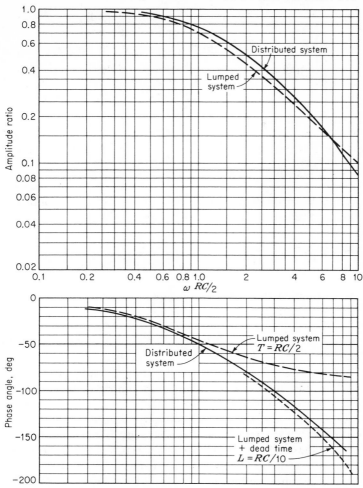

FIGURE 5-16   Frequency response of a distributed system.

the resistances in the fluids near the wall, and there is no need to consider the wall as a distributed system.

5-12. SYSTEMS WITH SIDE CAPACITIES

For all the systems considered so far, the output lags the input by an angle that gradually increases with increasing frequency. However, when the signal at an intermediate point in a series of elements is the output of interest, the downstream elements are side capacities and the phase angle for the output may exhibit local minima and maxima. Consider an insulated tank used to smooth temperature fluctuations in a feed stream. The wall has

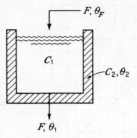

F, θ_F

C_1

C_2, θ_2

F, θ_1

FIGURE 5-17 A tank with
an appreciable
side capacity.

appreciable heat capacity, but the wall resistance is negligible compared with the liquid-film resistance. The heat-balance equations for tank and wall are

$$C_1 \frac{d\theta_1}{dt} = F(\theta_F - \theta_1) - UA(\theta_1 - \theta_2)$$

$$C_1 s \theta_1 = \frac{\theta_F - \theta_1}{R_1} - \frac{\theta_1 - \theta_2}{R_2} \qquad (5\text{-}33)$$

$$C_2 \frac{d\theta_2}{dt} = UA(\theta_1 - \theta_2) = \frac{\theta_1 - \theta_2}{R_2}$$

$$\qquad\qquad\qquad\qquad\qquad\qquad (5\text{-}34)$$

or $\quad (T_2 s + 1)\theta_2 = \theta_1$

The above equations are in the same form as those for a bulb in a thermowell, and the effective time constants T_a and T_b are obtained from Eq. (3-76).

$$\frac{\theta_2}{\theta_F} = \frac{1}{R_1 C_1 R_2 C_2 s^2 + (R_1 C_1 + R_2 C_2 + R_1 C_2)s + 1} = \frac{1}{(T_a s + 1)(T_b s + 1)}$$

$$\qquad\qquad\qquad\qquad\qquad\qquad\qquad\qquad\qquad\qquad (5\text{-}35)$$

The equation relating tank and wall temperatures is then used to get the desired transfer function,

$$\frac{\theta_1}{\theta_F} = \frac{\theta_2}{\theta_F} \times \frac{\theta_1}{\theta_2} = \frac{T_2 s + 1}{(T_a s + 1)(T_b s + 1)} \qquad (5\text{-}36)$$

The time constant T_2 is always smaller than T_a and larger than T_b. If T_a and T_b are quite different, the phase lead contributed by T_2 makes the overall phase curve go through a minimum.

For any kettle, reboiler, or exchanger, the outside wall acts as a side capacity for storage of heat. The wall capacity is particularly important for systems containing gas or steam. An article by Day (6) gives the frequency response of the pressure in a steam jacket to changes in steam flow; there is a minimum in the phase curve which is predicted by the theoretical analysis. The response of the top plate of a distillation column may also show unusual phase and amplitude curves because the lower plates act as side capacities which interact with the top plate.

PROBLEMS

1. Plot a Bode diagram for a control system with time constants of 10, 10, and 2 min. Determine the maximum overall gain and the critical frequency.

2. Calculate the critical frequency and maximum controller gain for a system with one time delay and one time constant for several values of the ratio L/T. Plot your results in a form suitable for accurate interpolation. Also plot $K/K_{\max}$ for a phase margin of $30°$.

3. A temperature control system has process time constants of 20 min and 5 min. The control valve and the thermometer bulb both have time constants of 10 sec. A 1-psi change in controller output changes the controlled flow 25 gpm from the normal value of 200 gpm. The process temperature is 175°F for 200 gpm and 174°F for 210 gpm. A bulb with a range of 80°F is used.

a. Calculate the maximum overall gain.

b. Calculate the maximum controller gain.

4*a*. Calculate the effective time constants in the transfer function relating input flow and level for two open tanks connected in series. Both tanks are vertical cylinders 4 ft in diameter and 10 ft high. The outlet of the first tank is connected to the bottom of the second tank, and the bottom of the second tank is 2 ft below the bottom of the first tank. The normal levels are 8 ft in the first tank and 7 ft in the second tank, and the normal flow is 50 cfm

b. Plot the frequency response of the level in both tanks to changes in flow to the first tank.

5. The open-loop frequency response of a process is given below.

Frequency, cpm	Amplitude	Phase lag, deg
0.1	10.0	5
0.5	9.8	30
1	9.2	65
2	7.1	140
4	2.2	290
8	0.80	470
16	0.23	800
32	0.06	

a. There is a time delay of at least 0.028 min in this process (a sampling delay). Can the behavior of the system be approximated by a time delay plus one or two other parameters?

b. By what factor could controllability be improved by elimination of the 0.028-min sampling delay?

6. The main elements in a concentration control loop are a proportional controller (negligible time constant), a control valve with a time constant of 20 sec, a flash evaporator with a time constant of 60 sec, and a time delay of 30 sec in the sampling line.

a. What is the natural frequency of the system?

b. What is the maximum controller gain if full span of the instrument corresponds to 1.5 weight per cent?

Pressure on valve, psi	Flow of heating medium, lb/min	Concentration, wt %
13	100	8.3
10†	150†	7.0†
6.5	200	6.0
3	230	5.4

† Normal values.

7. The frequency response of a process is to be determined experimentally. If the process is first-order, how many cycles are required before the amplitude ratio is within 5 per cent of the steady-state value? How many minutes are required if the time constant is 10 min?

8. Calculate the maximum controller gain for the following system:

$$\text{Valve: } T = 10 \text{ sec, } K_v = 3$$

$$\text{Second-order process: } \zeta = 0.2, \, \omega_n = 0.5 \text{ sec}^{-1}, \, K_p = 2$$

$$\text{Time delay: } L = 3 \text{ sec}$$

How much would the controllability be changed by reducing the time delay to 2 sec?

9. A control loop has four first-order elements with time constants of 10, 2, 1, and 0.1 min.

a. What are the critical frequency and the maximum overall gain?

b. If the controller gain is set at half the maximum value, calculate the phase margin.

c. Show that, for certain combinations of four time constants, the phase margin can be less than 15° if the controller gain is half the maximum value.

10. Calculate the frequency response for a two-capacity model of a distributed system, and compare with the exact response shown in Fig. 5-16.

REFERENCES

1. Thaler, G. J., and R. G. Brown: "Analysis and Design of Feedback Control Systems," 2d ed., p. 152, McGraw-Hill Book Company, New York, 1960.
2. Wilts, C. H. "Principles of Feedback Control," p. 22, Addison-Wesley Publishing Company, Inc., Reading, Mass., 1960.
3. Del Toro, V., and S. R. Parker: "Principles of Control System Engineering," p. 334, McGraw-Hill Book Company, New York, 1960.
4. D'Azzo, J. J., and C. H. Houpis: "Control System Analysis and Synthesis," pp. 218, 227, McGraw-Hill Book Company, New York, 1960.
5. Farrington, G. H.: "Fundamentals of Automatic Control," p. 128, Chapman and Hall, Ltd., London, 1951.
6. Day, R. L.: in "Plant and Process Dynamic Characteristics," p. 29, Butterworth & Co. (Publishers), Ltd., London, 1957.

6 FREQUENCY RESPONSE OF CONTROLLERS

In the examples of the previous chapter, the controller response was not included on the Bode diagrams, because a proportional controller has a constant gain and no phase lag out to quite high frequencies. If a controller has either derivative or integral action along with proportional action, both the phase angle and the effective gain of the controller change with frequency and the controller response must be considered in calculating the maximum gain and critical frequency for the system. The exact frequency response depends on the construction of the controller, particularly for pneumatic instruments which use interacting, nonlinear components in the derivative and integral circuits. This chapter deals mainly with ideal controllers, those whose response is given by the theoretical equations for proportional-integral, proportional-derivative, and proportional-integral-derivative control. Fortunately, the controller settings and pertinent frequencies generally are in the range where the difference between the ideal and actual controller response is small. Actual frequency-response data for the controller would of course be preferable, but such data are rarely published by the instrument manufacturers, and also the exact instrument response depends on the length of tubing or the volume of the bellows to which the instrument is connected.

6-1. PROPORTIONAL-INTEGRAL CONTROLLER

The transfer function for a proportional-integral controller is obtained from the basic equation

$$p = K_c e + \frac{K_c}{T_R} \int e \, dt \tag{6-1}$$

$$\frac{P}{E} = K_c \left(1 + \frac{1}{T_R s}\right) = \frac{K_c(T_R s + 1)}{T_R s} \tag{6-2}$$

The frequency response is found by substituting $s = j\omega$ in Eq. (6-2) and getting the length and angle of the resulting vector (Fig. 6-1).

$$\frac{P}{E} = K_c \left(1 + \frac{1}{j\omega T_R}\right) = K_c \left(1 - \frac{j}{\omega T_R}\right) \tag{6-3}$$

The actual gain of the controller for sinusoidal inputs is

$$\text{Gain} = K_c \left(1 + \frac{1}{\omega^2 T_R^2}\right)^{1/2} = \frac{\text{output amplitude}}{\text{input amplitude}} \tag{6-4}$$

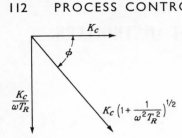

FIGURE 6-1 Vector diagram for proportional-integral controller.

At low frequencies the gain approaches $K_c/\omega T_R$, and at high frequencies the gain is K_c. On a logarithmic plot, these straight lines describing the limiting behavior intersect at $\omega = 1/T_R$, which is the corner frequency for integral action or the reset corner. The phase angle is given by

$$\text{Phase angle} = -\tan^{-1}\frac{1}{\omega T_R} \qquad (6\text{-}5)$$

The phase angle is $-90°$ for very low frequencies, $-45°$ at the reset corner, and approaches zero at high frequencies. Typical frequency-response curves are shown in Fig. 6-2. The gain or

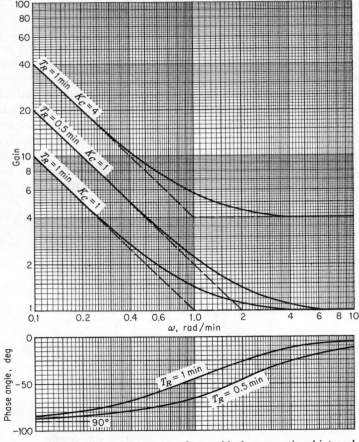

FIGURE 6-2 Frequency response of an ideal proportional-integral controller.

output-input ratio is used for this plot rather than the amplitude ratio as defined in Chap. 5, since the zero frequency gain is infinite. Increasing the nominal gain K_c (the dial setting) displaces the gain curves upward and does not affect the phase angle; so presenting only the curve for $K_c = 1.0$ would have been sufficient. Increasing the reset time, which means decreasing the reset rate, shifts both the gain and phase curves to the left.

The easiest way to obtain the frequency response for any values of gain and reset time is to use the curves for $K_c = 1$ and $T_R = 1$ in Fig. 6-1 and consider the abscissa to be ωT_R. The general plots for a first-order lag (Fig. 5-3) can also be used, since they are just mirror images of the curves in Fig. 6-2. For example, for $T_R = 5$ min and $\omega = 0.4$ rad/min, $\omega T_R = 2$, and the controller phase angle is $-27°$, from Fig. 6-2. A value of 2 for ωT_R corresponds to $\omega T = \frac{1}{2}$ in Fig. 5-3b, and the corresponding lag is $-27°$. The controller gain at $\omega T_R = 2$ is $1.12K_c$ from Fig. 6-2, which corresponds to $1/0.89$, where 0.89 is the amplitude ratio from Fig. 5-3a at $\omega T = \frac{1}{2}$.

EFFECT ON SYSTEM STABILITY

The phase lag introduced by the use of integral action adds to the lags contributed by the process elements and thus lowers the critical frequency of the system. Since the amplitude ratio of the process is usually decreasing with increasing frequency, the maximum controller gain tends to be lowered by this decrease in critical frequency. The fact that the actual controller gain at low frequencies is now greater than K_c, as shown by Fig. 6-2, also contributes to lower values of the maximum gain setting. The lower critical frequency and the lower value of K_c make the response of the control system more sluggish, which is undesirable. If a very large value of reset time is used, the effect on process stability is negligible but a long time is required to eliminate the offset resulting from a load change. The optimum reset time is determined by balancing the need for quick elimination of offset against the desire for high critical frequency and maximum gain.

A common rule is to set the reset time equal to the ultimate period for the process with just proportional control.†

$$T_{R,\text{opt}} = P_u \qquad (6\text{-}6)$$

The reasoning behind this rule is that the reset time should be decreased until the reset action starts to have an appreciable effect on the frequency response of the system. When the reset time equals the ultimate period, the value of $\omega_c T_R$ is $(2\pi/P_u)T_R$ or 2π. The phase lag contributed by integral action at the former critical frequency is then 9°, or 5 per cent of 180°.

† The value recommended by Ziegler and Nichols (1) is $P_u/1.2$; the slight difference is not important to the argument presented here.

For most processes, this extra phase lag in the controller decreases the critical frequency and maximum gain only 10 to 20 per cent. Therefore, going to a higher reset time, say, 3 to $10P_u$, would have little effect on the frequency response of the system, but it would make the recovery from load changes much slower. Using a quite low reset time, say, 0.1 to $0.3P_u$, would make the controller phase lag fairly large, and the critical frequency and maximum gain would be quite a bit smaller. The optimum reset time is not the same for all processes but depends on the slopes of the frequency-response plots, as brought out in the more detailed discussion of optimum settings in Chap. 9.

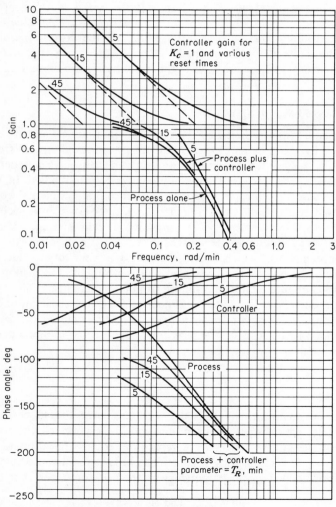

FIGURE 6-3   Bode diagram for Example 6-1.

Example 6-1

Show the effect of various reset times on the stability and critical frequency for a process with time constants of 10, 5, and 2 min (same ratios as for Example 5-1 and Fig. 5-11).

The Bode diagram for the process is Fig. 6-3.   For proportional control only, $\omega_c = 0.41$ rad/min.

$$K_{max} = (K_c K_p)_{max} = \frac{1}{0.085} = 12$$

$$P_u = \frac{2\pi}{0.41} = 15.3 \text{ min}$$

Choose $T_R = 5$, 15, and 45 min for proportional-integral control.   To include controller response on the Bode diagram, a gain setting $K_c = 1$ is used.   The actual controller gain (for $K_c = 1$) times the amplitude ratio for the process elements can

Table 6-1

T_R min	$\dfrac{T_R}{P_u}$	ω_c rad/min	A.R. process	Controller gain $K_c = 1$	Overall A.R.	$(K_c K_p)_{max}$
∞	...	0.41	0.085	1	0.085	12
45	3	0.39	0.095	1	0.095	11
15	1	0.34	0.13	1.02	0.13	8
5	⅓	0.22	0.29	1.37	0.40	2.5

be considered an overall amplitude ratio, and the reciprocal of this ratio is the maximum value of the product $K_c K_p$.   Table 6-1 shows that using a reset time equal to the period decreases the permissible gain setting by 30 per cent and lowers the critical frequency by 15 per cent.   Using $T_R = 5$ lowers K_c by 80 per cent and the frequency by 50 per cent, which is too much of a penalty to pay for increased reset rate.   With a reset time of 45 min, the frequency and gain are almost unchanged, but it would take too long to eliminate offset.

As should be apparent from Example 6-1, the optimum reset time can be determined only to within a factor of about 2 from the frequency-response diagram.   Lowering the reset time tends to compensate for lowering the gain, and, as shown in Chap. 9, there are fairly broad minima in the graphs of integrated error versus T_R or K_c.

6-2. PROPORTIONAL-DERIVATIVE CONTROLLER

The transfer function for a proportional-derivative controller comes from the equation

$$p = K_c e + K_c T_D \frac{de}{dt} \tag{6-7}$$

$$\frac{P}{E} = K_c(1 + T_D s) = K_c(1 + j\omega T_D) \tag{6-8}$$

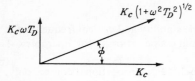

FIGURE 6-4   Vector diagram for proportional-derivative controller.

As shown by Fig. 6-4, the controller gain is K_c at very low frequencies and $K_c\omega T_D$ at very high frequencies. The asymptotes intersect at the derivative corner, where $\omega T_D = 1$, and at this frequency the actual gain is 1.41 times the nominal gain. Thus adding either integral or derivative action makes the actual gain of the controller greater than the nominal gain for certain frequencies. The advantage of using derivative control comes from the phase lead of the controller, which varies from zero at low frequencies to 90° at high frequencies.

$$\text{Gain} = K_c(1 + \omega^2 T_D^2)^{1/2} \qquad (6\text{-}9)$$

$$\text{Phase angle} = \tan^{-1} \omega T_D \qquad (6\text{-}10)$$

The controller response for typical settings is shown in Fig. 6-5. Increasing the derivative time shifts both the phase curve and the gain curve to the

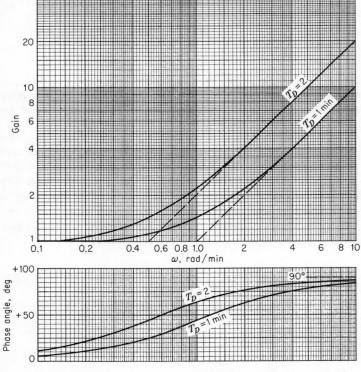

FIGURE 6-5   Frequency response of an ideal proportional-derivative controller.

left. The gain and phase for any values of T_D are easily obtained from the curves for $T_D = 1$ by considering the abscissa to be the product ωT_D. Increasing the gain just displaces the gain curve upward by the factor K_c and does not change the phase curve.

SELECTING THE DERIVATIVE TIME

A reasonable value of the derivative time can be chosen by examining the open-loop frequency response of the process. For a process that consists of a series of first-order elements, the critical frequency would be increased 1.5-fold to three fold by having the controller contribute 45° phase lead at the critical frequency. This corresponds to $\omega_c T_D = 1$, and the actual controller gain for this case would be $1.4K_c$. The maximum permissible gain would be two- to five fold higher than with just proportional control, since the lower amplitude ratio for the process would more than compensate for the 1.4-fold higher controller gain. If the derivative time were chosen so that $\omega_c T_D = 4$, the controller would contribute 77° phase lead and a still higher critical frequency would be obtained. However, the actual controller gain would be about $4K_c$, and so the maximum value of K_c might be lower than before. A high value of K_c is desirable to minimize the offset following load changes and to ensure rapid corrective action for disturbances. The maximum value of K_c will generally be obtained when the controller phase lead is 40 to 60°, or when $\omega_c T_D = 1$ to 2. If 45° lead is used, the critical frequency ω_c is the frequency at which the process lag is $180 + 45$, or 225°, and the derivative time is then set equal to $1/\omega_c$.

The derivative time can also be chosen as a fraction of the ultimate period for the process with only proportional control. For a three-mode controller, Ziegler and Nichols (1) suggested using $T_D = P_u/8$. If the period with proportional control is 1.5-fold greater than the period with proportional-derivative control, this recommendation corresponds to $T_D = \dfrac{1.5}{8} \dfrac{2\pi}{\omega_c}$, or $T_D \omega_c = 1.18$. (In this discussion, ω_c is the critical frequency of the process plus the proportional-derivative controller, and P_u is the period when only proportional control is used.)

The following example shows the effect of derivative time on critical frequency and maximum gain for a typical process. Note that four time constants are considered in this example, since instability can theoretically be avoided for a three-capacity process just by making the derivative time greater than the smallest time constant.

Example 6-2

A process has four first-order lags with time constants of 5, 2, 2, and 1 min. Calculate the derivative time that will permit the highest controller gain to be used.

The frequency response of the process is shown in Fig. 6-6. For proportional control, the critical frequency is 0.48 rad/min, and the amplitude ratio is 0.18. A

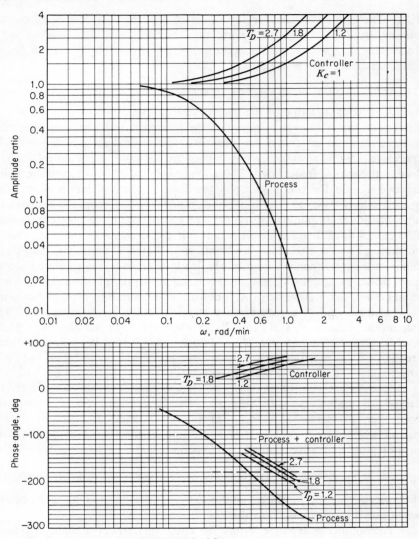

FIGURE 6-6   Bode diagram for Example 6-2.

derivative time of 2 min would give 45° phase lead at $\omega_c = 0.5$ rad/min, and derivative times of 1.2, 1.8, and 2.7 min are chosen for calculation. A controller gain $K_c = 1$ is used to simplify plotting the controller response. The maximum gain is obtained as shown below.

For $T_D = 1.2$ min

$$\omega_c = 0.74 \text{ rad/min (at 180° lag for process + controller)}$$

Process A.R. $= 0.07$

$$G_c = \text{controller gain} = 1.33K_c$$

$$(K_cK_p)_{max} = \frac{1}{(0.07)(1.33)} = 10.7$$

The maximum permissible gain is obtained by using a derivative time of about 2 min, which corresponds to $\omega_cT_D = 2$, or 60° phase lead for the controller.   The use of derivative action permits twice as high a gain to be used and also doubles the

Table 6-2

T_D, min	ω_c, rad/min	A.R.	G_c	$(K_cK_p)_{max}$	ω_cK_{max}
0	0.48	0.18	K_c	5.5	2.6
1.2	0.74	0.070	$1.33K_c$	10.7	7.9
1.8	0.88	0.042	$1.9K_c$	12.5	11.0
2.7	1.0	0.030	$2.8K_c$	11.8	11.8

critical frequency, and the net result is approximately a fourfold improvement in controllability.   To maximize the product ω_cK_{max}, a derivative time of about 3 min would be used.

SIGNIFICANCE OF BODE-PLOT SLOPES

The improvement resulting from the addition of derivative action depends on the slopes of the phase and amplitude curves near the critical frequency.   If the phase curve has a relatively low slope, the phase lead of the controller causes a large shift in critical frequency.   If at the same time the amplitude curve is steep, even a small increase in critical frequency produces a large increase in the maximum gain.   For processes containing only first-order elements with a wide range of time constants, both the above conditions exist, and derivative action improves the controllability several-fold.   As an extreme example, consider a process with time constants of 100, 50, 1, and 0.5 sec.   At the former critical frequency, each of the two largest time constants contributes 85 to 90° lag, and the lag changes only slightly with frequency.   The smaller time constants contribute 5 to 10° lag, and these values also change only slightly with increasing frequency.   Therefore, adding derivative action with 60 to 70° phase lead would give quite a large increase in both critical frequency and maximum controller gain.   However, most pneumatic controllers cannot give such large phase leads, and the benefits of derivative action are less than calculated for ideal controllers.

Derivative action has relatively little effect if the phase curve is steep and the amplitude curve almost horizontal, which is characteristic of processes with a large time delay.   For such processes, the advantage (if any) of adding derivative control comes primarily from the slight increase in critical frequency.

Example 6-3

The major elements in a process are a first-order lag with a time constant of 2 min and a time delay of 5 min. Calculate the critical frequency and maximum gain if proportional control is used. How much would derivative action improve the performance of the control system?

Calculations show 154° lag at 0.4 rad/min and 189° lag at 0.5 rad/min (see Fig. 6-7).

At $\omega = 0.48$, $\omega T = 0.96$, $\phi = 44°$, A.R. $= 0.72$ (Fig. 5.3).

$$\omega L = 0.48(5) = 2.4 \qquad \phi_L = 57.3(2.4) = 138°$$

$$\phi_{total} = 44 + 138 = 182°, \text{ close enough to } 180°$$

$$\omega_c = 0.48 \text{ rad/min} \qquad K_{max} = \frac{1}{0.72} = 1.39$$

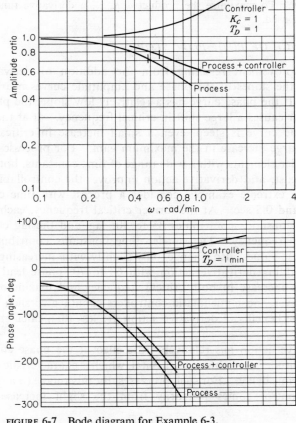

FIGURE 6-7   Bode diagram for Example 6-3.

Try $T_R = 2$ min for derivative action; since $\omega T_R = \omega T$, the phase lead of the controller cancels the phase lag of the time constant and the critical frequency depends just on the time delay.

$$\omega_c = \frac{180}{57.3(5)} = 0.63 \text{ rad/min}$$

$$K_{max} = \frac{1}{(\text{A.R.}_{\text{process}})(G_c)} = 1.0$$

Table 6-3. Summary of Results for Various Derivative Times

T_D, min	ω_c, rad/min	$\omega_c T_D$	G_c	Process A.R.	K_{max}	$\omega_c K_{max}$
0	0.48	...	...	0.73	1.39	0.67
0.5	0.52	0.26	1.03	0.69	1.41	0.73
1	0.56	0.56	1.15	0.66	1.32	0.74
1.5	0.60	0.90	1.35	0.64	1.16	0.70
2	0.63	1.26	1.61	0.62	1.0	0.63
3	0.67	2.01	2.27	0.60	0.73	0.49

Using a maximum value of $\omega_c K_{max}$ as a criterion, the optimum derivative time is about 1 min, which corresponds to $\omega_c T_D = 0.5$. The controllability is only 10 per cent greater than with just proportional control, which is hardly enough improvement to justify adding derivative action. If the derivative time were adjusted to give 60° lead, which was optimum for the previous example, control would be worse than with no derivative action at all.

6-3. PROPORTIONAL-INTEGRAL-DERIVATIVE CONTROLLER

An ideal three-mode controller is one whose output is the sum of the proportional, integral, and derivative signals, as shown by the following equations:

$$p = K_c e + \frac{K_c}{T_R} \int e \, dt + K_c T_D \frac{de}{dt} \tag{6-11}$$

$$\frac{P}{E} = K_c \left(1 + \frac{1}{T_R s} + T_D s \right) \tag{6-12}$$

If the derivative time is much less than the reset time, the controller response can be obtained by using the general solution for proportional-integral control for low frequencies and the solution for proportional-derivative control for high frequencies. Thus the gain curve would have a slope of -1 at low frequencies, 0 at intermediate frequencies, and $+1$ at high frequencies.

The phase curve would start at $-90°$, have three inflection points (at the reset corner, the derivative corner, and a point between these corners), and approach $+90°$ at high frequencies. When the derivative time is about the same as or greater than the reset time, the phase and gain must be calculated from the complete equation.

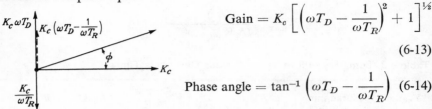

$$\text{Gain} = K_c\left[\left(\omega T_D - \frac{1}{\omega T_R}\right)^2 + 1\right]^{\frac{1}{2}}$$

(6-13)

$$\text{Phase angle} = \tan^{-1}\left(\omega T_D - \frac{1}{\omega T_R}\right) \quad (6\text{-}14)$$

Figure 6-8 shows the response curves for three ratios of derivative time to reset time. The gain curves are symmetrical about the frequency $\omega = 1/(T_D T_R)^{\frac{1}{2}}$, and the phase angle is zero at this frequency. When the ratio T_D/T_R is 1 or less, the gain curves can be approximated by the straight-line

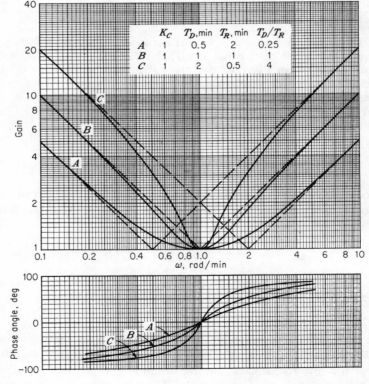

	K_C	T_D, min	T_R, min	T_D/T_R
A	1	0.5	2	0.25
B	1	1	1	1
C	1	2	0.5	4

FIGURE 6-8   Frequency response of an ideal proportional-integral-derivative controller.

asymptotes which start at the reset and derivative corners and a horizontal line (at a value K_c) between these corner frequencies.   A large value of the ratio T_D/T_R puts the derivative corner to the left of the reset corner, and for most frequencies the gain is much greater than the nominal gain K_c.   Near the intersection of the asymptotes, the gain drops rapidly to K_c.   Large values of T_D/T_R are rarely used; as is shown later, some controllers become unstable if the derivative time is made greater than the reset time, and even for ideal controllers the sharp dip in the gain curve would usually be undesirable, since the optimum gain setting for the control system might be very sensitive to shifts in system parameters.

6-4. CONTROLLERS WITH LIMITED ACTION

For an ideal controller with reset and derivative actions, the gain approaches infinity as the frequency approaches either zero or infinity.   The gain of an actual controller is limited to a maximum value by the feedback system within the controller, and the phase shifts at high and low frequencies also differ from those for an ideal controller.   The mechanism of a typical pneumatic controller is shown in Fig. 6-9, and the transfer functions for this controller are compared with those for an ideal controller in the next section.

The responses with only two control modes are considered first; the reset valve is closed to obtain proportional-derivative control, and the derivative value is bypassed to obtain proportional-integral control.   The derivative and reset valves are assumed to be linear resistances so that a time constant

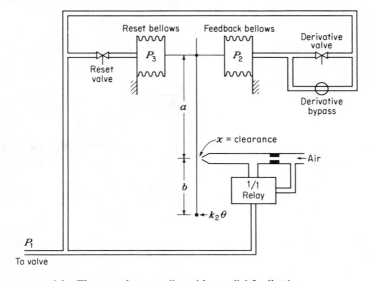

FIGURE 6-9   Three-mode controller with parallel feedback.

can be used to describe the response of P_2 and P_3 to changes in P_1.

$$C \frac{dP_2}{dt} = \frac{P_1 - P_2}{R_2}$$

$$C \frac{dP_3}{dt} = \frac{P_1 - P_3}{R_3}$$

(6-15)

$$\frac{P_2}{P_1} = \frac{1}{T_D s + 1} \qquad T_D = R_2 C$$

$$\frac{P_3}{P_1} = \frac{1}{T_R s + 1} \qquad T_R = R_1 C$$

(6-16)

The output pressure is a linear function of the clearance between the nozzle and the baffle, and the initial value is taken as zero for the transformed equation.

$$P_1 = -k_1 x$$

(6-17)

The negative sign is not important. A reverse-acting relay could be used to make P_1 increase with increases in x. The clearance x depends on the displacement caused by the input signal θ and the displacement resulting from any difference in pressures in the feedback and reset bellows, which are assumed to have the same spring constant.

$$x = k_2 \theta \frac{a}{a + b} + k_3 (P_2 - P_3) \frac{b}{a + b}$$

(6-18)

For proportional-integral control, the derivative valve is bypassed so that $P_1 = P_2$. Combining Eqs. (6-16) to (6-18) gives

$$P_1 = -k_1 k_2 \frac{a}{a + b} \theta - k_1 k_3 \frac{b}{a + b} \left(P_1 - \frac{P_1}{T_R s + 1} \right)$$

(6-19)

$$\frac{P_1}{\theta} = \frac{-k_1 k_2 a}{a + b} \frac{T_R s + 1}{T_R \left(1 + \frac{k_1 k_3 b}{a + b} \right) s + 1}$$

(6-20)

For comparison with the response of an ideal controller, Eq. (6-20) can be put in the form

$$\frac{P}{\theta} = \frac{K_c A (T_R s + 1)}{T'_R s + 1}$$

(6-21)

where $K_c = \dfrac{-k_2 a}{k_3 b}$

$T'_R = T_R (1 + A)$

$A = k_1 k_3 \dfrac{b}{a + b}$

The term A is usually about 100, and T'_R is therefore two orders of magnitude greater than T_R, the nominal reset time.   For high frequencies, $T'_R s + 1 \cong T'_R s$, and the approximate transfer function becomes the same as for an ideal proportional-integral controller.

$$\frac{P}{\theta} \cong \frac{K_c A(T_R s + 1)}{T'_R s} \tag{6-22}$$

and since $T'_R \cong A T_R$,

$$\frac{P}{\theta} \cong \frac{K_c(T_R s + 1)}{T_R s} \qquad \text{for } \omega \gg \frac{1}{T'_R}$$

At very low frequencies, the terms $T_R s + 1$ and $T'_R s + 1$ are both 1.0, and the controller has a gain of $K_c A$ and zero phase shift.   (The phase lag goes through a maximum between the corner frequencies $\omega = 1/T'_R$ and $\omega = 1/T_R$.)   The terms A and T'_R vary with K_c, since changing the gain setting involves moving one of the pivot points to change the ratio a/b.   The maximum phase lag therefore varies with both the reset and gain settings and is always less than 90°.   Typical response curves are shown in Fig. 6-10.

For proportional-derivative control, the reset valve in Fig. 6-9 is closed, and the displacement of the upper end of the baffle arm depends only on P_2.

$$x = k_2 \theta \frac{a}{a+b} + k_3 \frac{b}{a+b} P_2 \tag{6-23}$$

$$P_1 = -k_1 k_2 \theta \frac{a}{a+b} - k_1 k_3 \frac{b}{a+b} \frac{P_1}{T_D s + 1} \tag{6-24}$$

The final form of the equation is

$$\frac{P_1}{\theta} = K_c \frac{A}{1+A} \frac{T_D s + 1}{T'_D s + 1} \tag{6-25}$$

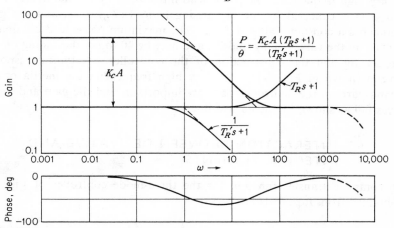

FIGURE 6-10   Proportional-reset controller with limited action.

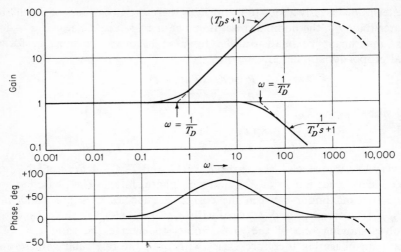

FIGURE 6-11   Proportional-derivative controller with limited action.

where K_c and A are given in Eq. (6-21) and

$$T'_D = \frac{T_D}{1 + A}$$

At low frequencies, $T'_D s + 1$ is about equal to 1; so the approximate transfer function is the same as for an ideal controller.

$$\frac{P_1}{\theta} = K_c(T_D s + 1) \qquad \text{for } \omega \ll \frac{1}{T'_D} \qquad (6\text{-}26)$$

At very high frequencies, the phase lead introduced by the numerator term $(T_D s + 1)$ is canceled by the lag arising from $1/(T'_D s + 1)$, and the gain approaches a constant value of $K_c A$. The maximum phase lead obtainable depends on the values of T_D and T'_D and may be 40 to 60° depending on the gain setting and the construction of the controller. A typical response curve is shown in Fig. 6-11. At very high frequencies the inertia of the moving parts in the controller becomes important, and the gain and phase curves fall as shown by the dashed lines.

6-5. INTERACTION BETWEEN DERIVATIVE AND INTEGRAL ELEMENTS

The complete transfer function for the three-mode controller of Fig. 6-9 is obtained from Eqs. (6-16) to (6-18).

$$P_1 = -k_1 k_2 \frac{a}{a + b} \theta - k_1 k_3 \frac{b}{a + b}\left(\frac{P_1}{T_D s + 1} - \frac{P_1}{T_R s + 1}\right) \qquad (6\text{-}27a)$$

Considerable rearrangement gives an equation which is similar to the equation for an ideal controller except for the limited action terms in the denominator.

$$\frac{P_1}{\theta} = \frac{K_c \dfrac{A}{B}\left(\dfrac{T_D s}{I} + 1 + \dfrac{1}{T_R I s}\right)}{\left(\dfrac{T_D s}{IB} + 1 + \dfrac{1}{T_R I B s}\right)} \tag{6-27b}$$

where $I = 1 + \dfrac{T_D}{T_R}$ = interaction factor

$$K_c = \frac{-k_2 a}{k_3 b}$$

$$A = k_1 k_3 \frac{b}{a + b}$$

$$B = 1 + A\left(\frac{T_R - T_D}{T_R + T_D}\right)$$

Because B is much greater than 1, the denominator terms become important only at very low and very high frequencies, where they act to limit the controller gain and bring the phase angle to zero. For intermediate frequencies the response is that of an ideal controller except that the gain, derivative time, and reset time differ from the dial settings, which are K_c, T_D, and T_R. The effective reset time is the dial setting times the interaction factor I, and the effective derivative time is the dial setting divided by I. The effect of interaction on the controller gain is even larger, since the gain is increased by a factor of about $(T_R + T_D)/(T_R - T_D)$.

$$\frac{P_1}{\theta} \cong K_c \frac{T_R + T_D}{T_R - T_D}\left(\frac{T_D s}{I} + 1 + \frac{1}{T_R I s}\right) \quad \text{at intermediate frequencies} \quad (6\text{-}28)$$

As T_D approaches T_R, the actual gain approaches infinity, according to Eq. (6-28). [The gain is really $K_c A$ when $T_D = T_R$ and becomes infinite at slightly higher values of T_D, i.e., when $B = 0$ in Eq. (6-27b).] When T_D is greater than T_R, the denominator terms containing B become negative, indicating unstable behavior. The instability arises because the pressure in the reset bellows changes more rapidly than the pressure in the feedback bellows. If a change in input θ causes a small increase in P_1, the pressure difference $P_2 - P_3$ *decreases* and causes a large further increase in P_1, rather than the canceling effect normally achieved by the feedback system. The block diagrams in Fig. 6-12 may help explain the operation of the controller. The signs of the original diagram were changed to show more clearly that P_2 is a negative feedback signal and that P_3 is positive feedback. The system is stable only when the negative feedback predominates, which means that T_D must be less than T_R.

In using controllers with this type of interaction, the ratio T_D/T_R is usually one-quarter or less. The Ziegler-Nichols recommendation (1) is

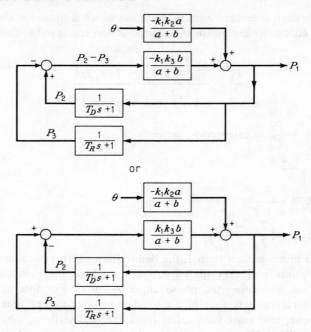

FIGURE 6-12   Block diagram for an interacting controller.

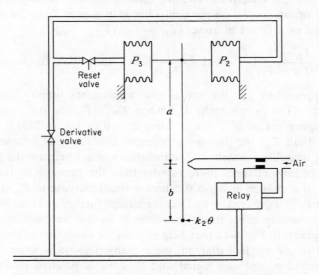

FIGURE 6-13   Three-mode controller with series feedback.

that $T_D/T_R = 1.2/8 = 0.15$.   With an interaction factor of 1.15, the actual gain is $1.35K_c$, the derivative time is $0.87T_D$, and the reset time is $1.15T_R$. For frequencies between the reset corner and the derivative corner, the response is about the same as for an ideal controller (Fig. 6-8).   The most important effect of the limited action is the lowering of the phase lead at frequencies near or above the derivative corner.   The maximum phase lead may be only 30 to 60°, and the maximum value is a function of the dial settings and the mechanical features of the instrument.

The instability arising from too much derivative action can be avoided by putting the derivative and integral resistances in series, as shown in Fig. 6-13.   The two resistances and capacities form a dead-end interacting system, and the equations relating P_1, P_2, and P_3 follow from Eq. (3-76).

$$\frac{P_3}{P_1} = \frac{1}{T_1 T_2 s^2 + (T_1 + T_2 + R_1 C_2)s + 1} = \frac{1}{T_D T_R s^2 + (2T_D + T_R)s + 1}$$

(6-29)

$$\frac{P_2}{P_1} = \frac{T_R s + 1}{T_D T_R s^2 + (2T_D + T_R)s + 1}$$

(6-30)

The transfer function for the controller is obtained by combining these equations with Eqs. (6-17) and (6-18).

$$\frac{P_1}{\theta} = \frac{K_c A I \left(\dfrac{T_D s}{I} + 1 + \dfrac{1}{T_R I s} \right)}{(A + I) \left[\dfrac{T_D s}{A + I} + 1 + \dfrac{1}{T_R (A + I)s} \right]}$$

(6-31)

where K_c and A are given by Eq. (6-27b) and

$$I = 1 + 2 \frac{T_D}{T_R}$$

The interaction factor I is greater than with parallel feedback, but the controller is stable for all derivative times.   Using the high derivative times permitted by this or similar arrangements is said to be helpful for starting up batch processes (2).

The interaction factors for several controller designs are given by Young (3), along with plots of the phase lead for some commercial instruments.   The interaction can be eliminated by generating proportional, derivative, and integral signals separately (in a more complex instrument) and then combining the signals, but there has been little demand for a pneumatic instrument of this type.   If the exact response of the controller is critical, an electronic instrument with practically ideal response could be used.

6-6. NONLINEAR EFFECTS

The response of pneumatic controllers with integral or derivative action is slightly nonlinear, because the values of the derivative and reset resistances vary with the pressure. If the air flow through the restriction is completely laminar and if the pressure drop is small, the mass rate of flow is directly proportional to the pressure drop times the density. The resistance in pounds per square inch per standard cubic foot per minute is therefore inversely proportional to the absolute pressure, and the effective reset and derivative times vary inversely with the pressure. If, instead, the air flow follows the orifice equation, the mass rate of flow varies with the square root of (pressure × density). The effective time constants of the controller then vary with only the square root of the absolute pressure but vary appreciably with the type and magnitude of the input signal.

Frequency-response tests of a Taylor Transcope controller showed the derivative and reset times to be proportional to about the -0.9 power of the absolute pressure, indicating practically laminar flow. The change in reset time was confirmed by making a *step* change in error and noting that the response curve was about 40 per cent steeper at high pressures (28 psia) than at low pressures (20 psia). However, this change in time constants would be important only for special cases; the derivative and reset times can generally be changed 1.5-fold with little noticeable effect on the system response, and the dial settings on controllers are often in error by 10 to 50 per cent in any case.

Further tests of the Transcope controller showed only a few per cent change in time constants for a several-fold change in amplitude of the input signal. This is consistent with the measured pressure effect; if the flow varies with the 0.9 power of the pressure drop, the resistance depends only on the 0.1 power of the pressure drop. In some controllers, capillary tubes of variable length or variable cross section are used, rather than valves, to make the flow more nearly proportional to the 1.0 power of the pressure drop. No matter how "linear" such devices are to changes in ΔP, the nonlinearity because of changes in absolute pressure still exists.

SATURATION EFFECTS

If the error signal is large, the controller output may reach the maximum or minimum value corresponding to a wide-open or a closed control valve. The controller is said to be saturated, since an increase in error has no further effect on the controller action. Saturation is a nonlinearity which makes the normal definitions of gain and phase lag invalid. However, the essential effect is to decrease the controller gain as the input amplitude increases. The system therefore has excess stability for large disturbances if the controller is tuned for small disturbances. The harmful effect of saturation is that the peak or the steady-state errors resulting from large load changes are greater than predicted from the linearized analysis or the performance for small load changes. Saturation is more likely to occur with cascade control than with simple control schemes because the combined controller gains can be quite high.

6-7. ELECTRONIC CONTROLLERS

Since 1960 several new types of electronic controllers have been developed to compete with pneumatic instruments for process applications. Electrical transmission lags are negligible, and electronic controllers can give better performance than pneumatic instruments for systems with very long transmission lines. The advantage would be small for most systems, since electronic control eliminates the transmission lag but not the valve lag; in fact the lag of electrically operated valves is generally greater than that of air-operated valves. For fast response the electrical signal is usually converted to an air pressure to operate a standard pneumatic valve or valve positioner. Other minor advantages of electronic controllers are that the gain, reset time, and derivative time are more nearly constant over the operating range, and there is no dead band caused by friction or backlash.

Major advantages of electronic control for certain installations are operability at low temperatures (ice can plug pneumatic lines) and compatibility with electrical sensing devices or computers. The main disadvantage of electronic control is the need for explosion-proof equipment in hazardous areas.   If these considerations do not apply, the choice between electronic and pneumatic control will probably be decided by the relative costs of installation and maintenance and the reliability of the instruments.

The heart of all electronic controllers is a high-gain operational amplifier connected between two electrical networks, input and feedback (4).   The networks can be formed by using two resistors or two capacitors, as shown in Fig. 6-14. The current drawn by the amplifier is negligible, and the amplifier gain usually exceeds 1,000, so that the junction point A is almost at ground potential. The output voltage of the amplifier must therefore be opposite in polarity to the input voltage.

Consider the effect of an increase in input voltage to the resistor circuit.   The input current increases, and if the output voltage remains the

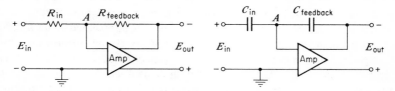

figure 6-14   Simple electronic control circuits.

same, the voltage at A will increase.   A slight increase in the amplifier input means a large change in amplifier output, and the output voltage drops until the amplifier input is again almost zero.   Since the current is the same in both resistors and the potential at A is practically zero, the ratio of the voltages is equal to the ratio of the resistances,

$$\frac{-E_{\text{out}}}{E_{\text{in}}} = \frac{R_{\text{feedback}}}{R_{\text{input}}}$$

Thus the gain of the circuit depends just on the ratio of the resistances, which can be accurately determined, and not on the gain of the amplifier itself.   The controller gain can be changed by changing the value of either resistance.

When capacitors are used, a change in input voltage leads to an input current as the capacitor is charged.   The imbalance in current is detected by the amplifier, and the amplifier almost immediately changes the output voltage to make the feedback current equal to the input current.   Thus the feedback capacitor receives the same charge as the input capacitor.   The charge is the capacitance times the voltage, and so the ratio of the voltages is equal to the inverse ratio of the capacitances.

$$-\frac{E_{\text{out}}}{E_{\text{in}}} = \frac{C_{\text{input}}}{C_{\text{feedback}}}$$

The gain of the circuit can be adjusted by varying one of the capacitances or by using a potentiometer to make the feedback voltage only a fraction of the amplifier output.

To obtain integral or reset action, a resistor is placed in parallel with the input capacitor (see Fig. 6-15).   As long as there is an input signal (a difference between the set point and the process variable) there is some input current and the amplifier continues to charge the feedback capacitor, which changes the output voltage.   To add derivative action requires a separate amplifier or a special circuit such as that shown in Fig. 6-15.

The schematic diagram and transfer function for a Taylor three-mode controller are presented here as one example of a modern electronic controller. This controller has direct currents in the range 1 to 5 ma as input and output signals.   (Other instruments have such ranges as 0 to 5 ma and 4 to 20 ma; an industry-wide standard may be agreed upon some day, as was done for pneumatic controllers.)   As shown in Fig. 6-15, the input current is passed through a 6,000-ohm resistor, and the voltage is subtracted from a set-point voltage to form the input voltage E_1.   The proportional response is changed by stepwise adjustments of the feedback capacitor bC_2, and the reset rate is adjusted in steps by changing the input resistance R_1.   Derivative action (Pre-Act) is obtained by putting a resistance R_2 in the feedback circuit and adding the capacitor $(1 - b)C_2$ to make the derivative time T_D independent

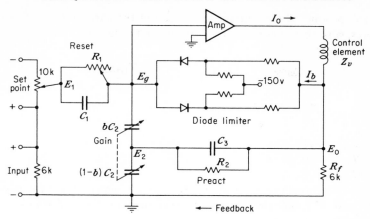

FIGURE 6-15 Simplified diagram of Taylor Transcope electronic controller, model 70RF.

of changes in gain.   The small capacitor C_3 presents excessive variation in controller output when noisy signals are present.   The diode limiters are provided to prevent the output current from dropping below 0.5 ma or exceeding 5.5 ma.

The following analysis of the Taylor controller is condensed from unpublished notes by N. B. Nichols (see also Ref. 5 and Ref. 8, Chap. 13). The amplifier gain is about 1,500, and so E_g is practically zero.   The amplifier current is normally less than 0.001 μa.   The transformed equations for the system are

$$C_1 s E_1 + \frac{E_1}{R_1} = -bC_2 s E_2 \tag{6-32}$$

$$(1 - b)C_2 s E_2 = C_3 s (E_o - E_2) + \frac{E_o - E_2}{R_2} - bC_2 s E_2 \tag{6-33}$$

$$E_o = [I_o - I_b - (1 - b)C_2 s E_2 - bC_2 s E_2]R_f \tag{6-34}$$

The output current I_o flows through the control element or other load, Z_v, and I_b is the small current flowing into the diode biasing circuit.   Combining Eqs. (6-32), (6-33), and (6-34) leads to the transfer function relating I_o and the input deviation E_1.

$$-\frac{I_o - I_b}{E_1/R_f} = \frac{C_1}{bC_2}\left(1 + \frac{1}{R_1 C_1 s}\right) \frac{R_2(C_2 + C_3)s + 1}{R_2 C_3 s + 1}$$

$$\times \frac{1}{1 + R_f C_2 s \dfrac{R_2 C_3 s + 1}{R_2(C_2 + C_3)s + 1}} \tag{6-35}$$

The product $R_f C_2$ is about 0.06 sec, and the current I_b is a few micro-amperes, and so the controller equation for the normal frequency range is approximately

$$-\frac{I_o}{I_{in}} = K_C \left(1 + \frac{1}{T_R s}\right) \frac{T_D s + 1}{T_D' s + 1}$$

where $K_C = \dfrac{C_1}{bC_2}$

$T_R = R_1 C_1$
$T_D = R_2(C_2 + C_3)$
$T_D' = R_2 C_3$

For this controller, C_2 is about 20 times C_3, which makes the maximum phase lead 65° and the maximum gain increase from derivative action twentyfold.

PROBLEMS

1. Calculate the maximum controller gain for a process with these characteristics:

$$1 \text{ first-order lag, } K_{P_1} = 2, \, T = 10 \text{ min}$$

$$6 \text{ equal first-order lags, } K_{P_2} = 1, \, T = 1 \text{ min each}$$

$$1 \text{ control valve, } K_v = 1.5, \, T_v = 2 \text{ sec}$$

If derivative action were added, about what derivative time should be used? How much improvement in controllability would result from adding derivative action? Can you cite an example where the effect of derivative action is much greater?

2a. Calculate the maximum controller gain for the following system:

$$\text{Valve: } T = 10 \text{ sec, } K_v = 3$$

$$\text{Second-order process: } \zeta = 0.2, \, \omega_n = 0.05 \text{ sec}^{-1}, \, K_p = 2$$

$$\text{Time delay: } L = 3 \text{ sec}$$

b. If reset action were added to the controller, with a reset rate of $1.2/P_u$, where P_u is the period at the ultimate sensitivity, what phase lag would be added and what would be the maximum controller gain?

3. Show that a system with four time constants and an ideal three-mode controller can have 180° phase lag at more than one frequency.

4. The approximate transfer function for a controller is

$$G_c = \frac{K_c(T_D s + 1 + 1/T_R s)}{T_D' s + 1 + 1/T_R' s}$$

Plot the maximum phase lead as a function of the ratio T_D/T_D'.

REFERENCES

1. Ziegler, J. G., and N. B. Nichols: Optimum Settings for Automatic Controllers, *Trans. ASME*, **64**:759 (1942).
2. Caldwell, W. I., G. A. Coon, and L. M. Zoss: "Frequency Response for Process Control," chap. 19, McGraw-Hill Book Company, New York, 1959.
3. Young, A. J.: "An Introduction to Process Control System Design," chap. 12, Longmans, Green & Co., Inc., New York, 1955.
4. Tucker, G. K.: User Guide for the Electronic Controller, *Automatic Control*, **11**(8):11 (August, 1959).
5. Taylor Instrument Companies: *Booklet* 11 A 100, November, 1960.

7 FREQUENCY RESPONSE OF CLOSED-LOOP SYSTEMS

The last two chapters showed how the frequency response of an open-loop system could be used to determine the stability limits of the closed-loop system and to predict the optimum controller settings. The open-loop analysis also gives the critical frequency of the system, which is a measure of the speed of response to step changes in set point or load. The frequency response of the closed-loop system is the response to sinusoidal changes in set point or load, and this additional information is often helpful in judging the performance of the control system. If the major load changes are cyclical fluctuations, the frequency response of the closed loop is more important than the transient response, especially since the error for certain frequencies is greater than if no control whatever is used. The closed-loop frequency response can also be used to estimate the peak error following step changes in load, and combining this information with the frequency and decay ratio predicted from the open-loop analysis gives a fairly accurate representation of the transient response. A further reason for studying the closed-loop frequency response is to predict the behavior of multiloop systems. With cascade control the maximum gain and critical frequency for the main loop are determined from a Bode plot which includes the response of the inner closed loop along with the response of the other elements.

7-1. CHANGE IN SET POINT

The simplest case to consider is the response of a system with no measurement lag to changes in set point, as shown in Fig. 7-1. The closed-loop transfer function was given in Chap. 4,

$$\frac{\theta}{\theta_c} = \frac{G}{1 + G} \tag{7-1}$$

For the case where there are three time constants and a proportional controller,

$$G = \frac{K}{(T_1 s + 1)(T_2 s + 1)(T_3 s + 1)} \tag{7-2}$$

$$\frac{\theta}{\theta_c} = \frac{K}{T_1 T_2 T_3 s^3 + (T_1 T_2 + T_1 T_3 + T_2 T_3)s^2 + (T_1 + T_2 + T_3)s + 1 + K} \tag{7-3}$$

By letting $s = j\omega$ in Eq. (7-3) and clearing the complex fraction, the amplitude and phase angle could be calculated for various frequencies. However, it is much easier to obtain G by combining the amplitudes and phase angles of the individual elements and then to get $1 + G$ and $G/(1 + G)$ by using vectors.

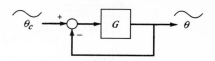

FIGURE 7-1 Frequency response of closed-loop system for servo-operation.

Example 7-1

Obtain the closed-loop response θ/θ_c for the system of Fig. 5-11 which has time constants of 10, 5, and 2 sec. Use an overall gain $K = 6$.

The values of gain and phase are obtained as shown below. Note that, to divide vectors, the amplitudes are divided and the denominator phase angle is subtracted from the numerator phase angle.

ω, rad/sec	0.1	0.2	0.41	0.6
A.R., Fig. 5-11	0.67	0.33	0.085	0.034
ϕ	$-82°$	$-130°$	$-180°$	$-203°$
G	$4.0\underline{/-82°}$	$2.0\underline{/-130°}$	$0.51\underline{/-180°}$	$0.20\underline{/-203°}$
$1 + G$	$4.45\underline{/-64°}$	$1.56\underline{/-100°}$	$0.49\underline{/0°}$	$0.83\underline{/5°}$
$\dfrac{G}{1 + G}$	$0.90\underline{/-18°}$	$1.28\underline{/-30°}$	$1.04\underline{/-180°}$	$-0.24\underline{/-208°}$

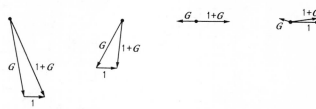

The closed-loop response is sketched in Fig. 7-2.

The general features of the closed-loop response are evident from Eq. (7-1) and the vectors in Example 7-1. At very low frequencies the closed-loop response is $K/(1 + K)$, since the time constants contribute no damping and no phase lag. The factor $K/(1 + K)$ is also the steady-state value of the output following a unit step change in set point, as was shown in Chap. 4. At a frequency somewhat less than the critical frequency there is a resonant peak similar to that obtained with an underdamped second-order system (see Figs. 5-7 and 5-8). If the controller gain is half the maximum, the resonant peak is about two to three times the value at low frequencies. Using a controller gain closer to the maximum gives lower damping coefficients, and the resonant peak becomes even higher. At the critical frequency, the gain is $(K/K_{max})/(1 - K/K_{max})$ or (overall gain)/(1 - overall

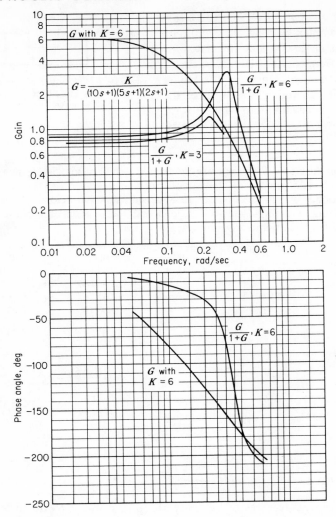

FIGURE 7-2   Closed-loop frequency response for Example 7-1.

gain), which is usually about 1.   The gain curve approaches that for the open loop at very high frequencies because as G goes to zero $1 + G$ goes to unity.

The phase lag for the closed loop is less than that for the open loop at all frequencies up to the critical frequency.   The curves intersect at this point and then approach each other again at very high frequencies.   With increasing controller gain, the phase curve becomes steeper at the critical frequency: the effect of increased gain or decreased damping coefficient is similar to that shown in Fig. 5-8 for second-order processes.

The maximum gain of the closed-loop response, which is called M_p, can be used as a criterion of system performance. A high value of M_p means a large error for sinusoidal inputs near the resonant frequency, and, what is generally more important, it indicates a large overshoot in the transient response. For the design of servomechanisms the recommended value of M_p is 1.4 ± 0.2 (1), which corresponds to a damping coefficient of about 0.4 for a simple second-order process. For process control systems, the recommended controller gain is closer to the maximum value, and M_p values of 2 to 3 would be typical. For the system of Example 7-1, a steady-state gain $K = 6$ ($K_{max} = 12.7$) would be used to get a favorable transient response, one with a decay ratio of about $\frac{1}{4}$ or a damping coefficient of 0.2 to 0.25. To get $M_p = 1.4$ would require a K of 3.5, or only one-fourth the maximum value. Different recommendations for servomechanisms and for process control systems are not surprising; a large overshoot may be intolerable in controlling a machine or a missile, but the peak overshoot is not so important as the error integral for most chemical processes.

Since the function $G/(1 + G)$ is a fairly simple one, generalized plots can be prepared to eliminate the need for combining vectors. The Black-Nichols chart of Fig. 7-3 is a semilog plot of open-loop gain versus open-loop phase angle with superimposed contours of closed-loop gain M and closed-loop phase angle α. The open-loop response is plotted on the chart for various frequencies and a smooth curve drawn through the points. The closed-loop gain and phase angle for each frequency are obtained by interpolation, or if only the peak gain is needed, the M contour which would just be tangent to the open-loop curve is selected. The curves in Fig. 7-3 are symmetrical about the 180° line. The closed-loop response for large phase lags can therefore be obtained by counting back from 180°, for example, by using the 170° line for an open-loop phase lag of 190°.

7-2. CHANGES IN LOAD

The closed-loop frequency response for changes in load can be readily calculated from the appropriate transfer functions. There is no general solution comparable to the Black-Nichols chart, which holds only for set-point changes, though this plot can be used to help get the response to load changes. The response to load changes depends on the location of the load disturbance. For a process with three time constants, load changes might enter the loop at three places, as shown in Fig. 7-4.

For a load disturbance before the first time constant

$$\frac{\theta}{L_1} = \frac{G_{L_1}}{1 + G} = \frac{K_1 K_2 K_3}{(T_1 s + 1)(T_2 s + 1)(T_3 s + 1)} \frac{1}{1 + G} \tag{7-4}$$

However, G_{L_1} differs from G only by a constant, and the transfer function

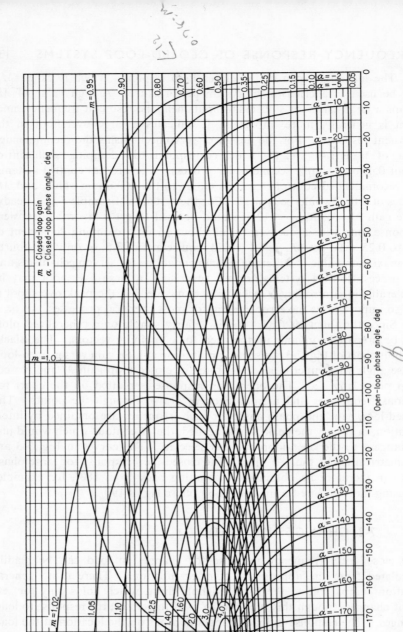

FIGURE 7-3 Black-Nichols chart. (Courtesy of the Taylor Instrument Companies.)

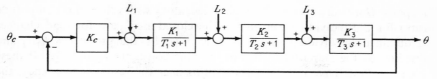

FIGURE 7-4   Frequency response of closed-loop system to load changes.

can be written in shorter form.

$$G = K_c G_{L_1} \tag{7-5}$$

$$\frac{\theta}{L_1} = \frac{1}{K_c} \frac{G}{1 + G} = \frac{1}{K_c} \frac{\theta}{\theta_c} \tag{7-6}$$

The term $G/(1 + G)$ is the response to changes in set point, which can be determined from the Black-Nichols chart.   The term $1/K_c$ merely shifts the gain curve by a constant factor, and the height of the resonant peak relative to the zero-frequency gain stays the same.

For load changes at L_2 in Fig. 7-4, the closed-loop transfer function is

$$\frac{\theta}{L_2} = \frac{G_{L_2}}{1 + G} = \frac{K_2 K_3}{(T_2 s + 1)(T_3 s + 1)} \frac{1}{1 + G} \tag{7-7}$$

This transfer function can also be put in a form that includes the term $G/(1 + G)$,

$$G = \frac{K_c K_1}{T_1 s + 1} G_{L_2} \tag{7-8}$$

$$\frac{\theta}{L_2} = \frac{T_1 s + 1}{K_c K_1} \frac{G}{1 + G} \tag{7-9}$$

The easiest way to obtain the frequency response for load changes at L_2 is to use the Black-Nichols chart to get $G/(1 + G)$, divide the gains by the term $K_c K_1$, and combine the result with the gain and phase lead corresponding to the linear lead $(T_1 s + 1)$.   Usually, the phase lag for load changes has little significance, and only the gain is calculated.

For load changes at L_3, or just before the last process element, the transfer function can also be expressed in two ways,

$$\frac{\theta}{L_3} = \frac{G_{L_3}}{1 + G} = \frac{K_3}{T_3 s + 1} \frac{1}{1 + G} \tag{7-10}$$

$$G = \frac{K_c K_1 K_2}{(T_1 s + 1)(T_2 s + 1)} G_{L_3} \tag{7-11}$$

$$\frac{\theta}{L_3} = \frac{(T_1 s + 1)(T_2 s + 1)}{K_c K_1 K_2} \frac{G}{1 + G} \tag{7-12}$$

Although Eq. (7-12) looks more complex, it is easier to use than Eq. (7-10) in getting the gain for load changes.   For most problems, the open-loop response G will already have been obtained, along with values of the amplitude ratio for the individual elements. The closed-loop gain for load changes is obtained by dividing the value $G/(1 + G)$ by the product $K_c K_1 K_2$ and dividing by the amplitude ratios of the time constants T_1 and T_2. If these time constants have low amplitude ratios near the resonant frequency, the peak gain may be much larger than the low-frequency gain, as is shown in the following example.

Example 7-2

Obtain the closed-loop gain for load changes at three different points in the

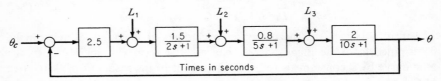

FIGURE E7-2   Diagram for Example 7-2.

following system (same as Example 7.1):

$$K_c = 2.5 = \tfrac{1}{2}K_{c,\,\text{max}} \qquad K = 6, K_{\text{max}} = 12.7$$

The closed-loop response to set-point changes was given in Fig. 7-2, which shows a zero-frequency gain of $\tfrac{6}{7}$ and a resonant peak of 3.0 at $\omega = 0.32$ rad/sec.
From Eq. (7-6),

$$\frac{\theta}{L_1} = \frac{1}{K_c}\frac{\theta}{\theta_c} = \frac{1}{2.5}\frac{\theta}{\theta_c}$$

At low frequencies, $\theta/L_1 = 0.86/2.5 = 0.34$; peak gain $= 3.0/2.5 = 1.2$.
From Eq. (7-9),

$$\frac{\theta}{L_2} = \frac{2s + 1}{(2.5)(1.5)}\frac{\theta}{\theta_c}$$

At low frequencies, $\theta/L_2 = 0.86/3.75 = 0.23$.
At $\omega = 0.2$, $2s + 1 = 1 + j(0.4) = 1.08$, or from Fig. 5-3,

$$\text{A.R.} = 0.92 \qquad \text{at } \omega T = 0.4$$

$$\frac{\theta}{L_2} = \frac{1.08}{3.75}(1.3) = 0.37$$

A peak gain of 0.95 is reached at $\omega \cong 0.33$. From Eq. (7-12),

$$\frac{\theta}{L_3} = \frac{(2s + 1)(5s + 1)}{(2.5)(1.5)(0.8)}\frac{\theta}{\theta_c}$$

At low frequencies $\theta/L_3 = 0.86/3.0 = 0.29$. At $\omega = 0.2$,

$$2s + 1 = 1.08$$

$$5s + 1 = 1.414$$

$$\frac{\theta}{L_3} = \frac{(1.08)(1.414)}{3}1.3 = 0.66$$

A peak gain of 2.2 is reached at $\omega \simeq 0.35$. The complete curves are shown in Fig. 7-5. The ratio of peak gain to zero-frequency gain is

$$R_1 = 3.5$$

$$R_2 = 4.1$$

$$R_3 = 7.6$$

The general features of the response to load changes deserve further comment. At low frequencies the closed-loop gain is $K_L/(1 + K)$, which is the same as the steady-state value after a step change in load, and the numerical values of course depend on the type and location of the load change. For Fig. 7-5, the values of K_L are $K_{L_1} = 1.5(0.8)(2) = 2.4$, $K_{L_2} = 1.6$, and $K_{L_3} = 2.0$; so the limiting values of θ/L do not fall in sequence.

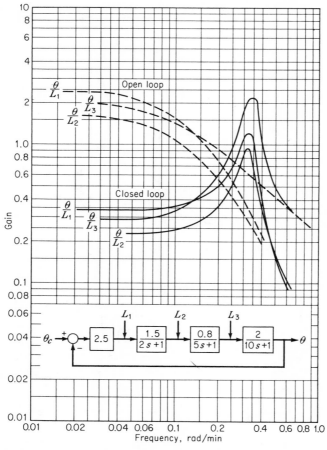

FIGURE 7-5   Closed-loop gain for Example 7-2.

Values greater than 1 could even be obtained for θ/L_3 if K_{L_3} were large enough; this would not mean poor control, since the open-loop gain would be much larger. The open-loop curves approach K_L at low frequencies; so for all cases the control system reduces the low-frequency error by a factor $1 + K$.

For frequencies near the resonant frequency the closed-loop gain exceeds the open-loop gain, which means that the error is greater than if no controller were used. The ratio of the peak gain to the zero-frequency gain increases as the load is moved closer to the end of the process. For load changes at L_1, the disturbance is damped by all three time constants before reaching the output, whereas a load change at L_3 is damped by only one time constant. The magnification of certain disturbances by the controller is not always a cause for concern, since the major load changes

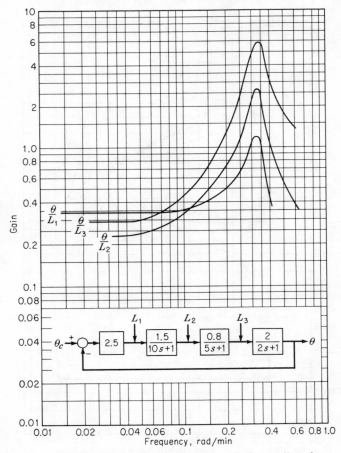

FIGURE 7-6   Closed-loop gain for a system with the smallest time constants at the end.

are usually step changes, ramp changes, or random fluctuations.   If periodic disturbances are expected, as from a piston pump or a previous control system, the control loop should be designed to have a critical frequency much greater than or much less than the frequency of the disturbance. For frequencies much higher than the critical frequency, the closed-loop gain is still slightly greater than the open-loop gain, but the error is small in any case.   Thus the main function of the controller in regulator operation is to compensate for low-frequency changes in load or for noncyclical changes; for frequencies greater than about half the critical frequency, the controller actually magnifies the effect of the disturbance.   Decreasing the controller gain lowers the peak gain but makes the control poorer for low-frequency disturbances.   The curves of Fig. 7-5 are typical of those for systems at recommended controller settings, except that the curves would approach zero at very low frequencies if reset action were added to the controller.

The difference between the response curves for various loads becomes greater if the smallest time constants are at the end of the system.   In Fig. 7-6, the 10-sec time constant is put first and the 2-sec time constant placed at the end.   The response θ/L_3 has a relative peak height of 20, or three times the value in Fig. 7-5.   Fortunately, the major time constant is usually the last one in the system (except for the measuring device), and load disturbances are rarely magnified to this extent.

7-3. USE OF CLOSED-LOOP RESPONSE TO PREDICT TRANSIENT RESPONSE

Most transient-response curves are damped oscillations that can be characterized by four parameters: the height of the first peak, the frequency, the decay ratio, and the steady-state value.   The steady-state value can be calculated exactly from the process gain and the controller characteristics. The frequency of the damped oscillations can be predicted from the open-loop or closed-loop frequency response.   The decay ratio can be estimated from the gain margin or phase margin, and a value of $\frac{1}{4}$ is typical.   A method of estimating the height of the first peak would permit the transient response to be predicted without the major effort of performing the inverse transformation.   Schemes for calculating the transient response from the frequency response have been presented (2), but the work involved is still much greater than that required to obtain the frequency response.   The method presented here is based on a correlation between the peak error and the peak gain of the frequency response, which is easy to use and is generally accurate to within 10 per cent.

The similarity between the frequency response of a closed-loop system and a second-order process has already been noted, and the transient-response curves for typical control systems and underdamped second-order processes are also of the same shape.   For a second-order process, the

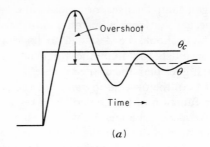

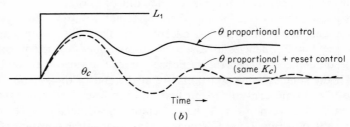

FIGURE 7-7   Transient response to changes at the beginning of the system.   (a) Set-point change proportional control;   (b) load change.

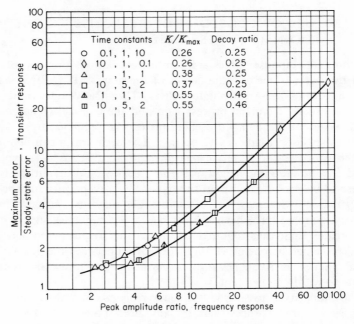

FIGURE 7-8   Correlation between transient response and frequency response.

maximum amplitude ratio and the overshoot are functions of ζ, the damping coefficient. This suggests that the effective damping coefficient for the closed-loop system can be obtained from the relative peak height of the frequency-response curve and this value of ζ used to predict the overshoot from the equation for second-order systems.

Consider first the response to set-point changes, and take Fig. 7-2 as an example. For $K = 6$, the peak gain θ/θ_c is 3, the zero-frequency gain is 0.86, and the relative peak height is 3.5. A second-order system has a peak amplitude ratio of 3.5 when the damping coefficient is 0.14 [see Eq. (5-16) or Fig. 5-7]. For a unit change in input, the overshoot for the second-order system is 0.66 when $\zeta = 0.14$, as shown by Fig. 3-19. The predicted overshoot for the actual process is 0.66 if the final value is 1.0. For a unit change in set point, the final value is $K/(1 + K)$, or 0.86, and the predicted peak value is $0.86(1 + 0.66) = 1.43$. A relative peak height of 1.67, which results from using $K = 3$, corresponds to $\zeta = 0.32$ and an overshoot of 0.35 relative to the final value. The overshoot does not change much with gain, since the maximum overshoot for a second-order system is 1.0, corresponding to $\zeta = 0$. For most processes the overshoot is generally between 0.3 and 0.7 for unit changes in set point.

For changes in load that occur at the beginning of the process, the closed-loop response θ/L_1 has the same shape as the response θ/θ_c, and with just proportional control the transient-response curves for set-point changes and load change should be similar. For the example of Fig. 7-2, a unit load change at L_1 results in a steady-state value of 0.34, and the peak error is $0.34(1.66) = 0.56$. If integral plus proportional control is used, the overshoot cannot be based on the final value, since the final offset is zero. However, for the recommended controller settings, the reset action has hardly any effect on the height of the first peak, and the predicted overshoot can be based on the final value that would result if only proportional control were used. The predicted responses for the cases discussed so far are sketched in Fig. 7-7.

If the load change occurs after the first process element, the peak error may be more than twice the final value (see Figs. 4-6 and 4-8), and the concept of an equivalent second-order system cannot be used. However, moving the load change toward the end of the process increases both the overshoot of the transient response and the peak gain of the frequency response. Figure 7-8 shows that a general correlation exists between the two responses, at least for systems with three time constants. To allow for the effects of the constants K_c, K_L, and K, the correlation is based on the amplitude ratio of the frequency response, i.e., the peak gain relative to the zero-frequency gain. Similarly, the peak error of the transient response is divided by the steady-state error that would exist with just proportional control, which is $K_L/(1 + K)$ for a unit change in load. The transient-response peaks for the plotted points were obtained by using an analog computer: data for two systems were taken from published curves (3). The upper curve is for cases where the transient response shows a decay ratio of 0.25, generally considered to be ideal. The lower curve is for a decay ratio of 0.46, obtained by using gains closer to the maximum. Increasing the gain for a given system increases both the peak amplitude ratio and the peak error in the transient response; but the change in the amplitude ratio is relatively greater. As the controller gain approaches the maximum, the amplitude ratio approaches infinity but the peak error in the transient response approaches K_L or a lower value.

The three points plotted for each system in Fig. 7-8 were obtained for load changes introduced before the first, second, and last time constant, and the maximum error increases in that order. Thus for the 10, 5, 2 system of Fig. 7-6, unit load changes at the three points result in maximum errors of 1.5, 2.7, and 4.4 times $K_L/(1 + K)$. The transients for the 2, 5, 10 system of Fig. 7-5 were not calculated, but the peak error would be estimated as follows: With $K_c = 2.5$ (0.5 times the maximum), the peak amplitude ratio is 7.6 for load changes at L_3. If a controller gain of half the maximum were assumed to give a decay ratio of 0.25, the predicted maximum error would be $2.8K_L/(1 + K)$. The decay ratio would actually be about 0.35 for $K_c = 2.5$, to judge from the results for the 10, 5, 2 system, and a better estimate of the peak error is $2.5K_L/(1 + K)$. The error in the first prediction would not often be significant. If necessary, the controller gain needed for a decay ratio of 0.25 can be calculated by using the damped frequency-response method (4).

Example 7-3

Predict the peak error for a unit load change to the system of Fig. 4-20 (a 1-min time constant, a 1-min time delay, and a controller gain of 1.2).

Obtain the open-loop response, and use the Black-Nichols chart to get the peak value of θ/θ_c.

ω, rad/min	$\omega T = \omega L$	Open loop				Closed loop	
		ϕ_T	ϕ_L	ϕ	A.R.	G	$G/(1 + G) = \theta/\theta_c$
1.5	1.5	57°	86°	143°	0.55	0.66	1.08
1.6	1.6	58°	92°	150°	0.53	0.63	1.12
1.7	1.7	59°	97°	156°	0.50	0.60	1.17
1.8	1.8	61°	103°	164°	0.48	0.58	1.21
1.9	1.9	62°	108°	170°	0.46	0.55	1.17

$$\frac{\theta}{L} = \frac{G_L}{1 + G} = \frac{K_L}{1.2e^{-s}}\frac{G}{1 + G}$$

$$\text{Peak amplitude ratio for } \frac{\theta}{L} = \frac{1.21}{1.2}\, 2.2 = 2.22$$

From Fig. 7-8,

$$\text{Predicted } \theta_{max} = 1.4 \frac{1}{2.2} = 0.64$$

$$\text{Actual } \theta_{max} = 0.68 \quad \text{from Fig. 4-20}$$

7-4. EFFECT OF MEASUREMENT LAG

The Black-Nichols chart gives only the measured response to set-point changes, and the measurement lag has been neglected in the previous examples. If the measuring device has one or two first-order elements (a thermometer bulb or a bulb in a well), the actual output amplitude is greater than the measured output. If the measuring device is an underdamped second-order element such as a manometer, the measured amplitude may be greater than the actual amplitude of the output fluctuations.

The actual output response for set-point changes is obtained from the transfer function

$$\frac{\theta}{\theta_c} = \frac{G}{1 + GG_m} \qquad (7\text{-}13)$$

The Black-Nichols chart can be used to obtain $GG_m/(1 + GG_m)$ and the result divided by G_m.

$$\frac{\theta}{\theta_c} = \frac{GG_m}{1 + GG_m} \frac{1}{G_m} \qquad (7\text{-}14)$$

If the measurement lag is significant, the actual output for load changes is

$$\frac{\theta}{\theta_L} = \frac{G_L}{1 + GG_m} = \frac{GG_m}{1 + GG_m} \frac{G_L}{GG_m} \qquad (7\text{-}15)$$

Example 7-4

The components of a control loop for an air heater are a valve, the heater, and a temperature bulb, which have time constants of 5, 20, and 100 sec, respectively, and a proportional controller. It is suggested that the response of the system would be improved by using a thermowell to double the bulb time constant, since this would permit a higher controller gain to be used. Calculate the main characteristics of the actual temperature response to load changes when $T_m = 100$ and when $T_m = 200$ sec.

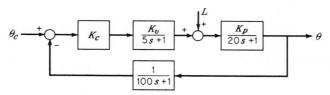

FIGURE E7-4a Diagram for Example 7-4.

Original system: Pick K_c to allow 30° phase margin.

ω, rad/sec	5ω	20ω	100ω	Phase lag, deg	A.R.
0.06	0.3	1.2	6	$17 + 50 + 81 = 148$	$0.95 \times 0.64/6 = 0.101$
0.07	0.35	1.4	7	$19 + 55 + 82 = 156$	$0.94 \times 0.58/7 = 0.078$

150° lag at $\omega = 0.062$, A.R. $= 0.095$. Use $K = 1/0.095 = 10.5 = K_c K_v K_p$.

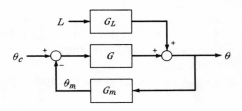

FIGURE E7-4b   Diagram for Example 7-4.

Assume $K_v = K_p = 1$ for convenience.

$$\frac{\theta}{L} = \frac{G_L}{1 + GG_m}$$

$$\frac{\theta}{L} = \frac{GG_m}{1 + GG_m} \frac{G_L}{GG_m}$$

$$\frac{\theta}{L} = \frac{\theta_m}{\theta_c} \frac{(100s + 1)(5s + 1)}{K}$$

$[K_p/(20s + 1)$ cancels out.]

ω	GG_m	$\dfrac{GG_m}{1 + GG_m} = \dfrac{\theta_m}{\theta_c}$	$(100s + 1)(5s + 1)$	$(\theta/L)K$
0.06	$1.06\underline{/-148°}$	1.83	6/0.95	11.7
0.07	$0.82\underline{/-156°}$	1.95	7/0.94	14.6
0.08	$0.54\underline{/-163°}$	1.05	8/0.92	9.1

Max relative peak height $\cong \dfrac{15}{K}(1 + K) = 16.4$

Predicted transient peak for unit load change $= 5.5 \dfrac{K_L}{1 + K}$ from Fig. 7-8

$$= 5.5 \frac{1}{11.5} = 0.48$$

Modified system, $T_m = 200$ instead of 100. At $\omega = 0.057$, $\phi = -150°$, A.R. $= 0.055$.   Use $K = 18.2$ for 30° phase margin.

ω	GG_m	$\dfrac{GG_m}{1 + GG_m}$	$(200s + 1)(5s + 1)$	$\dfrac{\theta}{L}K$
0.057	$1/\underline{-150°}$	1.9		
0.06	$0.92/\underline{-152°}$	1.92	12/0.95	24.3
0.07	$0.71/\underline{-160°}$	1.93	14/0.94	28.8
0.08	$05.5/\underline{-166°}$	1.1	16/0.92	19.1

$$\text{Max relative peak height} \cong 29\,\frac{19.2}{18.2} = 31$$

$$\text{Predicted transient peak} = \frac{10}{19.2} = 0.52$$

Summary	Peak error	Frequency of damped oscillations
$T_m = 100$	0.48	$\omega = 0.062$
$T_m = 200$	0.52	$\omega = 0.057$

Increasing the measurement lag increases both the period and the peak error by about 10 per cent, and the control is slightly poorer than before, though the *measured* error would be cut almost in half.

PROBLEMS

1. A flow control system has a valve time constant of 10 sec, a process time constant of 0.5 sec, and a measurement time constant of 1.0 sec.

   a. What is the maximum permissible value of K?

   b. Plot the actual and measured closed-loop frequency response for load changes occurring after the control valve if the controller gain is 0.3 and 0.5 times the maximum value.

2. A proportional controller is used for a system with three equal time constants.   Using the closed-loop frequency response to predict the peak error, sketch the transient response to a load change just after the first element.   Use an overall gain of 4.4, and compare your prediction with Fig. 4-10.

3. Compare the closed-loop frequency response for a pure dead-time process with that for a system with three time constants.   Use half the maximum gain for both cases.   If possible, give a physical explanation for the difference in shape.

4. A process has time constants of 1, 2, 5, and 20 sec.   The gain, integral, and derivative settings are to be chosen by using Eq. (9-3) or (9-2).   Compare the closed-loop gain for sinusoidal load changes at the start of the process for proportional-integral control and proportional-integral-derivative control.

REFERENCES

1. Brown, G. S., and D. P. Campbell: "Principles of Servomechanisms," p. 107, John Wiley & Sons, Inc., New York, 1948.
2. Del Toro, V., and S. R. Parker: "Principles of Control Systems Engineering," p. 352, McGraw-Hill Book Company, Inc., New York, 1960.
3. Brown, G. S., and D. P. Campbell: "Principles of Servomechanisms," p. 332, John Wiley & Sons, Inc., New York, 1948.
4. Caldwell, W. I., G. A. Coon, and L. M. Zoss: "Frequency Response for Process Control," p. 106, McGraw-Hill Book Company, Inc., New York, 1959.
5. Caldwell, W. I., G. A. Coon, and L. M. Zoss: "Frequency Response for Process Control," p. 267, McGraw-Hill Book Company, Inc., New York, 1959.

8 COMPLEX CONTROL SYSTEMS

The performance of a control system is determined by the nature of the process, the characteristics of the controller, and the location and magnitude of the disturbances. Sometimes the performance of a simple feedback control system can be improved considerably by making relatively minor changes, such as reducing a time delay or one of the smaller time constants, using a positioner to improve the valve response, or adding derivative action to the controller. If the response is still unsatisfactory because of large uncontrolled load changes, more complex control schemes can be considered.

8-1. TYPES OF CONTROL SCHEMES

One approach is to use additional controllers to regulate the flows, temperatures, and compositions that are load variables or inputs to the process. For example, the flow rate and temperature of the feed to a distillation column are often controlled by separate feedback systems to make it easier to achieve close control of product composition. If the different feedback systems are completely separate, they can be designed by conventional methods and need not be considered here. The interactions that occur when two control systems have elements in common can be serious and will be discussed later.

Feedforward action can be added to the control system for cases where the major load variable can be measured but not controlled. This variable might be the flow rate or concentration from a previous process or perhaps the ambient temperature. By using changes in the load variable to adjust the output of the main controller, the upsets from changes in this variable can be anticipated and made less severe. The transfer function for load changes must be accurately known and nearly constant to achieve the full benefits of feedforward compensation. The scheme is usually restricted to correcting for only one load variable, though a simple computer could be used to take account of changes in two or three inputs.

Frequently the best way of using an additional controller to decrease upsets is to use the scheme called "cascade control." The output of the primary controller is used to adjust the set point of a secondary controller, which in turn sends a signal to the control valve. The process output is fed back to the primary controller, and a signal from an intermediate stage of the process is fed back to the secondary controller. The main advantage of cascade control is that the performance is better for all types of load changes. For disturbances that enter near the beginning of the system, the secondary controller starts corrective action before the process output shows any deviation, and the error may be 10- to 100-fold lower than with a single controller. For disturbances that enter the last element of the process, the error integral may be reduced about 2- to 5-fold because the cascade system has a higher natural frequency.

There are several other schemes involving the use of two or more interconnected instruments. A pressure transmitter may be used to adjust the set point of a temperature controller for an evaporator or distillation column. Similarly, the flow of stream A may be controlled at a value proportional to the measured flow of stream B. These two schemes should be called "ratio control" or "proportioning control" and the term "cascade control" saved for cases where there are two complete feedback loops, one within the other. Examples of selector control systems include two controllers that are connected to a single valve to provide override action for safety and a single controller used to distribute flow among several pipes in a manifold (1). In this chapter, the main emphasis is on cascade control because it is the most widely used of the above systems; feedforward compensation and interacting control systems are treated briefly. Examples of ratio control, selector control, and other special systems are given by several authors (1, 2, 3).

8-2. EXAMPLES OF CASCADE CONTROL

Cascade control is often used for temperature control of chemical reactors because a slight deviation in temperature may lead to poor product quality and sometimes to a runaway reaction. Figure 8-1 shows a typical system for a jacketed reactor. The main controller acts on the difference between the set point and the measured reactor temperature, and the output is a pressure signal to the set-point bellows of the secondary controller. (In most newer pneumatic controllers, the set point is changed by changing the air pressure in a bellows rather than by moving a mechanical linkage.) The secondary controller adjusts the flow of coolant to keep the jacket temperature at the value determined by the primary controller. The main lags come from the thermal capacities of the jacket fluid, the reactor wall, and the kettle fluid. To simplify the diagram and the following qualitative discussion, these lags are shown as separate blocks, though they actually form an interacting system.

Consider the effect of a sudden increase in cooling water temperature, which corresponds to a load change L_2 in Fig. 8-1. On the assumption that the jacket fluid is well mixed, the jacket temperature θ_2 starts to rise immediately, and within a few seconds the control valve opens to increase

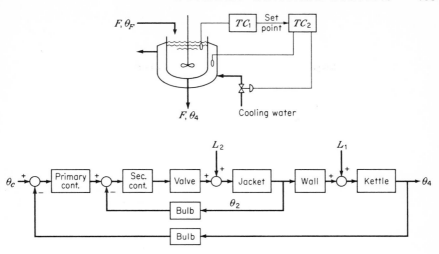

FIGURE 8-1   Cascade system for reactor temperature control.

the flow of water.   The fluctuations in jacket temperature damp out after
a few cycles, and the period of these oscillations depends primarily on the
natural frequency of the inner loop.   The fluctuations in kettle temperature
are much smaller than the jacket temperature fluctuations because of the
damping action of the wall and the kettle.

If there were no secondary controller, the change in water temperature
would not be detected so quickly, because the signal would have to pass
through four time constants in series, including a distributed resistance in
the wall, before reaching the controller.   By the time corrective action was
started, the jacket temperature would be appreciably higher and the mass
of warmer water accumulated in the jacket would lead to a large deviation
in reactor temperature.   Both the maximum deviation and the period would
be greater than with cascade control.

Changes in the flow rate or temperature of the feed stream to the kettle
are represented by L_1 in Fig. 8-1.   The initial effects of such changes are
detected almost immediately by the temperature bulb and the main controller,
but the effect of the corrective action is delayed by the other lags in the
control system.   The maximum deviation is considerably greater than for
corresponding changes at L_2, and it may be only slightly smaller than when
a single controller is used.   However, the period of oscillation is always
smaller when cascade control is used, since the inner feedback loop tends
to reduce the lag in the jacket system, and so the error integral following a
change in L_1 is also reduced by adding cascade control.

Cascade control is especially effective if the inner loop is much faster
than the outer loop and if the main disturbances affect the inner loop first.
When a flow rate is the manipulated variable in a single-loop control system,

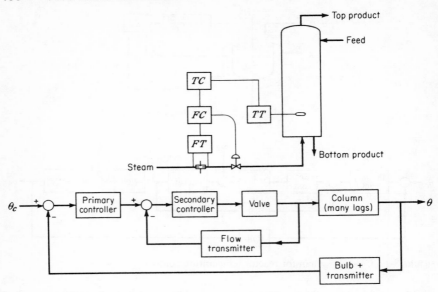

FIGURE 8-2 Cascade control system for a stripping column.

using a secondary loop to control the flow almost eliminates the effects of pressure changes, which would otherwise cause unwanted changes in flow rate. Figure 8-2 shows a stripping column where the temperature near the bottom is used to adjust the set point of a steam-flow controller. When the feed volatility decreases, the temperature in the column increases and the primary controller calls for a lower steam rate. The response of the system to changes in feed composition is only slightly improved by using cascade control, since the lags in the column are much larger than those in the flow control system. The response to changes in steam pressure or column pressure is very much better with cascade control; changes in flow rate are corrected very rapidly by the fast flow control loop, and a change in flow lasting only 2 to 3 sec would probably have too small an effect on column temperature to be noticed with a normal recorder.

When there are several lags ahead of the main process lag, the correct choice of secondary variable may not be obvious. A system discussed by Ziegler (3) consists of a column and reboiler, and the temperature a few plates above the reboiler is the main variable (Fig. 8-3). The secondary variable could be steam flow, pressure on the steam side of the exchanger, or vapor flow from the reboiler. Controlling the steam flow could eliminate only the effects of changes in steam supply pressure or condensing pressure. The time constants for the steam-flow system are an order of magnitude smaller than the other process lags; so making the flow response still faster by a closed-loop system has a negligible effect on the response of the

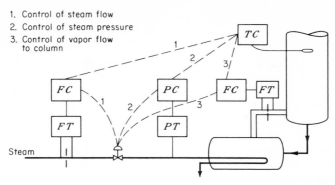

1. Control of steam flow
2. Control of steam pressure
3. Control of vapor flow to column

Steam

FIGURE 8-3   Cascade control of column temperature.

column reboiler system.   Using a secondary flow controller is said to be rarely worthwhile in this case (3, 4), probably because the major disturbances enter further on in the system.   Also, the conventional orifice meter and flow controller do not give a constant mass rate of flow if the absolute pressure changes, and a constant reading on the recorder would therefore not mean a constant heat input.   If control of the steam flow seems desirable, the pressure in the steam header should be regulated or one of the newer mass flowmeters used.

The steam pressure in the exchanger is a measure of the condensing temperature and to some extent a measure of the wall temperature (if the steam film resistance is not large).   The time constant for the metal walls[†] is usually several seconds, and by including this in the inner loop the critical frequency for the outer loop is significantly increased.   However, keeping the steam pressure constant does not ensure a constant vapor rate, which is really the important variable.   By using the vapor flow as a secondary variable, the inner loop would include the lag on the liquid side of the re-boiler, which can be represented by a several-second time constant.   With the enlarged inner loop, disturbances such as changes in reboiler level, reboiler temperature, or heat-transfer coefficient would enter the inner loop and be more quickly corrected than with the other two schemes.   Of course the disturbances in steam supply pressure would not be corrected as rapidly as with the first scheme.

How many elements should be included in the inner loop depends on the frequency and magnitude of the load changes at various points.   A general rule is that the response of the inner loop should be much faster than the response of the outer loop; there is little benefit from cascade control if the sum of the time constants in the inner loop exceeds that for the outer loop (6).

† The interactions between the wall and the kettle fluid are neglected in this analysis. The article by Day (5) gives complete transfer functions for a reboiler.

8-3. CONTROLLER SETTINGS FOR CASCADE SYSTEMS

Predicting the best settings for the two controllers of a cascade system is not as difficult as it might seem.   The first step is to pick reasonable values for the secondary controller, based on the same rules that are used for single-loop systems.   A gain of half the maximum can be used if there are three or more time constants or a time delay in the loop.   If only the two largest time constants are known, a gain that gives about 30° phase margin should be used.   Stable automatic control of the whole system can be obtained with an unstable inner loop, but since the main loop may sometimes be switched to manual control, unstable inner loops are avoided.   Some authors recommend only proportional action for the secondary controller (2, 7).   The reason for omitting integral action is that the gain is usually large, and the slight offset resulting from load changes is eventually corrected by the primary controller.   As pointed out by Wills (6), integral action is worthwhile if the secondary loop has a low gain, as is often the case for flow control.   The reason for omitting derivative action is not so clear. Derivative action is undesirable for fast, noisy processes like flow and pressure control, but, for slow systems with three or four time constants and no time delay, derivative action can increase the critical frequency two- to threefold (see Example 6-2).   Derivative action in the secondary controller should therefore improve some cascade systems, though the improvement is small compared with that obtained by going from a single controller to a cascade system.

The settings for the main controller and the critical frequency for the entire system are found from a Bode plot.   For experimental measurements, the outer loop is opened after the main controller and a sinusoidal signal sent through the inner closed loop and the other elements in the outer loop.   To calculate the response, we just combine the responses of the outer loop elements with the gain and phase lag for $G/(1 + G)$, where G is the product of the transfer functions for the inner loop.   As shown in Chap. 7 (see Fig. 7-2), the phase lag for $G/(1 + G)$ is always less than that for G for angles up to 180°.   This means that the phase lag for the entire system will reach 180° at a higher frequency than if only one controller is used.

Example 8-1

Calculate the critical frequency and the maximum gain of the primary controller for the following cascade system.   Compare with the values for a single-loop

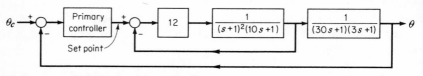

FIGURE E8-1   Diagram for Example 8-1.

system.   The secondary controller gain is set at 12, half the maximum value for the inner loop.   The measurement lags for both loops are negligible.   Let

$$G_2 = \frac{1}{(s + 1)^2(10s + 1)} \qquad \text{all times, min}$$

$$G_3 = \frac{1}{(30s + 1)(3s + 1)}$$

The values of $G/(1 + G)$ are obtained from the Black-Nichols chart (Fig. 7-3).

ω, rad/min	Single-loop system, A.R., ϕ, deg			Cascade system, A.R., ϕ, deg	
	G_2	G_3	$G_2 G_3$	$\dfrac{12 G_2}{1 + 12 G_2}$	$\dfrac{12 G_2 G_3}{1 + 12 G_2}$
0.1	$0.70/-55$	$0.29/-88$	$0.20/-143$		
0.15	$0.53/-73$	$0.21/-101$	$0.11/-174$		
0.2	$0.41/-85$	$0.14/-112$	$0.057/-197$		
0.4	$0.20/-120$	$0.053/-135$	....	$1.15/-25$	$0.061/-160$
0.5	$0.16/-133$	$0.037/-143$	....	$1.33/-32$	$0.049/-175$
0.6	$0.72/-143$	$0.027/-148$	....	$1.7/-50$	$0.046/-198$

Single-loop system	*Cascade system*
$\omega_c = 0.16$	$\omega_c = 0.53$
A.R. $= 0.093$	A.R. $= 0.048$
$K_{max} = 10.8$	$K_{max} = 21$
$K_{rec} = 5$	$K_{rec} = 10$

Using cascade control increases the critical frequency threefold and doubles the allowable gain for the main controller.   Both factors contribute to an increase in controllability, and the error integral $\int |e|\, dt$ following a small load change in the main loop would be reduced four- to sixfold.   There would be a much greater improvement for load changes entering the secondary loop because of the high gain in the secondary controller.[†]   The quantitative performance of cascade systems is discussed further in Sec. 8-4.

Note that the increase in critical frequency in Example 8-1 comes about because the phase curve for $G/(1 + G)$ is quite flat until the critical frequency for the inner loop is approached.   The curve would not be so flat if a lower gain were used in the secondary controller, and the critical frequency of the cascade system would not be so high.   This is why tight control (high gain) is often recommended for the secondary system.

Another way of explaining the effect of cascade control is to consider

[†] For a system with the same time constants, Franks and Worley (8) reported that cascade control gave an improvement factor of 7 for a primary loop disturbance and 350 for a secondary loop disturbance.   In their study, three-mode controllers were used in both loops, saturation effects were considered, and $\int t |e|\, dt$ was used as a criterion of performance.

the case where there is one dominant time constant in the secondary loop and two or three other lags in the main loop. For proportional control of a process with one time constant T, the system responds as a first-order system with a time constant $T/(1 + K)$. Cascade control thus reduces the time constant of the element in the inner loop, and if this element was the second largest time constant, there is a great improvement in controllability (9). If either the largest or the smallest time constant is reduced by making it the chief element in the inner loop, there is little effect of cascade control on the frequency and maximum gain, though there may still be a worthwhile improvement for disturbances entering the inner loop. If there were really only one time constant in the inner loop, its value could be reduced to zero by using infinite gain on the secondary controller. However, closing the loop around one element always introduces additional measurement and transmission lags, and these lags must be considered before choosing controller settings. If the process lags are very small, the extra lags associated with a secondary measurement might even cancel the potential advantage of cascade control.

In designing a cascade system, care should be taken to avoid nonlinearities arising from large signals circulating in the inner loop (10). With a very high gain on the secondary controller, a small change in the secondary variable or the set-point signal from the primary controller causes the secondary controller to saturate, and the valve goes to full open or closed. The system would probably still be stable, but its response would be poorer than predicted for a linear system.

When there is only one time constant in the primary loop (not counting those in the inner loop), the critical frequency of the system is close to that for the inner loop. At 180° phase lag for the open-loop system, the inner closed loop contributes at least 90° lag, which falls in the steep region of the phase curve (see Fig. 7-2). The following example shows that the primary controller gain may have to be reduced when cascade control is added under these conditions.

Example 8-2

Calculate the critical frequency and maximum gain for the primary controller for this system. The secondary controller gain of 6 is half the maximum for the inner loop.

The frequency response for a closed-loop system with time constants of 2, 5,

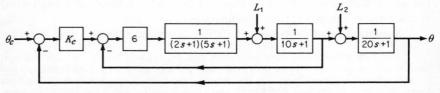

FIGURE E8-2   Diagram for Example 8-2.

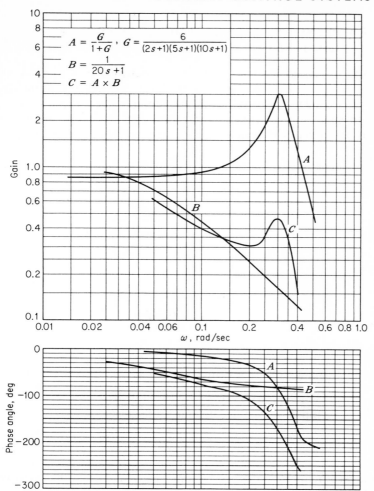

FIGURE 8-4   Bode diagram for cascade control system (Example 8-2).

and 10 sec was given in Fig. 7-2.   Combining this with the response for the 20-sec time constant gives the curves in Fig. 8-4.

$$\omega_c = 0.32 \text{ rad/sec} \qquad (180° \text{ lag})$$

$$\text{A.R.} = 0.42 \qquad (\text{curve } C \text{ at } \omega = 0.32)$$

$$K_{c, \text{max}} = \frac{1}{0.42} = 2.4 \qquad K_{\text{rec}} = 1.2$$

The gain is low because the critical frequency coincides with the resonant peak of the inner loop.

If only one controller were used, the critical frequency would be 0.15 rad/sec and the recommended gain would be 6.8/2, or 3.4.

8-4. RESPONSE OF CASCADE SYSTEMS TO LOAD CHANGES

The transfer functions for load changes are presented here to show more quantitatively the difference between the performances for disturbances entering the secondary loop and the primary loop. The expressions are too complex to make an analytical solution for a step change worthwhile, but much can be learned just from the steady-state response and the frequency response.

For load changes within the inner loop of Fig. 8-5

$$\theta_1 = K_1(\theta_c - H_1\theta_4) = -K_1H_1\theta_4 \qquad \text{for } \theta_c = 0 \tag{8-1}$$

$$\theta_2 = (\theta_1 - \theta_3H_2)G_2 \tag{8-2}$$

$$\theta_3 = (\theta_2 + L_2)G_3 = (\theta_1 - \theta_3H_2)G_2G_3 + L_2G_3$$
$$= \frac{\theta_1G_2G_3 + L_2G_3}{1 + H_2G_2G_3} \tag{8-3}$$

$$\theta_4 = G_4\theta_3 = -\frac{K_1H_1G_2G_3G_4\theta_4 + L_2G_3G_4}{1 + H_2G_2G_3}$$

$$\frac{\theta_4}{L_2} = \frac{G_3G_4}{1 + H_2G_2G_3 + K_1H_1G_2G_3G_4} \tag{8-4}$$

Equation (8-3) could have been written directly by applying the relationship

$$\frac{\text{Output}}{\text{Input}} = \frac{\text{feedforward transfer function}}{1 + \text{feedforward} \times \text{feedback}}$$

or

$$\frac{\theta_3}{L_2} = \frac{G_3}{1 + G_2G_3H_2} \qquad \frac{\theta_3}{\theta_2} = \frac{G_2G_3}{1 + G_2G_3H_2} \tag{8-5}$$

If the zero-frequency gains are 1.0 for all elements except the controllers, the steady-state solution is

$$\frac{\theta_4}{L_2} = \frac{1}{1 + K_2 + K_1K_2} \tag{8-6}$$

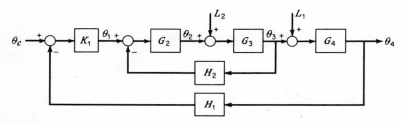

FIGURE 8-5   General diagram for cascade control.

When both gains are large, the offset varies inversely with the product of the gains. Going back to Example 8-1, the offset θ_4/L_2 is

$$\frac{1}{1 + 12 + 120} = 0.008$$

which is small enough to make integral action unnecessary. If the response curves have the same shape, the value given by Eq. (8-6) can be used to estimate the reduction in peak error that is possible with cascade control. For Example 8-1, the use of cascade control should reduce the error integral by a factor of about 70.

$$\frac{0.53 \text{ rad/sec}}{0.16 \text{ rad/sec}} \frac{1 + 12 + 120}{1 + 5} = 73$$

For disturbances entering the primary loop (L_1 in Fig. 8-5), the transfer function is

$$\theta_3 = \frac{\theta_1 G_2 G_3}{1 + H_2 G_2 G_3} \qquad \theta_1 = -K_1 H_1 \theta_4 \tag{8-7}$$

$$\theta_4 = (\theta_3 + L_1)G_4 = -\frac{H_1 K_1 G_2 G_3 G_4 \theta_4}{1 + H_2 G_2 G_3} + L_1 G_4$$

$$\frac{\theta_4}{L_1} = \frac{G_4(1 + H_2 G_2 G_3)}{1 + H_2 G_2 G_3 + K_1 H_1 G_2 G_3 G_4} \tag{8-8}$$

The steady-state solution comparable to Eq. (8-6) is

$$\frac{\theta_4}{L_1} = \frac{1 + K_2}{1 + K_2 + K_1 K_2} \tag{8-9}$$

When both gains are large, the offset varies inversely with K_1, which means that the gain of the inner loop does not help to reduce the offset and presumably has little effect on the peak deviation. The advantage of cascade control for primary disturbances comes mainly from the higher frequency, with some additional benefit if a higher gain can be used in the primary controller.

When the critical frequency of the cascade system is close to that of the inner loop, it is harder to predict the possible advantage of cascade control. In Example 8-2, the maximum gain for the primary controller was 2.4, compared with 6.8 for a single-loop system. The lower gain tends to offset the advantage created by the higher frequency and raises the question as to whether cascade control actually makes the response worse for primary loop disturbances. To help answer this question, the frequency response to load changes was calculated for Example 8-2 by using Eqs. (8-4) and (8-8).

Figure 8-6 shows the frequency response to secondary disturbances. The curve for cascade control is more peaked than for single-loop control, but both curves are enough like those for second-order systems to permit

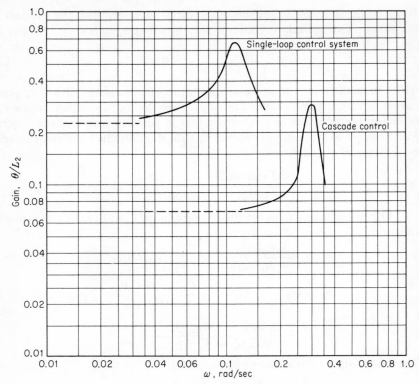

FIGURE 8-6   Frequency response for secondary load changes (Example 8-2).

Fig. 7-8 to be used in predicting the peak height of the transient response. The prediction method is outlined in Table 8-1.

Transient-response curves obtained with an analog computer for controller settings close to the optimum values are shown in Fig. 8-7. The reduction in peak error and period by the use of cascade control agrees

Table 8-1. Predicted Transient Response to Secondary-loop Disturbances

	Cascade system, $K_1 = 1.2$, $K_2 = 6$	Single-loop system, $K = 3.4$
Relative peak height from frequency response (Fig. 8-6)	$\dfrac{0.29}{0.07} = 4.15$	$\dfrac{0.67}{0.23} = 2.9$
$\dfrac{\text{Transient peak}}{\text{Potential offset}}$ (Fig. 7-8)	1.9	1.6
Potential offset for $L_2 = 1$	$\dfrac{1}{1 + 6 + 7.2} = 0.070$	$\dfrac{1}{1 + 3.4} = 0.23$
Predicted transient peak	0.13	0.37
Predicted period, sec	21	52

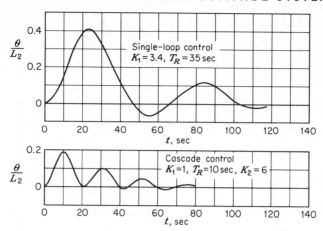

FIGURE 8-7   Step response for secondary load changes (Example 8-2).

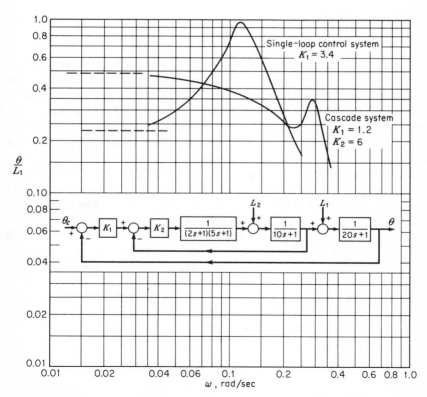

FIGURE 8-8   Frequency response for primary load changes (Example 8-2).

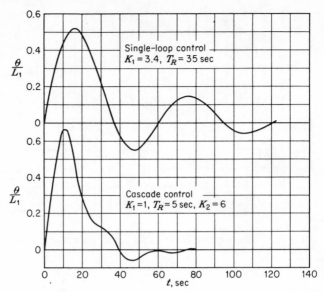

FIGURE 8-9   Step response for primary load changes (Example 8-2).

fairly well with the above predictions. (The primary controller gain was slightly lower for the transient tests than for the frequency-response calculation, but the difference is not significant for this approximate prediction.)

In Fig. 8-8, the frequency response of the same system to primary load changes is shown. The curve for the single-loop system has the familiar shape and shows a large peak gain because the disturbance enters just before the last time constant. The curve for cascade control is quite different, and the transient response cannot be predicted readily. However, the relatively low height of the resonant peak and the higher frequency suggest that cascade control will be somewhat better than single-loop control. The analog tests showed a slightly higher peak error with cascade control (Fig. 8-9), but the error integral was only half as great. The relatively low reset time was chosen after a few tests with other settings. The optimum reset time for this example is lower than the value usually recommended [Eq. (9-2)], because the phase curve is quite steep at 180° lag.

The benefits of cascade control for primary-loop disturbances are also shown in examples by Webb (11). Error reductions of two- to threefold were obtained mainly because of the increase in natural frequency.

8-5. FEEDFORWARD MODIFICATIONS

Feedforward control means the use of measured inputs to adjust the process conditions to give a desired output. Since it is difficult to measure all possible inputs and to predict their effect quantitatively, feedforward control is generally used along with feedback control. A typical scheme is shown in Fig. 8-10. This scheme has also been called "disturbance feedback" (2). Whether the additional path for information is considered as forward or backward is a matter for a terminology committee. The important feature is that the measured load variable is used to correct the output of the primary controller. The transfer functions are derived below for the system of Fig. 8-10.

$$\theta_2 = (\theta_1 + G_4L)G_2 \qquad \theta_1 = -\theta_3G_1 \qquad \text{if } \theta_c = 0 \qquad (8\text{-}10)$$

$$\theta_3 = (\theta_2 + LK_L)G_3 = -\theta_3G_1G_2G_3 + LG_2G_3G_4 + LK_LG_3 \qquad (8\text{-}11)$$

$$\frac{\theta_3}{L} = \frac{G_3(K_L + G_2G_4)}{1 + G_1G_2G_3} \qquad (8\text{-}12)$$

According to Eq. (8-12), perfect compensation for load changes can be achieved just by making $K_L + G_2G_4$ equal to zero. If G_2 represents a single time constant with the transfer function $K_2/(T_2s + 1)$, the ideal load compensator would be a proportional-derivative controller with the transfer function $(-K_L/K_2)(T_2s + 1)$. A gradual increase in load would cause a gradual decrease in θ_2 such that the combined signal to G_3 remained constant. (It is better to use a ramp change rather than a step change in load to evaluate the system, since a step change calls for infinite output from the derivative-action controller. A real controller has a limited response; so θ_3 would show some deviation after a step disturbance, and the value would be hard to calculate because of the controller nonlinearity.)

Though complete compensation is possible for anything but step changes in load, it is hard to achieve in practice because the transfer functions are not accurately known. For most process control systems, very careful measurements are required to get time constants and gains that are within 5 to 10 per cent of the true values. Furthermore, elements like heat exchangers, valves, and reactors are nonlinear, and the gain may change two-fold or more over the normal range of inputs. Most pneumatic controllers

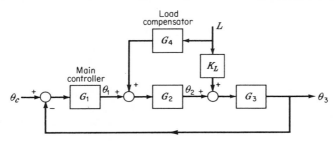

FIGURE 8-10   Feedforward modification of feedback control system.

are calibrated at only one point, and gains or time constants 10 to 20 per cent different from the dial settings are not uncommon. The controller error can be minimized by recalibration or by using a more accurate electronic controller, but the process uncertainties and nonlinearities are not so easily reduced. In a typical system, the imperfect compensation for load changes would probably make the net disturbance about 20 per cent of the actual load change. While a 5-fold reduction in error is appreciable, it is not so great as the 50- to 100-fold reductions possible with some cascade systems.

If there are two lags between the load compensator and the point where load changes enter the system (block G_2 in Fig. 8-10), perfect compensation would require a controller with first- and second-derivative action, so that

$$K_L + \frac{K_2 G_4}{(T_2 s + 1)(T_2' s + 1)} = 0$$

The difficulties in determining the system parameters and the effects of noise on second-derivative control have ruled out such devices for the present. By using a conventional instrument, the derivative time for the compensating controller can be set at the sum of the two time constants, as suggested by Young (2). The factor of improvement then depends on the ratio of the total lag in the final process elements to the sum of the lags in the first portion of the process. In the example given by Young (2), G_2 is two 1-min time constants and G_3 a 2-min time constant. Using just proportional compensation had hardly any effect, but proportional-derivative compensation reduced the error about threefold.

When the last time constant in the system is much larger than the others, the dynamic effects of the lags in the compensating circuit can be neglected and a single proportional controller would serve just as well as a two- or three-mode instrument. The success of the scheme then depends on the error in estimating K_2, K_4, and in particular K_L. For example, L might be the gas-flow rate to an absorber and K_L the gain in change in exit concentration per unit change in inlet flow rate, and K_L would certainly change with gas- and liquid-flow rates and with compositions. Note that, in Fig. 8-10, a separate block is included for K_L because the load signal to the compensating controller does not have the same units as the load signal to the final element G_3. In most other diagrams (Fig. 8-5, for example), L really represents LK_L, and a unit load change means a unit change in LK_L.

When should a feedforward modification be used instead of cascade control? If the main disturbance can be included in the inner loop, cascade control usually leads to the greater reduction in error. When the disturbance enters just before the last process element, feedforward compensation is somewhat better than cascade control if the last element has a much greater lag than the others. As we learn more about process dynamics and can compensate more exactly for load changes, feedforward compensation will be more widely used.   Process nonlinearities can be accounted for by using simple computers as compensating controllers, and the growing need for more accurate control should stimulate development of controllers which can do more than the basic three-mode controller, yet are much cheaper than the typical digital computer.

Example 8-3

A reactant stream is preheated from 70° to 160°F by steam condensing on the tubes of a shell-and-tube exchanger.   The flow rate varies from 20 to 30 gpm, and the fluctuations in output temperature are too great when a conventional feedback control system is used.   If the measured flow rate is used to correct the signal to the control valve, what settings should be used for the compensating controller? What reduction in error can be expected?
Additional data are as follows:

$$\text{Normal flow rate } F = 25 \text{ gpm}$$

$$\text{Normal steam temperature} = 250°F \text{ (30 psia)}$$

$$\text{Range of temperature bulb} = 150°F$$

$$\text{Steam supply pressure} = 80 \text{ psia}$$

$$\text{Equal-percentage control valve, } K_v = 4, \ T_v = 3 \text{ sec}$$

$$\text{Residence time of fluid in exchanger} = 20 \text{ sec}$$

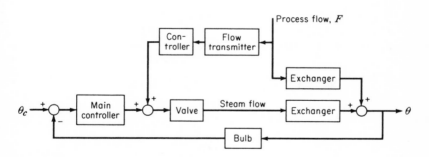

FIGURE E8-3a   Diagram for Example 8-3.

According to Thal-Larsen (Ref. 7, Chap. 11), the dynamic behavior of the exchanger can be approximated by the following transfer functions:

$$\frac{\theta}{Q} = \frac{\text{outlet temperature}}{\text{steam flow}} = \frac{K_1 e^{-Ls}}{(T_1s + 1)(T_2s + 1)}$$

$$\frac{\theta}{F} = \frac{\text{outlet temperature}}{\text{process flow}} = \frac{K_2 e^{-Ls}}{(T_1s + 1)(T_2s + 1)}$$

For this example, $T_1 \cong 20$ sec, $T_2 \cong 5$ sec, $L \cong 5$ sec.   Since a 1 per cent increase in steam flow increases the temperature rise by 1 per cent, or 0.9°F,

$$K_1 = \frac{0.9°F/150}{0.01} = 0.6 \frac{\% \text{ change in temperature}}{\% \text{ change in steam flow}}$$

A trial-and-error procedure is used to obtain K_2.

$$Fc_p(T_2 - T_1) = UA \, \Delta T_{1.m.} \qquad \Delta_1 = 180, \, \Delta_2 = 90, \, \Delta_{1.m.} = 130$$

Base case $UA = \dfrac{25c_p(90)}{130} = 17.3c_p \qquad [c_p, \text{Btu}/(\text{gal})(°F)]$

For $F = 30$, h is $\sim$16 per cent higher; assume that U is 8 per cent higher.

$$UA = 18.7c_p$$

Guess $T_2 = 154$, $\Delta_1 = 180$, $\Delta_2 = 96$, $\Delta_{1.m.} = 134$.

$$UA = \frac{30c_p(84)}{134} = 18.8c_p \qquad \text{(correct)}$$

$$K_2 = \frac{-6°F/150°F}{5/25} = -0.20 \frac{\% \text{ change in temperature}}{\% \text{ change in flow}}$$

For $F = 20$, h is $\sim$16 per cent lower; assume that U is 8 per cent lower.

$$UA = 15.9c_p$$

Guess $T_2 = 165°F$, $\Delta = 180, 85$; $\Delta_{1.m.} = 127$.

$$UA = \frac{20c_p(95)}{127} = 15.0c_p \qquad \text{(too low)}$$

Guess $T_2 = 168°F$, $\Delta = 180, 82$; $\Delta_{1.m.} = 125$.

$$UA = \frac{20c_p(98)}{124} = 15.7c_p$$

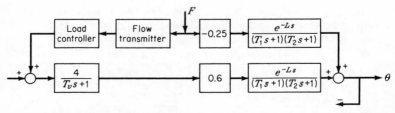

FIGURE E8-3b   Diagram for Example 8-3.

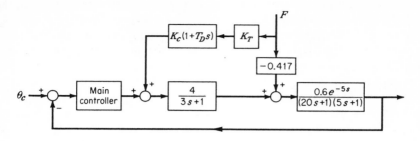

Diagram for Example 8-3.

By extrapolation, $T_2 = 169°F$.

$$K_2 = \frac{+\frac{9}{150}}{-0.2} = -0.30 \frac{\% \text{ change in temperature}}{\% \text{ change in flow}}$$

$$K_{2, \text{av}} = -0.25$$

The critical portion of the block diagram is shown in Fig. E8-3b. To fit the form of Fig. 8-8, the gain K_L must be $-0.25/0.6$, or -0.417.

The gain of the load controller K_C must be chosen to make

$$4K_C K_T = -0.417$$

Assuming that $K_T = 1$,

$$K_C = -0.104$$

If derivative action were used, the derivative time T_D should be 3 sec but almost as good results would be obtained with just proportional action, since the total exchanger lag is about 10 times the valve lag. The feedforward compensation should reduce the error by 80 to 90 per cent for small load changes but only by 60 to 70 per cent for large changes, because of the change in K_L with flow rate.

8-6. INTERACTING CONTROL SYSTEMS

When two or more related variables are regulated by separate feedback controllers, the control systems interact and the interaction affects the stability of both systems. Interactions arise when a controller is used to regulate the pressure in a tank and other controllers act to change the liquid level or the flow of gas into the system. An increase in liquid level compresses the gas in the tank; so the block diagram for the pressure control system must include level changes as a load variable. Increases in pressure tend to increase the flow of liquid from the tank, and so pressure changes are load changes for the level control loop. There is some degree of self-regulation in this example, since the higher pressure resulting from a sudden increase in level tends to increase the discharge flows of both liquid and gas. However, there are more lags to be considered in the combined system than in either system alone, and controller settings that are satisfactory for the individual loops may lead to cycling when both controllers are operating. The effect of the interaction is greatest when both loops have about the same critical frequency. Interaction exists when the pressure at the top of a distillation column and the reflux drum level are automatically controlled, but the pressure control loop is usually much faster than the level control system, and so the interactions are not serious.

An interesting example of pressure-level interaction in a vacuum evaporator is given by Johnson (12). The interaction was recognized only after plant tests showed much poorer performance than predicted by analog simulation of the separate control loops. Later measurements showed that a change of 0.1 in. of mercury vacuum corresponded to an 8.5-in. change in level and that slight changes in vacuum caused saturation of the level control system.

Interactions can arise when the relative humidity of air is controlled by spraying water into a chamber and the temperature is controlled by circulating the air over a heater. Interactions also occur when both top and bottom products of an absorber, distillation column, or other countercurrent-transfer device are independently controlled. A thorough analysis of such systems requires an analog controller, though an analytical solution is possible if the system is simplified to two or three time constants.

A more serious type of interaction occurs when an attempt is made to control too many variables. The density and boiling point of the product from an evaporator should not be separately controlled, since they both depend primarily on concentration. The feed rates and flow rates of all streams leaving a process obviously cannot be controlled, since a slight error in adjusting the set points leads to an accumulation or depletion of material. These and similar situations can usually be avoided by using common sense and the phase rule.

Example 8-4

Pressure and level controls are to be used on a continuous reactor. The tank is 6 ft in diameter and 10 ft high, and the normal liquid depth is 8 ft. The holdup times are 20 min for the liquid and 10 min for the unreacted gas. The operating pressure is 5 psig, and the gas is vented through a pressure control valve ($K_v = 5$) to atmospheric pressure. The level controller operates a valve in the liquid discharge line ($K_v = 3$), and the downstream pressure is 1 atm. The liquid density is 70 lb/ft^3, the level controller has a range of 10 ft, and the pressure controller has a range of 15 psi.

a. Draw a complete block diagram, and obtain the transfer functions for pressure and level.

b. Assuming a gain of 10 for both controllers and neglecting the lags in control valves and sensing devices, calculate the damping coefficient for the system.

Consider first the transfer functions for the separate loops, assuming turbulent flow through the control valves.

For liquid level

$$A\left(\frac{dh}{dt}\right) + h\left(\frac{\partial F_2}{\partial h}\right) = -x_2\left(\frac{\partial F_2}{\partial x_2}\right) \qquad \text{at constant } P, F_{in}$$

From Eq. (3-46)

$$\left(\frac{\partial F_2}{\partial h}\right) = \frac{\bar{F}_2}{2(\bar{h} + h_0)}$$

$$h_0 = \frac{5(144)}{70} = 10.3 \text{ ft}$$

$$T_4 = \frac{2A(\bar{h} + h_0)}{\bar{F}_2} = \frac{2(\bar{h} + h_0)}{\bar{h}} T_H$$

$$= \frac{2(8 + 10.3)}{8} (20) = 92 \text{ min}$$

$$\left(\frac{\partial F}{\partial x_2}\right) = K_{v2}\bar{F}_2 \qquad \text{change in flow/\% change in lift}$$

$$K_4 = \frac{-K_{v2}\bar{F}_2}{\partial F_2/\partial h} = -K_{v2}(2)(18.3) \qquad \text{ft/\% change in lift}$$

or

$$K_4 = -K_{v2} \tfrac{2}{10}(18.3) \qquad \text{\% change in depth/\% change in lift}$$

since

$$K_{v2} = 3 \qquad K_4 = -11.0 \frac{\text{\% change in depth}}{\text{\% change in lift}}$$

$$\frac{h}{X_2} = \frac{-11}{92s + 1}$$

For gas pressure changes at constant level, the time constant is given by Eq. (3-48).

$$T_2 = \frac{2\overline{\Delta P}}{\bar{P}} (T_H')$$

$$= 2\left(\frac{5}{19.7}\right)10 = 5.1 \text{ min}$$

$$K_2 = -\left(\frac{\partial F_1}{\partial x_1}\right)\left(\frac{\partial P}{\partial F_1}\right) = -(5\bar{F}_1)\left(\frac{2\overline{\Delta P}}{\bar{F}_1}\right)\frac{1}{15}$$

$$= \frac{-5(10)}{15} = -3.3 \frac{\text{\% change in pressure}}{\text{\% change in position}}$$

$$\frac{P}{X_1} = \frac{-3.3}{5.1s + 1}$$

The response of liquid level to changes in pressure is characterized by the same time constant as the response of level to changes in valve position.

$$\frac{h}{P} = \frac{K_4 K_5}{T_4 s + 1}$$

$$K_4 K_5 = \left(\frac{\partial h}{\partial P}\right)\frac{15}{10} = -(144/70)15/10 = -3.08 \frac{\% \text{ change in depth}}{\% \text{ change in pressure}}$$

or $$\frac{h}{P} = 0.28\left(\frac{-11}{92s + 1}\right)$$

The effect of level on pressure is a little different, since a sudden increase in level immediately increases the pressure and the pressure then decays to the original value.   The transfer function can be shown to be

$$\frac{P}{h} = (K_6 K_2)\frac{T_2 s}{T_2 s + 1}$$

$$K_6 K_2 = \left(\frac{\partial P}{\partial h}\right)_{\text{no flow}} = \frac{\bar{P}}{2} \text{ psia/ft} \qquad (2 \text{ ft of gas space})$$

$$= \frac{19.7/15}{2/10} = 6.6 \frac{\% \text{ change in pressure}}{\% \text{ change in depth}}$$

$$\frac{P}{h} = -2\left(\frac{-3.3}{5.1s + 1}\right)$$

The final block diagram is shown in Fig. 8-11.   With the dynamics of valve and transmitter neglected, the equations are

$$[(P_c - P)K_1 + K_6 T_2 sh]\frac{K_2}{T_2 s + 1} = P$$

$$[(h_c - h)K_3 + K_5 P]\frac{K_4}{T_4 s + 1} = h$$

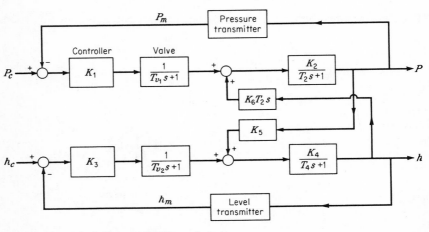

FIGURE 8-11   Block diagram for Example 8-4.

Combining and letting $h_c = 0$ gives

$$P[T_2T_4s^2 + (T_2 + T_2K_3K_4 + T_4 + K_1K_2T_4 - K_2K_6K_4K_5T_2)s$$
$$+ (1 + K_3K_4 + K_1K_2 + K_1K_2K_3K_4)] = K_1K_2(T_4s + 1 + K_3K_4)P_c$$

The gains and time constants are

$$\begin{aligned}
K_1 &= -10 & T_2 &= 5.1 \\
K_2 &= -3.3 & T_4 &= 92 \\
K_3 &= -10 \\
K_4 &= -11 \\
K_5 &= 0.28 \\
K_6 &= -2.0
\end{aligned}$$

For these values the quadratic term becomes

$$0.128s^2 + 1.0s + 1$$

which corresponds to $\zeta = 1.4$

or $$(0.85s + 1)(0.15s + 1)$$

This system is more stable than a single-loop system with the same two time constants, because the interactions are in the right direction and are not delayed by additional time constants. If there were additional lags before the interactions took effect, the system would be less stable.

PROBLEMS

1. The mator elements in a control system are, in order, time constants of 10, 5, and 60 sec and a time delay of 15 sec. How much would the critical frequency of the system be increased by including the 10- and 5-sec time constants in a secondary loop? A measurement time constant of 2 sec would be introduced by use of a secondary loop.

2. How large a change in L_1 or L_2 would be required to cause saturation of the primary or secondary controller of Example 8-2? Use the curves of Figs. 8-7 and 8-9, and assume that both controllers are initially at the middle of their range.

3. A first-order chemical reaction is carried out in a series of three stirred-tank reactors of equal size. The desired conversion is 75 per cent and the rate is changed by changing the flow of catalyst to the first tank. One of the major load variables is the concentration of reactant in the feed. Compare the probable performance of a conventional control system, a cascade system based on sampling the first and third tanks, and a feedforward-feedback system. The holdup time in the tanks is 20 min, and measuring the conversion or the feed concentration involves a sampling delay of 1 min.

4. Gas A is sparged into a stirred tank to react continuously with B, which is dissolved in an aqueous salt solution. A temperature bulb in the tank is used to control the flow of cooling water to the jacket. A differential pressure-type level

controller connected to the base of the tank adjusts the flow of liquid from the bottom of the tank.   The flow of gas A is automatically controlled to maintain 50 per cent conversion of B.   The tank pressure is controlled at 2 atm by throttling the gas exit stream, which contains unreacted A plus water vapor (the normal temperature is 110°C).   Draw a complete block diagram for this system, showing all interactions.

5. Derive the transfer function relating gas pressure and tank level for Example 8-4.

6. Draw a block diagram, and give the main transfer functions for cascade control of a continuous-flow chemical reactor.   The reactor temperature is used to adjust the set point of the jacket temperature controller, which regulates the input temperature of the cooling fluid.   Assume that the tank and the jacket are perfectly mixed.   Show how the interaction between jacket and kettle influences the time constants by making up a typical numerical example.

7a. Determine the transfer functions relating y and y_c for the interacting system shown.   Calculate the damping coefficient or the effective time constants if $T_1 = 8$ min, $T_2 = 4$ min, $K_1 = K_2 = 10$, $K_3 = 3$, and $K_4 = 5$.

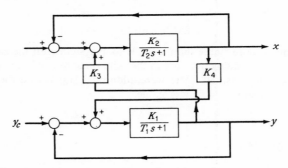

b. What happens if K_3 is negative or if both K_3 and K_5 are negative?

REFERENCES

1. Mamzic, C. L.: Basic Multiloop Control Systems, *ISA J.*, **7**(6):63 (June, 1960).
2. Young, A. J.: "An Introduction to Process Control System Design," p. 310, Longmans, Green & Co., Inc., New York, 1955.
3. Ziegler, J. G.: Cascade Control Systems, 1954 Bulletin, Texas A & M Symposium on Instrumentation.
4. Aikman, A. R.: How to Solve Analyzer Control Problems, *ISA J.*, **4**:364 (1957).
5. Day, R. L.: Dynamic Characteristics of the Distillation-column Reboiler, in Plant and Process Dynamic Characteristics, p. 29, Butterworth Scientific Publications, London, 1957.
6. Wills, D. M.: Cascade Control—Applications and Hardware, *Tech. Bull.* TX 119-1, Minneapolis-Honeywell Co., 1960.
7. Eckman, D. P.: "Automatic Process Control," p. 257, John Wiley & Sons, Inc., New York, 1958

8. Franks, R. G., and C. W. Worley: Quantitative Analysis of Cascade Control, *Ind. Eng. Chem.*, **48**:1074 (1956).
9. Gollin, N. W.: Cascade Control Systems, *Control Eng.*, **3**(7): 94 (1956).
10. Brown, G. S., and D. P. Campbell: "Principles of Servomechanisms," p. 275, John Wiley & Sons, Inc., New York, 1948.
11. Webb, P. U.: Reducing Process Disturbances with Cascade Control, *Control Eng.*, **8**(8): 63 (August, 1961).
12. Johnson, D. E.: Simulation and Analysis Improve Evaporator Control, *ISA J.*, **7**(7): 46 (July, 1960).

9 OPTIMUM CONTROL SETTINGS

The controller settings for problems in the earlier chapters were chosen by simple empirical rules which will be examined in more detail here. Methods of finding the best settings for existing plants or processes are discussed, as well as the problem of predicting the optimum settings at the design stage. A special section is included on processes with a large dead time, because such systems are very difficult to control and special controllers may be beneficial.

9-I. OPTIMUM SETTINGS FROM THE PLANT RESPONSE

CONTINUOUS-CYCLING METHOD

Determining by trial and error the optimum values of controller gain, integral time, and derivative time for a particular system would be very time-consuming because of the many possible combinations of settings. Values close to the optimum can be obtained quite easily by testing the closed-loop system with only proportional action on the controller. The integral time is set at infinity, the derivative time is set at zero or the lowest possible value, and the response of the system to a step change in set point is obtained for various gain settings. The gain setting that results in continuous cycling at a constant amplitude is the maximum gain $K_{c,\max}$. For controllers calibrated in sensitivity units $S = \text{psi/in.}$, the corresponding setting is the "ultimate sensitivity" S_u. The period of cycling at the maximum gain is called the "ultimate period" P_u. Usually only a few tests are needed to establish $K_{c,\max}$ and P_u, since the decay ratio of the first response curve shows whether the gain is close to or far from the maximum value (see Fig. 9-1).

The controller settings recommended by Ziegler and Nichols (1) are given below. For a variety of processes, these settings were found to give a decay ratio of about $\frac{1}{4}$, a period of oscillation close to the ultimate period, and a reasonable overshoot or peak error.

Proportional control $\qquad\qquad K_c = 0.5 K_{c,\max}$ $\qquad\qquad$ (9-1)

Proportional-integral control

$$K_c = 0.45 K_{c,\text{max}}$$

$$T_R = \frac{P_u}{1.2} \qquad (9\text{-}2)$$

Proportional-integral-derivative control

$$K_c = 0.6 K_{c,\text{max}}$$

$$T_R = \frac{P_u}{2.0} \qquad (9\text{-}3)$$

$$T_D = \frac{P_u}{8}$$

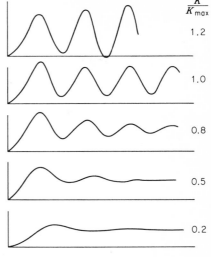

FIGURE 9-1   Typical response curves for proportional control.

Note that, with proportional-integral control, the recommended gain is 10 per cent lower than with only proportional control. Integral action makes the system less stable because of phase lag in the controller, as was shown in Chap. 6. The gain setting given by Eqs. (9-2) is actually 50 to 70 per cent of the gain at which instability would develop for the given reset time. The value of $K_{c,\text{max}}$ in these equations is based on tests with proportional control only and is not the true maximum gain setting for a system with other control actions.

When derivative control is added, the phase lead of the controller helps to stabilize the system and a higher gain and lower reset time are recommended. The settings given by Eqs. (9-3) do not reflect the large improvement that derivative control can make for some systems. For a process with several time constants, the maximum gain and the optimum gain may be doubled by the use of derivative action (1) (see also Example 6-2). On the other hand, derivative action has little effect on systems with a large dead time, and to avoid trouble from such cases, the general rules must be somewhat conservative.

DAMPED-OSCILLATION METHOD

In many plants, sustained oscillations for testing purposes are not allowable, and the ultimate frequency method cannot be used to secure the optimum settings. The following modification (2) is easy to follow and perhaps more accurate than the ultimate method. By using only proportional action and starting with a low gain, the gain is adjusted until the transient response of the closed loop shows a decay ratio of $\frac{1}{4}$. The reset time and derivative time are based on the period of oscillation, P, which is always greater than the ultimate period P_u.

For proportional-integral-derivative control,

$$T_R = \frac{P}{6}$$

(9-4)

$$T_D = \frac{P}{1.5}$$

With the derivative and reset times at the above values, the gain for $\frac{1}{4}$ decay ratio is again established by transient-response tests.

REACTION-CURVE METHOD

A second method given by Ziegler and Nichols (1) is based on the process reaction curve, the response of the open-loop system to a step input. The control loop can be opened at any point, but the usual place is between the controller and the valve.   With the controller on "manual operation," a step change is made in the pressure to the control valve.   The process output, recorded on the controller, is generally an S-shaped response, as shown in Fig. 9-2.   The controller settings are based on the maximum slope of the curve, N, and the effective lag L, the intercept of the maximum slope line with a horizontal line from the initial value.   The final value and the maximum slope are proportional to the size of the step change in input, and so the equations are based on the slope N divided by Δ psi/12, the fractional change in input.

Proportional control $$K_c = \frac{\Delta \text{ psi}/12}{NL}$$ (9-5)

Proportional-integral control

$$K_c = \frac{0.9(\Delta \text{ psi}/12)}{NL}$$

(9-6)

$$T_R = \frac{L}{0.3}$$

Proportional-integral-derivative control

$$K_c = \frac{1.2(\Delta \text{ psi}/12)}{NL}$$

$$T_R = \frac{L}{0.5}$$ (9-7)

$$T_D = 0.5L$$

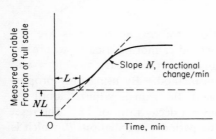

FIGURE 9-2   Process reaction curve.

The effective lag L is always greater than any true time delay which may exist in the system, though the values are almost the same if the system has a very large time delay. For this case, Eqs. (9-7) give a gain inversely proportional to the time delay, which is conservative, since the maximum gain actually approaches 1.0 as the time delay becomes very large. With one time constant much larger than the others and no time delay, L is small, and a very high controller gain can be used, as would be predicted from frequency-response analysis. The settings predicted by the reaction-curve method and the continuous-cycling method generally agree to within 10 to 50 per cent, as shown by many examples given in Ref. 3.

Frequency-response measurements could also be used to determine the maximum controller gain and the ultimate period in order to predict the optimum controller settings. However, the time required would be much greater than for the other methods, since several cycles must elapse at each frequency before the gain and phase lag reach constant values. Frequency-response tests are generally made only when simpler techniques of adjusting the controller have not proved satisfactory or when the frequency-response data are needed to determine system parameters for detailed controllability studies. The methods and equations presented here are intended to give only values close to the optimum controller settings. As shown in the next section, the optimum values are different for different types of processes with the same period and maximum gain. Also, errors in controller calibration and process nonlinearities influence the optimum settings. Fortunately, the response curves are not much affected by changes in controller setting near the optimum settings, and so the simple rules work fairly well. For accurate work, the recommended values should be taken as the starting point for a few extra tests to get the best value for all three dial settings.

9-2. CONTROLLER SETTINGS FOR MINIMUM ERROR INTEGRAL

If the complete transfer function of a process is known, the transient response can be calculated for various controller actions and the best settings determined by trial and error. The optimum settings are usually considered to be those which give a minimum error integral after a step change in set point or load. Ideal reset action is assumed so that the error eventually becomes zero and the error integral reaches a finite limit. Several workers have used the integral of the square of the error, $\int e^2\, dt$, as a criterion of control quality, either for convenience or to give more weight to large deviations. Others have used the integral of the time multiplied by the absolute error, $\int t\, |e|\, dt$, which weights deviations more heavily as time increases. The integral of the absolute value of the error, $\int |e|\, dt$, seems the best criterion for process control, since the penalty for poor control is generally a linear function of the error. The differences in the controller settings obtained by using these three criteria are generally small; the differences were less than 10 per cent for one case treated by Wills (4).

The controller settings selected to give the best response to a step input are not always the best settings for plant operation. The disturbances to the plant may be step changes, ramp changes, steady oscillations, or random fluctuations, and the optimum controller settings depend somewhat on how often these types of disturbances occur. When cyclic disturbances predominate, the controller gain should be adjusted to give a reasonable phase margin (30°) or a peak closed-loop gain of 1.5 to 2, rather than a minimum error integral following a step change. The difference in the best settings for step response and frequency response is significant when the time constants are widely spaced. For example, a system with time constants of 10, 5, and 0.5 min has a phase margin of only 16° and a peak closed-loop gain of

Table 9-1

Controller	Process				Optimum settings (4)		
$K_c\left(1 + \dfrac{1}{T_R s} + T_D s\right)$	$\dfrac{e^{-T_3 s}}{(T_1 s + 1)(T_2 s + 1)}$	$\dfrac{T_2}{T_1}$	$\dfrac{T_3}{T_1}$	$\dfrac{T_R}{P_u}$	$\dfrac{T_D}{P_u}$	$\dfrac{K_c}{K_{c,\max}}$
		1	0.1	0.8	0.25	
		0.1	0.1	1	0.2	0.54
				½†	⅛†	0.6†

† Ziegler-Nichols recommendations, Eqs. (9-3).

4 if $K = 0.5K_{\max}$, the optimum value for a step response (see Fig. 9-4). The phase margin is 30°, and the peak closed-loop gain is 2 at $K = 0.3K_{\max}$. Relatively low controller gains are also used when the signal to the controller contains much noise, as is true for most flow control systems (see Chap. 13). The optimum settings presented in the following paragraphs are based only on the step response, and the frequency response should also be considered before selecting controller settings for a process.

Optimum settings for a process with two time constants and a time delay were obtained by Wills (4); a step change in set point was made, and both $\int |e|\, dt$ and $\int t\, |e|\, dt$ were computed. In Table 9-1 the settings for minimum $\int |e|\, dt$ are presented in the same form as the recommendations of Eqs. (9-3). For these two cases, reset and derivative times given by Eqs. (9-3) are about half the optimum values, and the computed error integral using the settings of Eqs. (9-3) is 1.5 times the minimum value. These differences are not large, particularly since industrial controllers do not have ideal derivative or integral action. The phase lead is usually limited to 40 to 70°, which makes the best derivative setting less than that for an ideal controller.

Jackson (5) studied systems with three time constants and minimized the value of $\int e^2\, dt$ for a load disturbance entering just before the last time constant. Derivative action was not included, since a three-capacity system

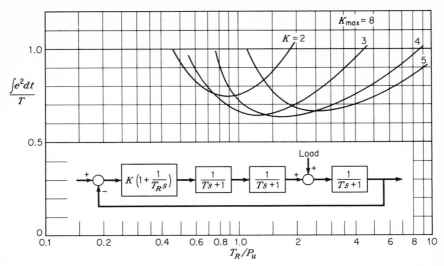

FIGURE 9-3   Effect of controller settings on the error integral.   [*Data of Jackson* (5).]

can in theory be stabilized at any gain by using enough derivative action. In agreement with earlier studies, Jackson found the minimum in the error curve to be quite flat, which means that settings near the optimum are almost as good as the exact values.   Data for a system with three equal time constants are plotted in Fig. 9-3.

The settings for minimum error integral are given in Fig. 9-4 as the ratios $K/K_{c,\text{max}}$ and T_R/P_u to make the results comparable with the predictions of the continuous-cycling method.   The optimum gain is between 0.47 and 0.56 times $K_{c,\text{max}}$, slightly greater than the value of $0.45K_{c,\text{max}}$ recommended by Ziegler and Nichols [Eqs. (9-2)].   The reset times are two to six times greater than the value of $P_u/1.2$ given by Eqs. (9-2).   Frequency-response analysis can explain why large values of T_R/P_u are desirable for systems with two nearly equal time constants and one very small time constant.   The phase-lag curve for such a system has a low slope near the critical frequency, and, with the reset time close to the ultimate period, reset action contributes enough phase lag to shift the critical frequency quite a bit.   Therefore a higher than normal reset time is used to minimize the shift in frequency (see Fig. 9-5).

The optimum controller settings for a given system depend on the location of the major load disturbance.   The settings in Fig. 9-4 are for a disturbance just before the last time constant, but the optimum settings would be almost the same for other disturbances, since the last time constant was the largest in the system.   If the last time constant were the smallest, the optimum gain would be lower and the optimum reset time greater than shown in Fig. 9-4.   Typical results taken from Jackson's article are given

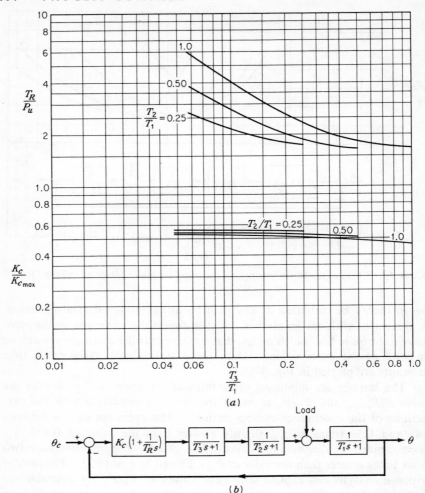

FIGURE 9-4   Optimum gain and reset settings for three-capacity processes.

in Fig. 9-6.   The first two cases are included as a reminder that the shape of the response curve depends only on the relative values of the lags in the closed loop and on the transfer function for load changes.   The order and magnitude do not affect the optimum gain.   The period and reset time are, of course, directly proportional to the size of the lags.

An earlier study by Hazebroek and van der Waerden (6) applied the error-squared criterion to systems with n equal time constants.   The disturbance was applied at the beginning of the system.   Figure 9-7 shows that K/K_{max} increases from 0.34 for $n = 3$ to 0.5 as n becomes very large.   The ratio T_R/P_u goes from 1.5 to 0.31 as n becomes large.   Relatively low reset

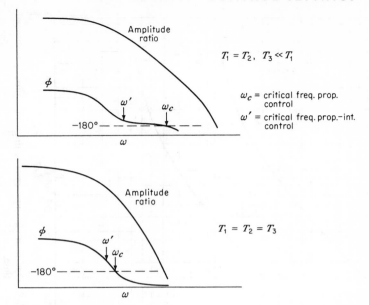

$$T_1 = T_2, \quad T_3 \ll T_1$$

ω_c = critical freq. prop. control

ω' = critical freq. prop.-int. control

$$T_1 = T_2 = T_3$$

FIGURE 9-5   Bode diagrams showing the effect of reset action on critical frequency.

times can be used when n is large, because the phase-lag curve is quite steep near 180° lag.   The phase lag of the controller has only a small effect on the frequency for 180° phase lag.   The differences between the values of K and T_R for $n = 3$ in Fig. 9-7 and $T_1 = T_2 = T_3$ in Fig. 9-4 may be due to the different disturbance used, or there may be computational errors because of the flat minimums encountered in the error integral.

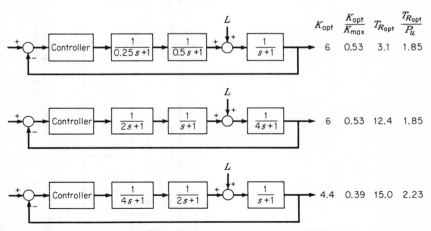

FIGURE 9-6   Effect of arrangement of time constants on optimum controller settings.

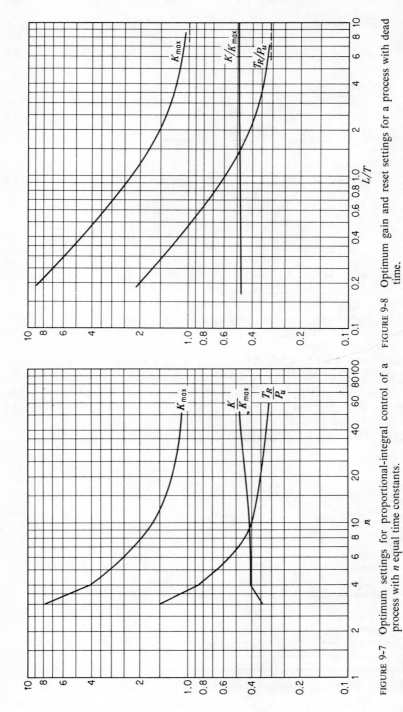

FIGURE 9-7 Optimum settings for proportional-integral control of a process with n equal time constants.

FIGURE 9-8 Optimum gain and reset settings for a process with dead time.

Optimum settings for a system with one time constant and one time delay (6) are plotted in Fig. 9-8.   The ratio T_R/P_u approaches 0.31 as the delay becomes very large, the same as the value recommended for a very large number of equal lags.   This is expected, since the frequency response of the two systems is about the same.

To summarize these studies, the gain for a minimum error integral with proportional-integral control is close to $0.5K_{c,\max}$ ($K_{c,\max}$ is based on proportional control only).   The optimum ratio of reset time to ultimate period varies over wider limits for different types of processes.   However, if we consider the weak dependence of the error integral on reset time, a reset time equal to the ultimate period is satisfactory for most systems.   The optimum reset time is greater than the period if the phase-lag curve has a low slope near the critical frequency, and the optimum reset time is 0.3 to 0.5 times the period for systems with a large time delay.

9-3. CONTROL OF PROCESSES WITH A TIME DELAY

In many process control systems there is an appreciable time delay or dead time which arises from the flow of material through pipelines or processing equipment.   Such time delays are sometimes called "transportation lags" or "velocity-distance lags."   A time delay may also be introduced by intermittent sampling devices, such as gas chromatographs, or by a digital computer in the feedback system.   With a true time delay of L sec, the output is absolutely constant for L sec after a change is made in the input. For a distributed resistance or several nearly equal time constants in series, there is no measurable change in output for a certain time after a change in input, and so these systems are said to have an effective time delay.

For systems with a large time delay, it is difficult to get good control with conventional two- or three-mode controllers.   The maximum controller gain is low, since the delay causes a large phase lag before the other lags contribute much damping.   The period of oscillation is large, at least twice the time delay, and so any deviations will not be reduced to zero until five to six times the time delay.   Many schemes have been presented for improving the control of processes with dead time, most of which involve use of a time-delay element in the controller (8 to 13).   In simple terms, a controller which uses the information that the process has L sec dead time should be able to make more intelligent corrections than a controller which receives nothing but the error signal.   The following examples show that the possible benefit of such schemes is less for a pure time-delay system than for a system having a time constant of the same magnitude as the delay.

Figure 9-9 shows a conventional control system for a process which has only a time delay.   (A long-pipeline reactor is a possible case.)   The maximum controller gain is 1.0, and the ultimate period is 20 min, twice the time delay.   The controller settings, $K = \frac{1}{2}K_{\max}$ and $T_R = 0.31P_u$, are

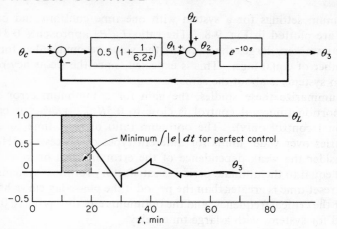

FIGURE 9-9 Proportional-integral control of a time-delay process.

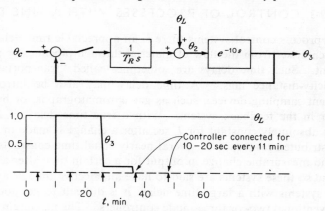

FIGURE 9-10 Step-by-step control of a time-delay process.

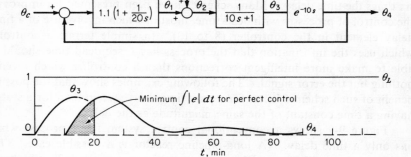

FIGURE 9-11 Proportional-integral control of a process with dead time.

the optimum values from Fig. 9-8.   A step change in load, which might be a change in impurity concentration or catalyst activity, is not detected for 10 min, and the controller action taken at this time is not noticed until $t = 20$.   The best that a feedback control system could do would be to reduce the error to zero at $t = 20$.   The minimum error integral would therefore be 10 for a unit load change, or $1.0L$ in general terms.   With proportional-integral control, the integral of the absolute error is about $1.5L$, only 50 per cent greater than for a perfect controller.   Thus the benefits of more complex feedback systems will not be very great for this process.   For large reductions in error, the load disturbances must be measured and feedforward compensation used.

Floating control or integral control could be used for the time-delay system, but based on Buckley's study (7), the minimum error integral is at least 50 per cent greater than that in Fig. 9-9.   Proportional control alone is out of the question, since the final offset would be two-thirds the load change with a gain of 0.5.

A very attentive human operator could cut the error integral for Fig. 9-9 to a minimum by making a step change in the process input when an error was noticed and waiting to see the result of this change before applying a further correction.   The principle of step-by-step control can be applied automatically by using controllers which sample the output at control intervals of T_c min and apply an almost instantaneous correction proportional to the error (8, 9).   A conventional integral or proportional-integral controller can easily be modified to act for a few seconds and then hold a constant output during the rest of the control interval (10).   For the pure time-delay process, the control interval should be slightly greater than the time delay to ensure stability, and the corrective action should be just great enough to bring the output back to zero if there is no further load change.   In practice, the process gain would not be known accurately enough to permit perfect compensation, and two or three adjustments would be needed to eliminate the offset, as shown in Fig. 9-10.   The error integral depends on how soon the intermittent sampling system notices an error.   For an error starting halfway between samples, the error integral is about 1.5 to $1.7L$, depending on the accuracy of the controller settings.   Judged by the error integral, control is no better than with a conventional system, though the error is reduced to zero a little sooner.

Figure 9-11 shows the transient response for conventional control of a system with a time constant equal to the time delay.   A typical example would be a stirred-tank reactor followed by a pipeline reactor.   The controller settings, taken from Fig. 9-8, are apparently a little conservative, since there is no overshoot.   The error integral $\int |e|\, dt$ is 18.2 min, or $1.82L$, as calculated just from the gain and integral time, since the error becomes zero only when $(K_c/T_R)\int e\, dt$ is 1.0 (the net change in controller output, which is needed to offset the load change, comes just from the integral

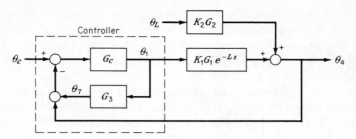

FIGURE 9-12   Block diagram of a special controller for processes with dead time.

action of the controller).   A perfect feedback control system would reduce the error to zero at $t = 20$, and the error integral would be $0.37L$.   However, a step reduction in θ_3 or θ_4 is impossible because of the capacity of the time constant and the tendency of the system to saturate for large inputs.   To reduce the error integral to 0.5 to 0.7L is a more realistic goal.

The control scheme suggested by Smith (11, 12) uses time-delay and time-constant units in the controller.   The purpose is to make the input to the controller the same as if the process had no time delay, which permits a much higher controller gain and reset rate to be used.   A general diagram is given in Fig. 9-12, using nomenclature to correspond to the specific example in Fig. 9-13.   The desired transfer function is

$$\frac{\theta_8}{\theta_1} = K_1 G_1 \tag{9-8}$$

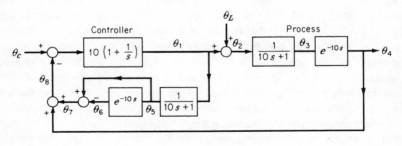

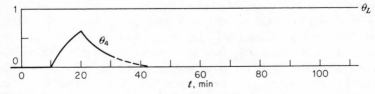

FIGURE 9-13   Modified control system for a process with dead time.   [*After* Smith (11, 12).]

where $\qquad \theta_8 = \theta_4 + \theta_7$

From Fig. 9-12,

$$\theta_8 = \theta_1 K_1 G_1 e^{-Ls} + \theta_1 G_3 = K_1 G_1 \theta_1 \qquad (9\text{-}9)$$

or $\qquad G_3 = K_1 G_1 (1 - e^{-Ls}) \qquad (9\text{-}10)$

Thus the extra controller elements G_3 must include the time constants of the process, $K_1 G_1$, in series with the pulse function $(1 - e^{-Ls})$.

For the system of Fig. 9-13, a gain of 10 and a reset time of 1.0 min were chosen as reasonable controller settings, considering the possibility of saturation for large changes and the difficulty of exactly matching the process parameters. The error integral is about $0.8L$, which is 0.4 to 0.5 the value obtained with conventional control.

The application of Smith's control scheme to an actual process was reported by Lupfer and Oglesby (14). The process has a time delay of 9.5 min and time constants of 13.1, 11.1, and 0.5 min. About a twofold reduction in error integral was obtained by using either the exact pulse function called for by Eq. (9-10) or a first-order approximation for the dead time.

PROBLEMS

1a. Sketch the reaction curve for a process with one time constant and one time delay. Use the Ziegler-Nichols method to obtain the optimum settings of a three-mode controller for the ratios $L/T = 0.1, 0.3, 1$, and 3.

b. Compare the settings from part a with those predicted from the maximum gain and the ultimate period.

2. Use frequency-response analysis to predict the optimum reset time for a process with time constants of 20, 10, and 0.1 min. Assume that the optimum corresponds to a maximum value for the product of the maximum gain and the critical frequency. Compare with the values predicted by Eqs. (9-2) and by Fig. 9-4.

3a. Plot the transient response of a process with time constants of 50 sec and 20 sec and a time delay of 10 sec. Use the reaction-curve method to obtain the recommended settings for a three-mode controller.

b. What is the phase margin for the recommended controller gain?

c. Prove that the effective lag does not depend on the size of the step change.

4. A proportional-integral controller is used on a pure time-delay process. Calculate the response to a step change in load if the controller gain is half the maximum value and the reset time is half the time delay. Calculate the integral of the absolute error, and compare with that for Fig. 9-9.

5. Determine the effective system parameters from the following transient response, and predict the critical frequency and maximum gain. Use Eqs. (9-3)

to secure the optimum settings for a three-mode controller. How do these values agree with those estimated directly from the reaction curve?

Time, min	Response, arbitrary units
0	0
1	0
2	0
3	4
4	10
5	19
6	27
7	35
8	41
9	45
30	50

6a. Sketch the response of a pure time-delay process with a step-by-step control system. Assume a control interval 1.1 times the time delay, and assume that the controller overcorrects by 50 per cent after each measurement.

b. Show the response for the same case, except that the control interval is 0.9 times the time delay.

REFERENCES

1. Ziegler, J. G., and N. B. Nichols: Optimum Settings for Automatic Controllers, *Trans. ASME*, **64**:759 (1942).
2. Taylor Instrument Companies: Instructions for Transcope Controller, *Bull.* 1B404, 1961.
3. Caldwell, W. I., G. A. Coon, and L. M. Zoss: "Frequency Response for Process Control," p. 188, McGraw-Hill Book Company, Inc., New York, 1959.
4. Wills, D. M.: Tuning Maps for Three-mode Controllers, *Control Eng.*, **9**(4): 104 (1962).
5. Jackson, R.: Calculation of Process Controllability Using the Error-squared Criterion, *Trans. Soc. Instr. Tech.*, **10**:68 (1958).
6. Hazebroek, P., and B. L. van der Waerden: Theoretical Considerations on the Optimum Adjustment of Regulators, *Trans. ASME*, **72**:309 (1950).
7. Buckley, P. S.: Automatic Control of Processes with Dead Time, in J. F. Coales (ed.), "Automatic and Remote Control," vol. I, p. 33, Butterworth & Co. (Publishers), Ltd., London, 1960.
8. Sartorius, H.: Deviation Dependent Step-by-step Control Systems and Their Stability, in A. Tustin (ed.), "Automatic and Manual Control," p. 421, Academic Press Inc., New York, 1952.
9. Oldenbourg, R. C.: Deviation Dependent Step-by-step Control . . . for Plants with Large Distance-velocity Lag, in A. Tustin (ed.), "Automatic and Manual Control," p. 435, Academic Press Inc., New York, 1952.
10. Hausner, M. J.: Get the Benefits of Sampled Data Control from Modified Process Controllers, *Control Eng.*, **8**(3):148 (March, 1961).

11. Smith, O. J. M.: A Controller to Overcome Dead Time, *ISA J.*, **6**(2):28 (February, 1959).
12. Smith, O. J. M.: Closer Control of Loops with Dead Time, *Chem. Eng. Progr.*, **53**:217 (1957).
13. Reswick, J. B.: Disturbance-response Feedback—A New Control Concept, *Trans. ASME*, **78**:153 (1956).
14. Lupfer, D. E., and M. W. Oglesby: Applying Dead-time Compensation for Linear Predictor Process Control, *ISA J.*, **8**:11, 53 (November, 1961).

10 CONTROL VALVES AND TRANSMISSION LINES

A control valve is the usual mechanism for adjusting the input of a process control system. A double-seat valve with a spring-diaphragm actuator or motor is shown in Fig. 10-1. The advantages of using two seats are that the pressure drop forces on the plugs almost cancel, and also the flow capacity is up to 30 per cent greater than for single-seat valves of the same size. However, a valve with two seats cannot be tightly closed and would not be used where positive shutoff is essential. With a single-seat valve there is an unbalanced force on the plug from the pressure drop across the valve, and the valve position depends on this force as well as on the force produced by the motor. The effect of valve pressure drop can be minimized by using either a more powerful motor or a valve positioner. A piston-type actuator with an integral positioner is shown in Fig. 10-2.

Both the steady-state and dynamic characteristics of the valve should be considered in the design of a process control system. The steady-state behavior depends mainly on the size and shape of the valve plug and on the pressure drop across the valve. The actuator design has a slight effect on the steady-state position, since a powerful actuator or a positioner can reduce the hysteresis caused by stem friction. The dynamic response of a valve depends primarily on the actuator and on the length of transmission line from controller to valve. The inertia of the moving valve stem and plug is usually negligible. This chapter deals only with the flow characteristics of a few common valve types and the dynamic response of pneumatic actuators and transmission lines. The mechanical features of these and many other types of valves, actuators, and positioners are described in Refs. 1, 2, and 3.

10-1. VALVE TYPES AND INHERENT CHARACTERISTICS

The flow through a sliding-stem control valve usually follows the orifice equation for incompressible fluids.

$$F = kA\sqrt{\frac{\Delta P}{\rho}} \qquad (10\text{-}1)$$

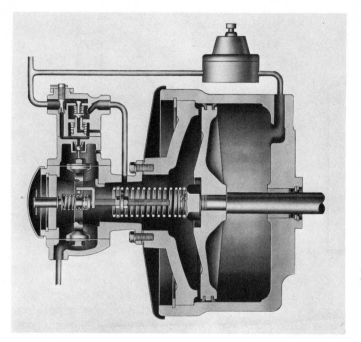

FIGURE 10-2   A piston-type actuator.   (*Courtesy of Conoflow Corp.*)

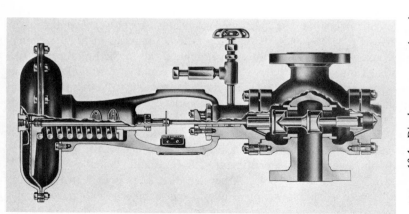

FIGURE 10-1   Diaphragm   control   valve.   (*Courtesy of   Mason-Neilan   Div., Worthington Corp.*)

195

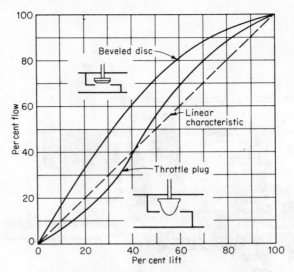

FIGURE 10-3   Valves with almost linear characteristics.
(*Data from Taylor Instrument Companies.*)

Control is achieved by moving the stem to vary the area for flow.   The gain of the valve, or change in flow for a given change in stem position, depends on the change in area with stem position, and also on the change in pressure drop with flow.   If there are resistances in series with the valve, the pressure drop across the valve decreases with increasing flow because of the increasing pressure drop across the other fixed resistances.

The manufacturers of control valves have standardized on a flow coefficient C_v to give the flow capacity of valves.   The flow coefficient is the flow of water in gallons per minute for a pressure drop of 1 psi across the wide-open valve.   The maximum flow rates for most other fluids and other pressure drops can be calculated by using Eq. (10-1).   Methods of sizing valves for viscous liquids, flashing liquids, or gases at high pressure drops are given in the manufacturers' catalogues and in the literature (5, 6). Values of C_v range from 40 to 60 for a 2-in. valve and vary roughly with the square of the nominal valve size.   The flows for various valve-stem positions are obtained from C_v and the valve characteristic, which is usually plotted as per cent of maximum flow versus per cent lift.

The inherent valve characteristic is the relationship between the flow and the stem position at a constant pressure drop across the valve.   The two common general characteristics are "linear" and "equal-percentage," as shown in Figs. 10-3 and 10-4.   A beveled-disc, or poppet, valve is just a globe valve with a sliding stem.   Full flow is achieved with a relatively short movement of the stem (⅛ to ¼ in. for a 1-in. valve), and so the terms "low-lift valve" and "quick-opening valve" are also used.   For up to 50 per cent of maximum flow the poppet valve has an almost linear character-istic, since the area of the cylindrical-shaped passage for flow is proportional to the lift.   At higher flows, the pressure drop through the valve body is significant, and the valve sensitivity decreases.

The throttle plug is shaped to give an almost linear characteristic over a greater lift than is possible with a poppet valve. (Parabolic, ratio, and throttle plugs have similar shapes.)   The flow area is the annular area between the plug and valve seat. A plug could be machined to give an exactly linear characteristic, but there is little to be gained by this, since the inherent characteristics are modified to a varying degree by the change in pressure drop across the valve.

The plug for an equal-percentage valve is generally a hollow cylinder with V-shaped ports on the side. The flow area increases more rapidly with lift as the valve opens. For an ideal equal-percentage valve, the sensitivity or gain would be directly proportional to the flow ($dF/dx = kF$), and the flow characteristic would be a straight line on a semilog graph. An actual valve has equal-percentage characteristics only over the upper portion of the range, as shown in Fig. 10-4.   At low lifts the flow through the clearance between the plug and the seat ring is appreciable, and the total flow does not change much with decreasing lift until the valve is almost seated.

The valve of Fig. 10-4 could be used to control flows down to 5 per cent of the maximum, which means that the "rangeability" for constant pressure drop is 20 to 1.   The rangeability is somewhat greater in large valves because

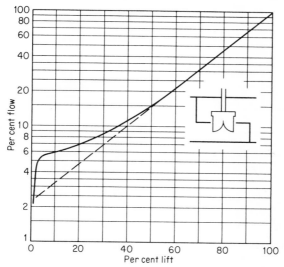

FIGURE 10-4   Inherent characteristic of an equal-percentage valve. (*Data from Taylor Instrument Companies.*)

of the relatively small leakage flow, but a rangeability greater than 50 to 1 is hard to achieve.   An equal-percentage valve can be made by using a shaped plug instead of a ported cylinder (4), but the rangeability is about the same.

10-2. EFFECTIVE VALVE CHARACTERISTICS

The flow characteristic of a valve in a process control system depends on the inherent characteristic and on the change in valve pressure drop with flow rate.   Consider the case where the control valve adjusts the flow to some process equipment and the total pressure drop across the valve, equipment, and piping is constant.   If the pressure drop across the valve is half the total at the normal flow rate, increasing the flow by 20 per cent increases the pressure drop in the line and equipment by about 1.4-fold, which leaves only 30 per cent of the total pressure drop across the valve. Therefore, to get a 20 per cent increase in flow, the valve lift must change more than is indicated by the inherent valve characteristic.

Figure 10-5 shows the effective characteristic for a beveled-disc valve with a normal pressure drop of half the total.   The curve was calculated

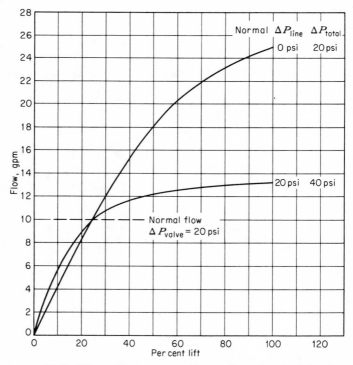

FIGURE 10-5   Effect of line pressure drop on flow characteristic of a beveled-disc valve.

from the data in Fig. 10-3 on the assumption that the pressure drop in the line and equipment varies with the square of the flow rate.† The effect of changing pressure drop makes the characteristic strongly nonlinear. At 10 per cent above the normal flow the valve gain is 0.12 gpm/per cent lift compared with 0.25 at 10 per cent below the normal flow.

The decrease in valve gain with increasing flow may make the process difficult to control when the load is changing. Suppose that the process gain does not change with flow and that the optimum overall gain is also constant. If the controller gain is set at half the maximum value when the flow is low and the valve gain high, the overall gain will be too low at high flows and the process response will be slower than desired. Tuning the controller for optimum performance at high flows would lead to instability at low flows.

If the process gain increases with flow rate, a valve with decreasing sensitivity is desirable. For flow control systems that use an orifice and a differential pressure transmitter, the gain of the measurement system is proportional to the flow, since the pressure drop varies with the square of the flow.

$$\Delta P = kF^2$$

$$\text{Gain} = \frac{d\Delta P}{dF} = 2kF$$

(10-2)

Thus, for a constant overall gain at given controller settings, the valve gain should be inversely proportional to the flow. Using a beveled-disc valve with half the total pressure drop across the valve overcorrects for the increase in gain of the flow process, since the valve gain changes twofold for only a 20 per cent change in flow. The normal valve pressure drop would have to be about three-fourths the total to make the overall gain almost constant.

The effect of line pressure drop on the characteristics of an equal-percentage valve is shown in Fig. 10-6. The gain increases with increasing flow for low flows, is almost constant over the intermediate range, and decreases as the flow approaches the maximum value. For the case where the valve pressure drop is normally half the total pressure drop, the gain changes only 20 per cent as the flow goes from 10 per cent below normal to 10 per cent above normal ($K_v = 0.20$ gpm/per cent flow at 9 gpm; $K_v = 0.16$ gpm/per cent flow at 11 gpm). For twofold changes in flow, the change in valve gain does become significant; the gain goes from 0.13 at 5 gpm to

† Sample calculations for Fig. 10-5 are:
Normal conditions, $F = 10$ gpm, $\Delta P_v = 20$ psi, $\Delta P_T = 40$ psi.
Valve with capacity of 25 gpm at 20 psi is chosen for an example.
$10/25 = 0.40$; so normal lift is 25 per cent, from Fig. 10-3.

At $F = 12$, $\Delta P_v = 40 - 20(12/10)^2 = 11.2$ psi

$$\text{Max flow at 11.2 psi} = 25\sqrt{11.2/20} = 18.8 \text{ gpm}$$

$$12/18.8 = 0.64$$

Get 64 per cent of maximum flow at 43 per cent lift, from Fig. 10-3.

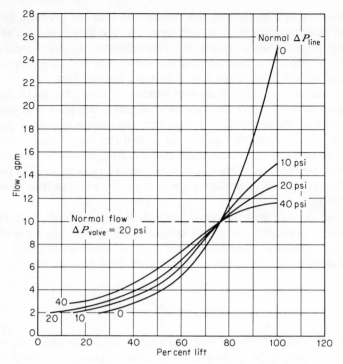

FIGURE 10-6   Effect of line pressure drop on flow characteristic of an equal-percentage valve.

0.20 at 10 gpm.   The change in valve gain is undesirable if the process gain is constant, but for many systems the process gain decreases with increasing flow, and an equal-percentage valve can make the overall gain almost constant over a wide range of flows.

The method of expressing the valve gain may need clarification.   For problems where only small load changes are considered, the change in flow per fractional change in lift is often divided by the normal flow to get a dimensionless valve gain.   For an equal-percentage valve with a constant pressure drop across the valve, the percentage gain is about 5 per cent change in flow per per cent change in lift, an easy number to remember. For the effective characteristic of Fig. 10-6, the percentage gain is 2 per cent change in flow per per cent change in lift at 10 gpm when half the pressure drop is across the valve.   To study the effect of large load changes, it is best to express both the valve gain and process gain by using absolute flow units (gallons per minute) to avoid uncertainty about the basis used for a percentage gain.   Since the valve characteristics are important mainly for large load changes, the term valve gain in this chapter refers to an absolute gain in gallons per minute fractional or per cent change in lift.

The flow characteristics in Figs. 10-5 and 10-6 are based on a constant pressure drop across the system. When a centrifugal pump is used, the head decreases with increasing flow and the decrease in gain with increasing flow is more pronounced than at constant total pressure drop. A similar effect is obtained when the control valve is placed in parallel with a fixed resistance to permit a smaller control valve to be used. The valve has to carry a higher percentage of the total flow at high flows, which decreases the change in total flow with lift.

10-3. SELECTING CONTROL VALVES

Some general principles governing selection of valves should now be apparent. If the possible load changes are small, say, less than 5 per cent, and the normal pressure drop across the valve is at least half the total, the valve characteristics have a minor effect on controllability. A poppet valve may be preferred for these cases because it is cheap and because the low lift permits a small motor to be used (7).

If large load changes are likely, the valve style, size, and pressure drop should be chosen so that the overall gain can be close to the optimum value for both high and low loads at one setting of the controller gain. When the process gain is constant, a poppet valve is suitable if nearly all the pressure drop occurs across the valve; if the valve pressure drop is small, an equal-percentage valve would be used to give an approximately linear effective characteristic. When the process gain decreases with increasing flow, an equal-percentage valve is recommended.

In all cases the valve should be sized to handle at least 20 per cent more than the anticipated maximum flow. The pressure drop across the valve may be a design variable, or it may already be fixed by other factors. A high pressure drop across the valve increases the range of controllable flow and makes it easier to obtain a desired flow characteristic, but a high pressure drop represents wasted energy. A valve pressure drop of at least one-half the total is recommended for small lines, though if load changes are very small, satisfactory control is possible with only one-fourth the total drop across the valve. For control of flows in large lines, it may be more economical to use a variable-speed drive on the pump than to throttle the flow with a control valve.

Example 10-1

The level in a tank is controlled by throttling the discharge stream, and the entire pressure drop occurs across the control valve. The flow to the tank varies over a threefold range.

a. What type of control valve would permit an almost constant overall gain?

b. Would the changing time constant of the tank influence the desired gain and selection of the valve?

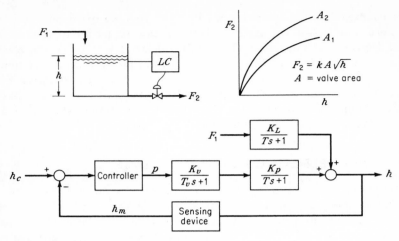

FIGURE E10-1   Diagram for Example 10-1.

a. The process gain K_p is the rate of change in level with change in flow,

$$K_p = \frac{\partial h}{\partial F_2} = \frac{2\sqrt{h}}{kA}$$

Since the level is controlled at the same value for all flows, the required valve area is directly proportional to the flow and K_p can be expressd in terms of the exit flow.

$$K_p = \frac{2\sqrt{h}}{F_2/\sqrt{h}} = \frac{2h}{F_2}$$

The valve gain is the rate of change of flow with change in either lift or air pressure, since the lift is directly proportional to the signal from the controller,

$$K_v = \frac{\partial F_2}{\partial x} = \frac{\partial F_2}{\partial p}$$

To make the product $K_p K_v$ constant, K_v must be proportional to F_2, which is the characteristic of an equal-percentage valve, but before deciding on a valve, the effect of changes in the system dynamics must be considered.

b. On the assumption that the time constant of the tank is much larger than any of the other lags, changing this time constant changes only the overall amplitude ratio of the system and not the critical frequency. A threefold increase in the flow lowers the tank time constant by a factor of 3 and lowers the maximum gain by the same amount. Therefore, the product $K_v K_p$ should be inversely proportional to F_2, which is possible if K_v is constant. *A linear valve is recommended.*

Example 10-2

A process stream is cooled from 150 to 100°F in a shell-and-tube exchanger. The outlet temperature is controlled by regulating the flow of cooling water. The water pressure is 70 psig, and the pressure drop across the exchanger is 60 psi at

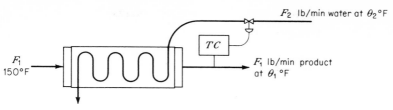

FIGURE E10-2 Diagram for Example 10-2.

a flow of 2,000 lb/min. The effects of water-flow rate and inlet-water temperature are shown below. How well would an equal-percentage value compensate for changes in process gain caused by changes in inlet-water temperature?

Normal values: $F_1 = 500$ $\theta_1 = 100°F$
$F_2 = 1,000$ $\theta_2 = 60°F$

F_2	Relative UA†	θ_1		
		For $\theta_2 = 50$	For $\theta_2 = 60$	For $\theta_2 = 70$
600	0.83	103.3		
800	0.93	98	103.5	
1,000	1	.....	100	
1,200	1.06	.....	97.4	
1,400	1.11	.....	95.5	101.3
1,600	1.15	.....	93.8	100.2

  † Calculated by assuming 40 per cent of the resistance on the water side at normal conditions.

The control valve is sized to pass 2,000 lb/min at 10 psi and full lift. The lift required for other flows is calculated by using the characteristic in Fig. 10-4 and assuming that $\Delta P = kF^2$.

F_2	ΔP valve, psi	F_2, max lift	F_2/F_2 max	% lift
600	65	5,100	0.12	42
800	60	4,900	0.16	52
1,000	55	4,700	0.21	59
1,200	48	4,370	0.27	66
1,400	41	4,170	0.33	71
1,600	32	3,580	0.45	79
2,000	10	2,000	1.0	100

For normal operation, the product of valve gain and process gain is

$$K_v K_p \cong \frac{200 \text{ lb/min}}{7\% \text{ lift}} \times \frac{3°F}{200 \text{ lb/min}} = 0.43°F/\% \text{ lift}$$

With water at 50°F, F_2 is about 700 lb/min, and the gain is

$$K_v K_p \cong \frac{200}{10\% \text{ lift}} \times \frac{5.3}{200} = 0.53°\text{F}/\% \text{ lift}$$

With water at 70°F, F_2 is about 1,600, and

$$K_v K_p \leq \frac{200}{8\% \text{ lift}} \times \frac{1.1}{200} = 0.14°\text{F}/\% \text{ lift}$$

Thus the equal-percentage valve compensates fairly well for the increased process gain at low flows, but it does not correct for the decrease in gain at high flow rates. If precise control is necessary for a wide range of loads, nonlinear elements could be incorporated in the controller, or the scheme of bypassing part of the process fluid might be considered.

These examples show that general rules for selecting valve characteristics must be used with caution if close control is required for widely varying loads. Because of changes in system dynamics, the optimum overall gain may change with changing load. Furthermore, the best characteristic for one type of load change may not be best for changes in another load variable. Additional examples of valve-selection procedures are presented in Refs. 6 to 10.

10-4. RESPONSE OF PNEUMATIC TRANSMISSION LINES AND VALVES

A pneumatic transmission line has resistance, capacitance, and inductance distributed along the length of the line, and the basic equations for signal transmission are the same as for an electrical transmission line. The problem in adapting available transmission-line theory to pneumatic lines is to obtain values of R, C, and L per unit length, with proper allowance made for the radial gradients of velocity and temperature. For small signals at low frequencies, the velocity profile is parabolic, and the flow is isothermal. At high frequencies the air in the center of the tube flows as a plug and expands adiabatically, and there are large velocity and temperature gradients near the wall. Rigorous solutions allowing for these effects have been presented by Iberall (11) and by Nichols (12), and frequency-response tests have shown the predictions to be accurate for small signals (± 0.1 psi) (11 to 13). However, for disturbances of 0.5 psi or greater, the attenuation and phase lag are generally greater than predicted because of turbulent flow in portions of the line. For standard ¼-in. (0.188 ID) tubing, the critical Reynolds number of 2,100 is reached at a pressure gradient of only 0.0046 psi/ft. This gradient is often exceeded in practice, at least for processes that are fast enough to make the transmission-line lag significant.

Several workers have presented simplified theories for pneumatic lines using empirical correction factors to allow for the greater frictional resistance at high velocities (13 to 15). The modified equations are still rather complex, and yet they may be in error by as much as 30 per cent for signals of medium size. The scatter of the data and the significant differences between the results of independent studies (14) probably arise because the transition from laminar to turbulent flow does not always occur at the same Reynolds number. Since the exact

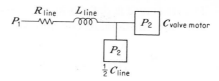

FIGURE 10-7   Lumped transmission line.

response is difficult to obtain, and since the transmission lag is small for most process control systems, approximate methods of obtaining the lag are often satisfactory.   The approximate frequency response can be obtained by inter-polating between published curves, or a lumped-parameter approach can be used, as described below.   If rough calculations show that the transmission lag is critical, a more rigorous treatment can be used or, better yet, experi-mental tests made with the actual components.

For low frequencies, the response of a distributed RC system is about the same as that of a first-order system with a time constant $RC/2$ (see Fig. 5-16).   Either half the capacity plus the total resistance or the total capacity plus half the resistance could be used in an approximate model if the system is dead-ended.   A transmission line has a capacitive load at the end (a valve motor or a small bellows), and so the model uses the total line resistance and half the line capacity combined with the load capacity.   The total inductance is included in series with the resistance, as shown in Fig. 10-7. The transfer function for this RLC system relates the pressures at the inlet and outlet of the line and is obtained from a material balance.

$$\frac{dQ}{dt} = \frac{P_1 - P_2 - L(d^2Q/dt^2)}{R}$$

(10-3)

where Q = volume of gas (STP) in capacity $C = CP_2$.

Transforming and then substituting for Q,

$$RsQ = P_1 - P_2 - Ls^2Q$$

$$P_2(RCs + LCs^2 + 1) = P_1$$

or

$$\frac{P_2}{P_1} = \frac{1}{sC(R + Ls) + 1}$$

(10-4)

To show that the inductance L is small for normal frequencies, the parameters R and L are evaluated for $\frac{1}{4}$-in. tubing at a mean pressure of 9 psig.   The resistance is derived from the Hagen-Poiseuille equation for streamline flow.

$$\frac{\Delta P}{x} = \frac{32v\mu}{D^2g_c} = \frac{128F\mu}{\pi D^4g_c} = \frac{128\mu}{\pi D^4g_c}\frac{F^*P^*}{P}$$

(10-5)

where F = volumetric flow rate

F^* = flow at standard conditions

The resistance is defined as the pressure drop per unit length divided by the flow rate, at standard conditions.

$$R = \frac{\Delta P}{xF^*} = \frac{128\mu}{\pi D^4 g_c} \frac{P^*}{P}$$

$$= \frac{128}{3.14} \frac{[1.21 \times 10^{-5} \text{ lb mass}/(\text{ft})(\text{sec})](\frac{1}{12} \text{ ft/in.})}{(0.188)^4 \text{ in.}^4 \, 32.2 \text{ lb mass, ft}/(\text{lb force})(\text{sec}^2)} \left(\frac{14.7}{23.7}\right)$$

$$= 6.3 \times 10^{-4} \frac{\text{lb force}/(\text{in.}^2)(\text{ft})}{\text{std. in.}^3/\text{sec}} \quad \text{or} \quad \frac{\text{psi/ft}}{\text{sci/sec}} \qquad (10\text{-}6)$$

The inductance comes from the inertia of the gas, which is proportional to the density, but since the flow in Eq. (10.3) is given at standard conditions, the inductance on this basis is independent of density.

$$A \frac{\Delta P}{x} = \left(\rho \frac{A}{g_c}\right)\left(\frac{d^2v}{dt^2}\right) = \rho \frac{A}{g_c} \left(\frac{1}{A} \frac{d^2Q}{dt^2} \frac{\rho^*}{\rho}\right)$$

Force per unit length = mass × acceleration   (10-7)

or

$$\frac{\Delta P}{x} = \frac{\rho^*}{g_c A}\left(\frac{d^2Q}{dt^2}\right)$$

$$L = \frac{\rho^*}{g_c} \frac{4}{\pi D^2}$$

$$= \frac{0.0748 \text{ lb/ft}^3}{32.2} \frac{4}{3.14(0.188)^2} \frac{1}{1,728} = 4.9 \times 10^{-5} \frac{\text{psi/ft}}{\text{sci/sec}^2} \qquad (10\text{-}8)$$

The inductance effect is proportional to the frequency, as shown by the term $R + Ls$ or $R + j\omega L$ in Eq. (10-4). For the above values of R and L, the inductance term ωL is only half the resistance at 1 cps and can be neglected at lower frequencies. The inductance is neglected except for the 50-ft line in the examples calculated for Fig. 10-8.

The isothermal capacitance of a unit length of line is the area divided by the standard pressure. For ¼-in. (³⁄₁₆ ID) tubing

$$C = \frac{A}{P^*} = \frac{\pi}{4}(0.188)^2 \times 12 \times \frac{1}{14.7} = 0.0224 \frac{\text{sci/ft}}{\text{psi}} \qquad (10\text{-}9)$$

According to Nichols (12), the flow should be isothermal up to 5 cps. The capacity of a fixed volume at the end of the line lies between V/P^*, the value for isothermal expansion, and $V/\gamma P^*$, the value for adiabatic expansion. ($\gamma = c_p/c_v = 1.4$ for air.)

The capacity of a diaphragm motor or bellows is greater than that of a fixed volume because of the increase in volume with pressure.

$$C = \frac{d}{dP}\left(\frac{VP}{P^*}\right) = \frac{V}{P^*} + \frac{P}{P^*}\left(\frac{dV}{dP}\right) \qquad (10\text{-}10)$$

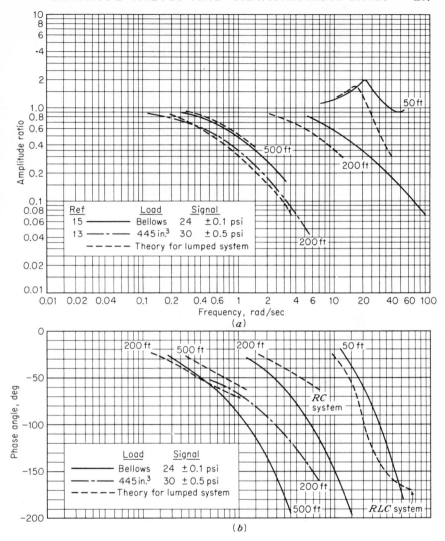

FIGURE 10-8   Frequency response of ¼-in. (³⁄₁₆ ID) tubing.

For a Foxboro No. 6 valve motor, the volume increases from 25 to 55 in.³ over the 3- to 15-psi range, and the isothermal capacity at 9 psi is

$$C = \frac{40}{14.7} + \frac{23.7}{14.7}\left(\frac{55-25}{15-3}\right) = 6.75 \text{ sci/psi}$$

Experimental frequency-response data for ¼-in. tubing are shown in Fig. 10-8. The bellows used to terminate the lines probably had a volume

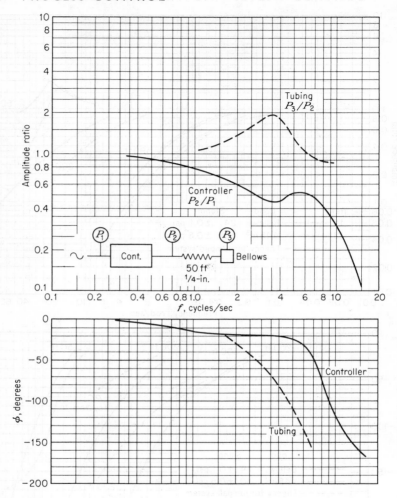

FIGURE 10-9   Frequency response of a Transcope pneumatic controller.
(*Courtesy of Taylor Instrument Companies.*)

of 1 to 4 in.³, much less than the tubing volume.   For dead-ended lines 500 ft or longer, the amplitude ratio is about the same as for the lumped *RC* system, but the phase lag is about 50 per cent greater.   Adding a time delay of 1 sec for each 1,000 ft of tubing would give a fairly good fit to the data.

For lines of moderate length which have a large capacity at the end, the first-order approximation is fairly good up to 70° phase lag, as shown by the data for a 200-ft line and a 445-in.³ load (13).   The predicted amplitude ratio is a little too low, probably because the isothermal capacity was used for the load.

Short lines show a resonant peak that could be a problem in systems with a high natural frequency.   However, most controllers and transmitters are designed to damp out these high-frequency oscillations.   Frequency-response data for a pneumatic proportional controller are shown in Fig. 10-9.   The controller input and output pressures were measured with strain gauge transducers, since the instrument bellows have appreciable lag at these frequencies.   The controller response depends on the diameter and length of the transmission line and the terminal volume, since controller and line form an interacting system.   With 50 ft of $\frac{1}{4}$-in. tubing and a small bellows, the amplitude ratio for this controller drops just about enough to cancel the peak caused by resonance in the transmission line, and the overall amplitude curve would be flat to about 5 cps.   With only 10 ft of tubing, this controller has a peak amplitude ratio of 1.6 at 4 cps (not plotted), but the resonant peak of a 10-ft line is at about 10 cps, and so there is little overlap.   The numbers given here and in Fig. 10-9 are intended only as a rough guide: if the performance of the control system at high frequencies is critical, tests should be carried out using the actual components.

The transient response of a long transmission line to a step upset is an S-shaped curve characteristic of distributed systems (see Fig. 10-10).   The apparent initial delay can be much greater than the length divided by the sonic velocity, because the product RC varies with the square of the length. The time for 63.2 per cent recovery, which should be about $RC/2$, can be used as an effective time constant for rough calculations of the frequency response.   Figure 10-11 shows data given by Bradner (16) for a 3- to 15-psi upset; the effective lags are about twice those given in Fig. 10-8 because of

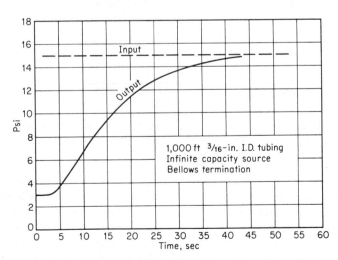

FIGURE 10-10   Step   response   of   pneumatic   tubing.   [*From Bradner* (16), *courtesy of Instrument Society of America.*]

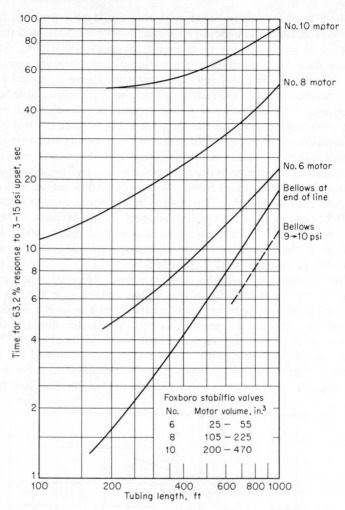

FIGURE 10-11   Transient response time for ¼-in. (³⁄₁₆ ID) tubing.
[*From Bradner* (16), *courtesy of Instrument Society of America.*]

the large upset used.   Even small step upsets will lead to turbulent flow for part of the test, since the initial pressure gradient is theoretically infinite.

The use of larger-diameter lines may be worthwhile when the transmission lag is large.   Since the resistance varies with D^{-4} and the line capacity depends on D^2, the effective time constant depends on D^{-2} for a long dead-ended line.   For ⅜-in. (0.305 ID) tubing, $RC/2$ is 2.63-fold less than for ¼-in. tubing.   If the capacity is mainly at the end of the line, a 6- to 7-fold reduction in time constant is possible by going to ⅜-in. tubing.

10-5. VALVE POSITIONERS

Both the static and the dynamic behavior of a pneumatic valve can be improved by using a valve positioner. The positioner consists of an input bellows, a feedback lever that is connected to the valve stem, and a relay that sends air to the actuator. A typical force-balance positioner is shown in Fig. 10-12. If the position of the valve stem differs from that called for by the input pressure, the pilot valve moves and air is sent to the actuator until the correct position is obtained. The positioner reduces the effects of stem friction and unbalanced pressure forces on the valve plug. It also increases the speed of response of the system, since the air entering the motor does not have to flow through a long transmission line. The response of the line plus bellows is practically the same as that of a dead-ended line. The response of the valve itself depends on the capacity of the relay and the volume of the motor. For different combinations tested by Mamzic (17), the phase lag reached 45° at frequencies of 2 to 20 rad/sec, corresponding to time constants of 0.05 to 0.5 sec.

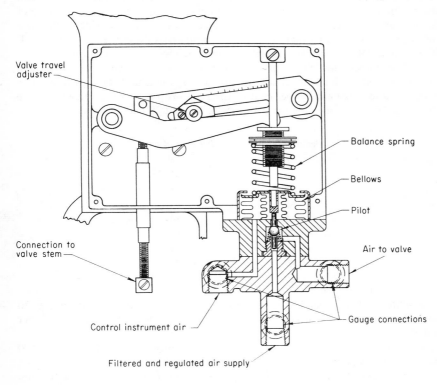

Valve travel adjuster

Balance spring

Bellows

Pilot

Connection to valve stem

Air to valve

Gauge connections

Control instrument air

Filtered and regulated air supply

FIGURE 10-12   A pneumatic valve positioner.   (*Courtesy of Moore Products Co.*)

PROBLEMS

1. A bypass arrangement is suggested for controlling a water stream between
the limits of 600 and 800 gpm. The
available pressure drop is 10 psi, independent of flow rate. Calculate the effective
valve characteristics for a 4-in. valve
($C_v = 150$) and for a 3-in. valve ($C_v = 90$) in the bypass line.

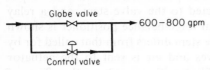

Globe valve
600–800 gpm
Control valve

2. A process stream is heated from about 80 to 120°F in a stirred tank. The
temperature is controlled by injecting live steam into the tank. What effective
valve characteristics are required to make the overall gain almost constant for the
following load changes?

   a. Variations in feed temperature from 60 to 100°F
   b. Twofold changes in flow rate of the process stream
   c. Changes in steam supply pressure of 20 to 50 psig

3. The temperature of an air stream is controlled by sending part of the stream
to a steam-heated exchanger and bypassing the rest of the stream through a control
valve. What type of valve should be used to give the best control over a range of
inlet air temperatures? Assume that the total flow of air is constant, that the
control point is 170°F, and that the exchanger could heat the entire stream to 200°F
with steam at 250°F. Neglect any change in dynamics of the exchanger.

4. The inherent flow characteristic of a typical butterfly valve is given below (2).
Calculate the effective characteristic for the case of a constant total pressure drop
across line and valve and 50 per cent of the total drop across the valve when the
valve is half open. Calculate the gain for 30, 50, and 70 per cent open.

Per cent open	Per cent of flow
20	15
30	25
40	40
50	65
60	80
70	87
80	92
90	96
100	100

5a. Moise (13) measured the response of pneumatic transmission lines at 30
psia. How would you correct these data to 24 psia for comparison with other
results?

   b. Would it be worthwhile to increase the transmission-line pressure to 100
psia for control systems with a large transmission lag?

6. A pressure control system has a tank with a 30-sec time constant, a pressure

transmitter with a 1-sec effective time constant, a control valve with a capacity of 10 sci/psi and 500 ft of line from the transmitter to the controller and from the controller to the valve.

  a. What is the critical frequency for this system?

  b. How much would the response be improved by using $\frac{3}{8}$-in. tubing instead of $\frac{1}{4}$-in?

  c. How much would the response be improved by using electronic transmission and control?

  7. An alternative arrangement for a lumped-transmission line would have the total line capacity at the middle of the line and the load capacity at the end. Derive the general transfer function for this system.   Take a 500-ft line ($\frac{1}{4}$ in.) and a fixed end volume of 100 in.^{3} as an example, and show how the frequency response compares with that for the system of Fig. 10-7.

  8. Calculate the ratio of effective capacity to isothermal capacity for a typical valve motor, assuming that air enters the motor at a constant temperature and that no heat is transferred from the air to the wall.

REFERENCES

 1. Beard, C. S.: "Control Valves," Instruments Publishing Co., Pittsburgh, Pa., 1957.
 2. Eckman, D. P.: "Automatic Process Control," p. 195, John Wiley & Sons, Inc., New York, 1958.
 3. Holzbock, W. G.: "Instruments for Measurement and Control," p. 317, Reinhold Publishing Corporation, New York, 1955.
 4. Wiedmann, J. A., and W. J. Rowan: Control-valve Plug Design, *Trans. ASME*, **78**:1367 (1956).
 5. Roth, G. L.: Factors in Selecting Valves for Compressible Flow, *Control Eng.*, **2**(12):46 (December, 1955).
 6. Valstar, J. E.: A Fresh Look at Selecting Control Valve Characteristics, *Control Eng.*, **6**:103 (March, 1959) (see also p. 123, April, 1959, and p. 119, May, 1959).
 7. Ziegler, J. G., and N. B. Nichols: Valve Characteristics and Process Control, *Instruments*, **22**(1):75 (1944).
 8. Milham, R., and A. R. Catheron: Valve Control of Liquid Flow, *Instruments*, **25**:596 (1952).
 9. Aikman, A. R.: Control Valves: Some Theoretical Considerations, *Trans. Soc. Instr. Technol.*, **5**:1 (1953).
10. Balls, B. W.: Control Valves: Application, *Trans. Soc. Instr. Technol.*, **5**:20 (1953).
11. Iberall, A. S.: Attenuation of Oscillating Pressures in Instrument Lines, *Bur. Standards J. Research*, **45**:85 (1950).
12. Nichols, N. O.: The Linear Properties of Pneumatic Transmission Lines, *ISA Trans.*, **1**(1):5 (1962).
13. Moise, J. C.: Pneumatic Transmission Lines, *ISA J.*, **1**(4):35 (1954).

14. Rohmann, C. P., and E. C. Grogan: On the Dynamics of Transmission Lines, *Trans. ASME*, **79**:853 (1957).
15. Samson, J. E.: Dynamic Characteristics of Pneumatic Transmission, *Trans. Soc. Instr. Technol.*, **10**:117 (1958).
16. Bradner, M.: Pneumatic Transmission Lag, *Instruments*, **22**(7):618 (1949).
17. Mamzic, C. L.: Improving the Dynamics of Pneumatic Positioners, *ISA J.*, **5**(8):38 (August, 1958).

11 THE DYNAMICS AND CONTROL OF HEAT EXCHANGERS

In the process industries, heat is transferred by radiation, by the mixing of hot and cold fluids, or, most frequently, by conduction through the walls of a heat exchanger. This chapter is an introduction to the response of tubular heat exchangers, a class that includes double-pipe exchangers, shell-and-tube exchangers, condensers, kettles with cooling coils, and tubular furnaces. For our purposes, the characteristic feature of these devices is that the capacities on one or both sides are distributed along the length of the exchanger. A jacketed kettle and a steam-heated reboiler belong to a simpler class, since the capacities on both sides are lumped, and the response can be derived from the treatment of interacting time constants in Chap. 3.

Because the exchanger parameters are distributed and interacting, the exact dynamic equations for an ordinary countercurrent exchanger are quite complex, and lengthy calculations are required just to get the open-loop frequency response. For exchangers with several passes or exchangers where large changes in velocity or physical properties occur, a digital computer would have to be used to determine the response. Although such a calculation takes only a few seconds of computer time, the expense of programming may not be worthwhile unless many similar cases must be investigated. In general, heat exchangers are fairly easy to control, and except where very close control is needed, simplified methods of dynamic analysis give answers that are accurate enough for the practicing engineer.

The approach taken here is to present exact transfer functions for a few of the simpler cases in order to show the parameters that determine the lags and to explain the resonance effects sometimes found with distributed systems. The solutions for these cases are then used to evaluate rules and models for obtaining the approximate response of more complex exchangers. The equations and approximate models for heat exchangers should also be useful in understanding the dynamics of countercurrent mass-transfer operations, such as absorption, extraction, and distillation, in packed or plate columns.

The measurement lag for temperature systems is given particular attention because it is generally greater than the lag in measuring flow, pressure, or level. When the measurement lag is the second largest time constant, as is often the case for temperature control loops, the quality of control is greatly improved by reducing the measurement lag. Furthermore, it is much easier to change the measuring lag by using a different element or a higher velocity in the exit line than to change the dynamics of the exchanger, which depend on the size and length of the tubes and the velocities in the exchanger.

11-1. DYNAMICS OF STEAM-HEATED EXCHANGERS

A relatively simple type of exchanger to analyze is one in which the temperature on one side of the wall is constant, as when a pure vapor is condensing on the outside of the tubes. A cooling coil in an agitated tank is also in this category, since the external fluid temperature is almost constant over the length of the coil. In a true countercurrent exchanger, the temperature on both sides varies with length, but when one fluid has a much higher flow rate than the other, the approximate behavior can be predicted by using the simpler equations for a constant temperature on one side.

Consider an exchanger in which steam is condensing on the outside of the tubes and liquid or gas is flowing in the tubes. The axial temperature profiles and radial gradients are pictured in Fig. 11-1. If the exit-fluid temperature is controlled by adjusting a valve in the steam line, the transfer functions needed for a stability analysis are P_s/X and θ/P_s or θ/θ_s. (The condensing temperature θ_s is directly related to the steam pressure P_s if there are no inerts in the system.) The steam pressure lags the valve opening X because of the vapor volume in the shell and the capacity of the shell wall. The shell and tube capacities really form an interacting system, but the interaction is ignored for the moment to concentrate on the transfer function θ/θ_s.

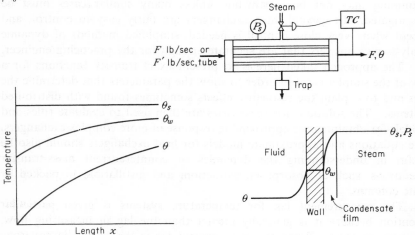

FIGURE 11-1   Temperature profiles for a steam-heated exchanger.

RESPONSE TO CHANGES IN STEAM TEMPERATURE

To obtain the transfer function θ/θ_s, the usual steady-state assumptions of no axial conduction, no backmixing, and constant fluid properties are made. It is also necessary to assume no wall resistance and no capacity in the condensate film, to limit the equations to second order. If significant, the wall resistance could be split and added to the two film resistances, and the condensate film capacity could be added to that of the wall.

The energy balance for the fluid stream is written for a length dx of a single tube.

$$M_f c_f \left(\frac{\partial \theta}{\partial t}\right) dx + F' c_f \left(\frac{\partial \theta}{\partial x}\right) dx = h_1 A_1 dx(\theta_w - \theta) \qquad (11\text{-}1)$$

or

$$T_1 \left(\frac{\partial \theta}{\partial t}\right) + v T_1 \left(\frac{\partial \theta}{\partial x}\right) = \theta_w - \theta$$

where $T_1 = \dfrac{M_f c_f}{h_1 A_1} = R_1 C_1$

$M_f =$ lb fluid/ft
$c_f =$ heat capacity of fluid, Btu/(lb)(°F)
$h_1 =$ inside coefficient, Btu/(sec)(ft²)(°F)
$A_1 =$ inside transfer area, ft²/ft
$v =$ velocity, fps $\left(v = \dfrac{F'}{M_f}\right)$

The energy balance for the wall is

$$M_w c_w \left(\frac{\partial \theta_w}{\partial t}\right) dx = h_2 A_2 dx(\theta_s - \theta_w) - h_1 A_1 dx(\theta_w - \theta) \qquad (11\text{-}2)$$

or

$$T_2 \left(\frac{\partial \theta_w}{\partial t}\right) = (\theta_s - \theta_w) - \frac{T_2}{T_{12}}(\theta_w - \theta)$$

where $T_2 = \dfrac{M_w c_w}{h_2 A_2} = R_2 C_2$

$$T_{12} = \frac{M_w c_w}{h_1 A_1} = R_1 C_2$$

$M_w c_w =$ wall capacity, Btu/(°F)(ft)
$h_2 =$ outside coefficient, Btu/(sec)(ft²)(°F)
$A_2 =$ outside transfer area, ft²/ft

Following Gould (1) and others (2, 3), the partial differential equations are converted to ordinary differential equations by taking the Laplace transform with respect to time. The variables θ and θ_w in the following transformed equations represent deviations from the normal values at any point along the exchanger:

$$T_1 s\theta + v T_1 \left(\frac{d\theta}{dx}\right) = \theta_w - \theta \qquad (11\text{-}3)$$

$$T_2 s\theta_w = \theta_s - \theta_w - \frac{T_2}{T_{12}}(\theta_w - \theta) \qquad (11\text{-}4)$$

Eliminating θ_w gives the first-order equation

$$\frac{v}{a}\left(\frac{d\theta}{dx}\right) + \theta = \frac{b}{a}\theta_s \tag{11-5}$$

where $a = \dfrac{(T_1 s + 1)(T_{12} T_2 s + T_{12} + T_2) - T_2}{T_{12} T_2 s + T_{12} + T_2}$

$\dfrac{b}{a} = \dfrac{1}{T_1 T_2 s^2 + (T_1 + T_2 + T_1 T_2/T_{12})s + 1}$

The solution of Eq. (11-5) for the boundary condition $\theta = 0$ at $x = 0$ is like the step response of a first-order system.

$$\frac{\theta}{\theta_s} = \frac{b}{a}(1 - e^{-ax/v}) \tag{11-6}$$

The term x/v is the time for fluid to flow through the tubes, which is a time delay L.

$$\frac{\theta}{\theta_s} = \frac{b}{a}(1 - e^{-aL}) \tag{11-7}$$

RESONANCE EFFECT

The term b/a can be factored to give equivalent time constants T_a and T_b to facilitate plotting the frequency response. Since the term a has s^2 in the numerator and s in the denominator, e^{-aL} is a vector with an ever-increasing phase lag and a length less than 1. The term $1 - e^{-aL}$ therefore shows regular fluctuations in amplitude and phase lag with frequency, which lead to resonant peaks in the overall frequency response. Figure 11-2 shows the calculated response for a typical exchanger, and similar resonance effects have been found in some experimental studies (4, 5, 6). The effect was probably not apparent in other studies, because the height of the peaks is quite small if the exit temperature is close to the steam temperature (see Example 11-1).

The resonance arises because the exchanger is forced in a distributed manner; i.e., the steam temperature is changed along the entire length of the exchanger and not just at one end. The peaks and valleys in the frequency response correspond to an integral number of half cycles in the exchanger. If the residence time corresponds to 1.5 cycles, some elements of the fluid are exposed to higher than normal wall temperatures for two-thirds of the time and to lower than normal temperatures for one-third of the time. For other elements, the proportions are reversed, and the difference in exposure tends to increase the amplitude ratio. With exactly one or more full cycles, all elements are exposed to a hotter than normal wall for half the time, and the amplitude ratio is relatively low. These two cases are shown in Fig. 11-3.

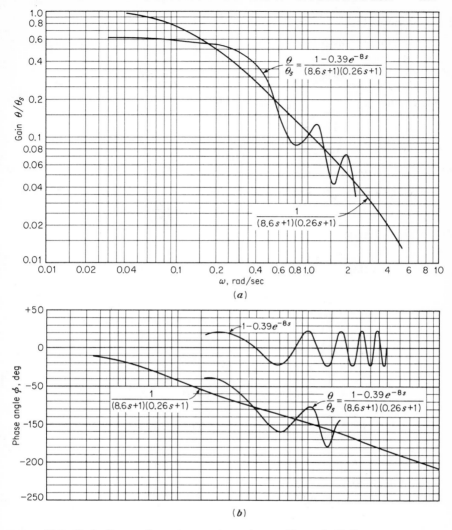

FIGURE 11-2   Bode diagram for a steam-water exchanger (Example 11-1).

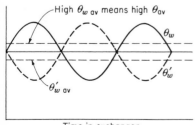

FIGURE 11-3   Explanation of resonance effect.

Example 11-1

Obtain the transfer function θ/θ_s for a steam-water exchanger with the following specifications:

Copper tubes, $\frac{3}{4}$ in. OD, 0.042 in. wall, 8 ft long.

For one tube,
$$A_1 = 0.174 \text{ ft}^2/\text{ft}$$
$$A_2 = 0.196 \text{ ft}^2/\text{ft}$$
$$S = 0.00242 \text{ ft}^2$$
$$M_w = 0.358 \text{ lb/ft}$$

Water velocity $v = 1$ fps

Average water temperature $= 150°F$

$$M_f = 0.00242 \times 62 = 0.150 \text{ lb/ft}$$

From Ref. 24 (p. 228),

$$h_1 = \frac{150[1 + 0.011(150)]}{(0.666)^{0.2}} = 430 \text{ Btu/(hr)(ft}^2)(°F)$$

$$h_2 = 2,000 \quad \text{(assumed)}$$

$$h_w = \frac{220}{0.042}(12) = 63,000 \quad \text{(neglect wall resistance)}$$

$$T_1 = \frac{M_f c_f}{h_1 A_1} = \frac{(0.150)(3,600)}{(430)(0.174)} = 7.2 \text{ sec}$$

$$T_2 = \frac{M_w c_w}{h_2 A_2} = \frac{(0.358)(0.094)}{(2,000)(0.196)}(3,600) = 0.31 \text{ sec}$$

$$T_{12} = \frac{M_w c_w}{h_1 A_1} = \frac{0.0337}{(430)(0.174)}(3,600) = 1.62 \text{ sec}$$

$$L = \frac{x}{v} = 8 \text{ sec}$$

From Eq. (11-5),
$$a = \frac{(7.2s + 1)(0.50s + 1.93) - 0.31}{7.2(0.50s + 1.93)}$$

$$= \frac{1.62(8.6s + 1)(0.275s + 1)}{13.9(0.259s + 1)}$$

$$a \cong 0.117(8.6s + 1)$$

$$e^{-aL} = e^{-8s-0.94} = 0.39e^{-8s}$$

$$\frac{b}{a} = \frac{1}{2.23s^2 + 8.89s + 1} = \frac{1}{(8.6s + 1)(0.26s + 1)}$$

$$\frac{\theta}{\theta_s} = \frac{1 - 0.39e^{-8s}}{(8.6s + 1)(0.26s + 1)}$$

The frequency response is plotted in Fig. 11-2.

In Example 11-1, the wall capacity contributed a time constant of 0.3 sec, which could be neglected for most control studies because of the larger lags in the measuring element, the control valve, and the fluid in the exchanger. A derivation neglecting the wall capacity leads to a time constant based on the overall coefficient.

$$\frac{\theta}{\theta_s} = \frac{1}{T's + 1}\left(1 - \frac{e^{-Ls}}{e^{L/T'}}\right) = \frac{1}{T's + 1}\left(1 - \frac{\Delta_2 e^{-Ls}}{\Delta_1}\right) \tag{11-8}$$

where $T' = \dfrac{M_f c_f}{U_1 A_1}$ $\dfrac{\Delta_2}{\Delta_1} = \dfrac{\theta_s - \theta_{exit}}{\theta_s - \theta_{inlet}}$

For Example 11-1, the single time constant T' is 8.6 sec, about the same as the larger time constant of the more exact solution.

Note that the time constants do not depend on the length of the exchanger or the time delay. Empirical correlations based on the time delay (7) may fit some of the data because the time delay and the major time constant are often about the same. If the exchanger of Example 11-1 were twice as long, or 16 ft, the exact transfer function would be

$$\frac{\theta}{\theta_s} = \frac{1 - 0.15e^{-16s}}{(8.6s + 1)(0.26s + 1)}$$

With the longer exchanger, the importance of the fluctuating term is decreased. The numerator of the above transfer function has an amplitude ratio between 0.85 and 1.15 and a phase angle between -9 and $+9°$. Fluctuations of this size would probably not be detected even in experimental studies.

The effective time constants for an existing system are often obtained from the response to a step input. However, the usual procedures for determining the largest or the two largest time constants from the transient response can give incorrect results for systems forced in a distributed manner. Consider a steam-water exchanger where the time delay L is about equal to T', the time constant based on the overall coefficient. After a step increase in steam temperature, the outlet temperature will gradually rise, and the final temperature would be reached in L sec if the capacity of the wall were negligible and in slightly more than L sec for a practical case. The time for 63 per cent response would be about 0.5 to 0.6L, and the frequency response calculated with this as the major time constant instead of T' would be appreciably in error. Of course the frequency response based on a single time constant T' would also be in error because of the wall capacity and the fluctuating term in Eq. (11-7) or (11-8), but these errors are small compared with a twofold error in T'. If L is very much greater than T', the time for 63 per cent response does approach T', as can be shown from Eq. (11-8) or by working Prob. 1.

RESPONSE TO LOAD CHANGES

For a complete system analysis the response to load changes must be considered. For the steam-water exchanger, possible disturbances include changes in inlet flow rate or temperature and changes in steam supply pressure. The response to a change in inlet temperature at constant steam temperature comes from the solution of Eq. (11-5) for $\theta_s = 0$, $\theta = \theta_F$ at $x = 0$.

$$G_{L_1} = \frac{\theta}{\theta_F} = e^{-aL} \tag{11-9}$$

When the wall capacity is small, the transfer function reduces to a gain multiplied by a time delay,

$$G_{L_1} = \frac{\theta}{\theta_F} \cong K_{L_1} e^{-Ls} \tag{11-10}$$

A decrease in feed temperature would not be noticed for L sec, but this does not mean that the error would be constant for L sec before control action took effect, as would be the case for a time delay in the main-loop transfer function G. A study of the closed-loop transfer function $G_{L_1}/(1 + G)$ indicates that a step change in θ_F would lead to a peak error of $K_{L_1}\theta_F$ followed by a damped oscillation at the natural frequency of the closed loop.

The response to changes in flow rate has received little attention in published studies, because an exact analysis is much more difficult than the analysis for temperature changes. If the flow is increased suddenly, there is an immediate change in the heat-transfer coefficient throughout the exchanger. However, the percentage change in overall coefficient is less than the percentage increase in flow, and so the exit temperature must drop, and this drop is spread out over a time interval of roughly L sec, the time for fluid to flow through the exchanger.

The following treatment of the response to flow changes is based on a study by Koppel (8), who considered only the case of negligible wall capacity. The first step is to linearize the relationship between the overall coefficient and the flow.

$$U = \bar{U}(1 + \beta r) \tag{11-11}$$

where $\bar{U}$ = normal coefficient

$r = \dfrac{\Delta F}{F}$ = change in flow/normal flow

The factor β is between 0 and 0.8 and depends on the fraction of the total resistance in the inside film. The heat balance for the fluid is

$$M_f c_f \left(\frac{\partial \theta}{\partial t}\right) + F'(1 + r)c_f \left(\frac{\partial \theta}{\partial x}\right) = \bar{U}A(1 + \beta r)(\theta_s - \theta) \tag{11-12}$$

or $\qquad T'\left(\dfrac{\partial \theta}{\partial t}\right) + \bar{v}T'(1 + r)\left(\dfrac{\partial \theta}{\partial x}\right) = (1 + \beta r)(\theta_s - \theta) \qquad$ (11-13)†

† T' is used to denote a time constant based on the overall coefficient.

Care must be taken at this point to separate the normal and fluctuating components of θ, it being remembered that $\bar{\theta}$ is a function of length.

$$\theta = \bar{\theta} + \Delta\theta$$

$$T'\left(\frac{\partial \Delta\theta}{\partial t}\right) + \bar{v}T'(1+r)\left(\frac{\partial \bar{\theta}}{\partial x} + \frac{\partial \Delta\theta}{\partial x}\right) = (1 + \beta r)(\bar{\theta}_s - \bar{\theta} - \Delta\theta) \quad (11\text{-}14)$$

Equation (11-14) can be solved (8), but the following simplification introduces little error and makes the main features of the result more apparent. Neglecting the terms $r(\partial \Delta\theta/\partial x)$ and $\beta r \, \Delta\theta$ and taking the Laplace transform leads to

$$(T's + 1)\,\Delta\theta + \bar{v}T'\left(\frac{d\,\Delta\theta}{dx}\right) + \bar{v}T'(1+r)\frac{d\bar{\theta}}{dx} = (1 + \beta r)(\bar{\theta}_s - \bar{\theta}) \quad (11\text{-}15)$$

Using the steady-state relationship $\bar{\theta}_s - \bar{\theta} = (\theta_s - \theta_0)e^{-L/T'}$,

$$(T's + 1)\,\Delta\theta + \bar{v}T'\left(\frac{d\,\Delta\theta}{dx}\right) = -(\theta_s - \theta_0)r(1 - \beta)e^{-L/T'} \quad (11\text{-}16)$$

The solution for $\Delta\theta = 0$ at $x = 0$ is the transfer function for flow-rate changes,

$$G_{L_2} = \frac{\Delta\theta}{r} = \frac{-(\theta_s - \theta_0)(1 - \beta)(1 - e^{-Ls})}{T's} = K_{L_2}\frac{1 - e^{-Ls}}{s} \quad (11\text{-}17)$$

The frequency response for flow changes has a fluctuating component similar to that for changes in steam temperature, because in both cases the exchanger is forced in a distributed manner. The term $(1 - e^{-Ls})/s$ approaches $1.0 \, \underline{/0°}$ at low frequencies, and the term $-(\theta_s - \theta_0)(1 - \beta)/T'$ is the steady-state gain for load changes. The closed-loop response to flow disturbances is similar to that for inlet temperature changes except that the peak error is less than $K_{L_2}r$.

LAG IN SHELL AND HEADERS

Two more factors that may have a significant effect on the exchanger response are the lags in the shell and the lags in the tube headers. The time constant for vapor holdup depends on the rate of condensation. With an overall coefficient of 200 or more, as obtained with liquid in the tubes, the vapor holdup time is 0.1 to 0.5 sec, which would usually be negligible. There is an appreciable energy-storage capacity in the shell wall, particularly for small exchangers. The heater used in one study (4) had an 8-in. shell, and the shell capacity was three times that of all the tubes.

The shell dynamics are really quite complex, because parts of the shell respond in different ways. The upper portion of a horizontal insulated shell has a very thin film of condensate and a time constant of 1 sec or less

$$\left[T_{sh} = \left(\frac{Mc_p}{hA} \right)_{shell} \right]$$ if the coefficient is 5,000 to 10,000 Btu/(hr)(ft²)(°F).

However, if the film evaporates completely following a decrease in steam pressure, the coefficient for the dry wall is orders of magnitude lower and the response becomes very slow (4). The lower portion of the shell is flooded with condensate from the tubes and might have an average time constant of 5 to 10 sec.

Fortunately, large changes in the shell response have only a small effect on the exchanger dynamics, since the shell capacity is in parallel with the capacities in the tubes and the fluid. The "exact" transfer functions presented by Catheron et al. (4) showed only a 30 to 50 per cent shift in the frequency-response curves for a 10-fold change in the assumed time constant for the shell. If only the approximate response is needed, the wall and shell capacities can be added to the fluid capacity to shift the frequency-response curves to the left by a factor that is usually about 1.5 to 2.0. The final transfer function for a one-time-constant model of a steam-liquid exchanger is given in Eqs. (11-18). The gain K_P is found from the valve characteristics, the valve pressure drop, and the exchanger characteristics by steady-state calculations.

$$\frac{\theta}{X} \cong \frac{K_P}{T''s + 1} \left(1 - \frac{\Delta_2}{\Delta_1} e^{-Ls} \right)$$

$$T'' = \frac{M_f c_f + M_w c_w + M_{sh} c_{sh}}{U_1 A_1}$$

(11-18)

If a little more effort is justified, a two-capacity model can be used to predict the wall effect, as outlined in Prob. 3. The main problem in an exact analysis is the difficulty in predicting h_{shell}.

The outlet header of a shell-and-tube exchanger causes a lag in fluid temperature, since the fluid temperature is usually measured in the exit pipe. With a multipass exchanger, there is some additional lag from each reversal chamber, and as shown by Morris (9), the cumulative effect of several such lags is appreciable. If the headers are well mixed, they act as time constants but several small time constants in series have almost the same effect as a time delay. Therefore, the holdup is probably equivalent to a time delay of 0.5 to 0.9 times the total holdup time in the headers, which could be a delay of 5 to 10 sec for a four-pass exchanger.

STEAM-AIR EXCHANGERS

The dynamic behavior of a steam heater becomes quite different when gas replaces liquid in the tubes. The capacity of the gas is small, and, in spite of the low coefficient, the time constant T_1 is generally smaller than that for liquids. Usually the gas capacity can be neglected compared with that in the metal walls. The response of the gas to changes in steam temperature is thus very rapid, but there is a large lag in steam temperature after changes in the valve position, as shown in Example 11-2. This shell lag is much greater than with liquid-steam exchangers, because the steam-flow rate is an order of magnitude lower with gas exchangers and a long time is required to heat the metal to a new temperature.

Example 11-2

Air is heated in a shell-and-tube exchanger by steam condensing on the outside of the tubes. Normal operating conditions are given below. Obtain the transfer

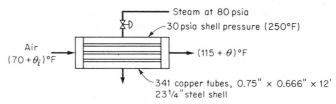

Steam at 80 psia
30 psia shell pressure (250°F)
Air (70 + θ_i)°F
(115 + θ)°F
341 copper tubes, 0.75" × 0.666" × 12'
23¼" steel shell

FIGURE E11-2   Diagram for Example 11-2.

functions θ/θ_s and θ/X.

$$\bar{v} = 30 \text{ fps} = \text{av velocity in tubes}$$

$$\bar{\rho}_1 = 0.1 \text{ lb/ft}^3 = \text{av density of air}$$

$$G = 10{,}800 \text{ lb/(hr)(ft}^2)$$

$$L = 0.4 \text{ sec} = \text{time delay}$$

$$A_1 = 0.174 \text{ ft}^2/\text{ft of tube}$$

$$S = 0.00242 \text{ ft}^2 \text{ per tube}$$

From Ref. 24 (p. 226),

$$h_1 = 0.0144 \, c_p \frac{G^{0.8}}{D^{0.2}} = \frac{0.0144(0.24)(1{,}680)}{0.56} = 10.3$$

If $h_2 \cong 1{,}000$,

$$U_1 \cong h_1 = 10.3 \text{ Btu/(hr)(ft}^2)(°\text{F})$$

$$A = nA_1 x = 341(0.174)(12) = 712 \text{ ft}^2$$

$$\ln \frac{\Delta_1}{\Delta_2} = \frac{UA_1}{Fc_p} = \frac{(10.3)(0.174)}{(26.1)(0.24)} = 0.286$$

$$\Delta_1 = 180°\text{F} \qquad \Delta_2 = 135 \qquad \Delta_{\text{l.m.}} = 157$$

$$T_1 = \frac{M_f c_f}{h_1 A_1} = \frac{(2.42 \times 10^{-4})(0.24)}{(10.3)(0.174)} \times 3{,}600 = 0.117 \text{ sec}$$

$$T_2 = \frac{M_w c_w}{h_2 A_2} = \frac{(0.358)(0.094)}{(1{,}000)(0.196)} + 3{,}600 = 0.617 \text{ sec}$$

$$T_{12} = \frac{M_w c_w}{h_1 A_1} = \frac{(0.358)(0.094)}{(10.3)(0.174)} \times 3{,}600 = 67.5 \text{ sec}$$

Since T_1T_2/T_{12} is negligible, Eq. (11-6) becomes

$$\frac{\theta}{\theta_s} = \frac{1 - e^{-aL}}{(0.12s + 1)(0.62s + 1)}$$

where

$$a = \frac{(71s + 1)(1.1s + 1)}{110(0.61s + 1)}$$

The above transfer function is mainly of theoretical interest, since the time constants are very small and, because of the short time delay, the resonant effects are significant only at frequencies above 1 cps. The major exchanger lag comes from the capacities in the tube wall, shell wall, and the vapor space on the shell side. For the following calculations, the total capacity is used, rather than the value per foot of tube.

Vapor capacity:
$$V = \frac{\pi}{4} (D_s^2 - nD_t^2)x = 23.1 \text{ ft}^3$$

$$M = \frac{23.1}{13.75 \text{ ft}^3/\text{lb}} = 1.68 \text{ lb steam}$$

$$C_v = \frac{M}{\bar{P}} \lambda \left(\frac{\partial P}{\partial \theta_s}\right) = \frac{1.68}{30 \text{ psi}} (924 \text{ Btu/lb})(0.53 \text{ psi/°F}) = 27 \text{ Btu/°F}$$

Tube capacity: $C_w = 0.358(0.094)(341)(12) = 138 \text{ Btu/°F}$

Shell capacity (⅜ in. thick): $V_{sh} = 2\pi(0.0312)(12) = 2.35 \text{ ft}^3 \text{ metal}$

$$C_{sh} = 2.35(500)(0.11) = 130 \text{ Btu/°F}$$

For a simple treatment, the shell and wall temperatures are assumed to be the same as the steam temperature. Heat transfer to the air is approximated by using the arithmetic-mean driving force. The steam flow depends only on the valve position, since the flow is critical. The energy balance for the shell and tubes is

$$(C_v + C_w + C_{sh}) \frac{d\theta_s}{dt} = (K_v \bar{F}_2 \lambda)X - UA\left(\theta_s - \frac{\theta_i + \theta}{2}\right)$$

For $\theta_i = 0$, the transformed equation is

$$Cs\theta_s = (K_v \bar{F}_2 \lambda)X - (UA)\theta_s + \frac{UA}{2} \theta$$

Neglecting the dynamics in the relationship between steam temperature and exit air temperature,

$$\theta = K_1 \theta_s \qquad K_1 = 0.25$$

$$\frac{C}{K_1} s\theta + UA\left(\frac{1}{K_1} - \frac{1}{2}\right)\theta = K_v \bar{F}_2 \lambda X$$

For $C = 295 \text{ Btu/°F}$, $UA = 2.04 \text{ Btu/(sec)(°F)}$,

$$\frac{\theta}{X} = \frac{(K_v \bar{F}_2 \lambda)/3.5UA}{165s + 1} = \frac{K_P}{165s + 1}$$

The process gain K_P is most accurately obtained from steady-state considerations. For $K_v = 5$, a 1 per cent change in valve signal changes the steam flow 5 per cent and the temperature rise 5 per cent, which is 2.25°F for this example.

11-2. COUNTERCURRENT EXCHANGERS

The partial differential equations for a countercurrent exchanger are Eqs. (11-1) and (11-2) plus corresponding equations for the shell fluid and the shell wall. Solutions have been presented by several investigators (1, 3, 10, 11), but the lengthy calculations required to secure numerical values of the frequency response would discourage most engineers. This has prompted a search for empirical equations or simple models to describe the approximate behavior.

The model recommended by Mozley (12) (his model II) uses a lumped capacity for both fluids and neglects the wall and shell capacities. Partially to account for the counterflow behavior, the rate of heat transfer is based on the arithmetic average of inlet and exit temperatures for both streams. The response of one stream to temperature changes of the other stream is given by the transfer function

$$\frac{\theta_{2,out}}{\theta_{1,in}} = \frac{4(T_2'/L_2)}{(T_1's + 2T_1'/L_1 + 1)(T_2's + 2T_2'/L_2 + 1) - 1} \tag{11-19}$$

where
$$T_1' = \frac{C_1}{UA} \qquad T_2' = \frac{C_2}{UA}$$

and L_1 and L_2 are the corresponding residence times.

A disturbing feature of Eq. (11-19) is that the effective time constants depend more strongly on the time delays L_1 and L_2 than on the time constants T_1' and T_2'. If L_1, L_2, T_1', and T_2' are all 10 sec, the effective time constants, from Eq. (11-19), are 5.0 and 2.5 sec. Changing T_1' and T_2' to 20 sec gives effective time constants of 5.0 and 3.3 sec; yet changing L_1 and L_2 to 20 sec changes the effective values to 10 and 3.3 sec. If the shell flow rate were high enough to make the shell temperature almost constant, the time delay in the tubes would not affect the major time constant [see Eq. (11-7)] and any effects of L_1 and L_2 for a normal countercurrent exchanger are expected to be small.

For one exchanger tested by Mozley, Eq. (11-19) described the dynamic behavior fairly well up to 150° phase lag, but this was probably a coincidence. When the time delays are equal and equal to the time constants T_1' and T_2', the model predicts effective time constants of $0.5T'$ and $0.25T'$. The actual response must be slower than this, since the major time constant would be T_1' if vapor were condensing in the shell and putting a flowing liquid in the shell should decrease the speed of response. The model probably predicts too large time constants when the time delays are large relative to the time constants and the driving force is quite small at one end of the exchanger.

Thal-Larsen (7) recommends an empirical correlation based on the residence time of the fluid in the exchanger, but the correlation was based mainly on data for steam-liquid exchangers and the only data given for a liquid-liquid exchanger did not fit very well.   Furthermore, the theory for steam-liquid exchangers shows that the time constant rather than the residence time should be the important parameter.

A method of analysis that promises to be quite accurate for all cases of counterflow or parallel flow was given by Paynter and Takahashi (13) and was used in the thesis by Hansen (14).   These workers expand the exact solutions to obtain the effective time constants and delays for lag-delay or root-lag models.   Although the method is easier than using the exact equation, it is too complex to be included here.

Hardly any of the studies of exchanger dynamics have included the response to flow-rate changes, though one of the flows would often be the controlled variable.   A theoretical solution would be very complex, since increasing the flow increases the heat-transfer coefficient as well as the driving force for heat transfer.   Some experimental data from a current study at Cornell are shown in Fig. 11-4.   The input was the water flow to the jacket of an 18.2-ft double-pipe exchanger, and the output was the exit temperature of the water in the tube, as measured with a bare thermocouple. The data shown are corrected for the slight time delay in the exit line, but not for thermocouple lag.   The calculated time constants are based on the overall coefficient with half the wall capacity added to each side.

Figure 11-4 shows that two time constants give a fair representation of the data and that the major lag is approximately T_1', the time constant based on the tube-side capacity and the overall coefficient.   The slight resonance at high frequencies is expected, since an increase in flow immediately changes the coefficient throughout the exchanger.   Thus the exchanger is being forced partly in a distributed manner, resulting in resonance similar to that found in steam-water exchangers (see Figs. 11-2 and 11-3).

The limited experimental data and a few theoretical frequency-response curves indicate that countercurrent exchangers can be adequately represented by two first-order lags.   A two-time-constant model ignores the resonance effects that are sometimes noticeable, and it is not much good for phase lags greater than 150°.   However, most exchangers have dead time in the headers and a large measurement lag, and so even a one-time-constant model for the heat-transfer section might be accurate enough for control calculations.   The problem then is to predict approximate values of the largest or perhaps the two largest effective time constants for those cases where a rigorous treatment is not justified.   Someone could make a real contribution by preparing exact solutions for a variety of conditions and correlating the effective time constants with parameters such as T, T', and L. Until such a correlation is presented, the following method can be used for rough calculations.

The major effective time constant T_a for a counterflow liquid-liquid exchanger can be taken as T_1' [Eq. (11-8)] if the temperature of the tube fluid is the output variable and the temperature or flow rate of the shell

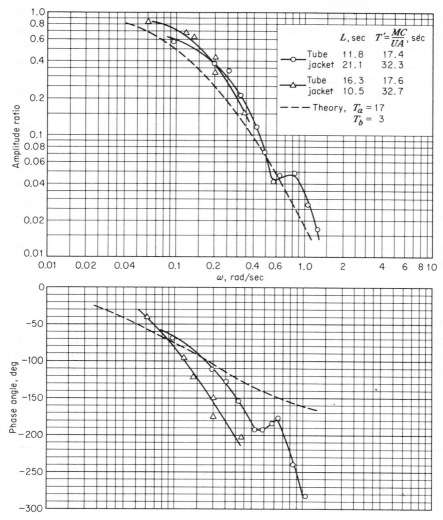

FIGURE 11-4   Frequency response of tube temperature to jacket flow rate for an 18-ft counterflow exchanger.

fluid the input variable.   One reason for using T_1' is that Eq. (11-8) would apply in the limit when the shell flow is very high and the shell temperature changes rapidly.   Equation (11-8) would also apply when the shell flow was the input variable and the main effect of a change in flow was the change in overall coefficient.   A change in coefficient would be felt at once through-out the exchanger, and the effect would be similar to a change in steam temperature for the steam-fluid exchanger.   The second effective time constant T_b may be as low as $0.05T_2'$ or as large as $0.5T_2'$, and a value of

$0.2T_2'$ is suggested as a compromise. In addition to the time constants T_a and T_b, the time delay in the headers and exit pipe must be included, and these delays may make the value of T_b unimportant. For gas-gas exchangers, the major lag comes from the capacity of the metal tubes and shell, and the response would be similar to that of a steam-air exchanger.

Multipass shell-and-tube exchangers have been the object of recent studies. Iscol and Altpeter (15) presented digital-computer programs for exchangers, and experimental tests showed their model to be accurate for temperature forcing of a 1-2 exchanger. Masubuchi (16) derived equations for exchangers with one shell pass and several tube passes, and he gives a few frequency-response curves to compare different exchanger arrangements. If the problem is to study the controllability of just one exchanger, it might be simpler to use an analog computer rather than a digital computer. Fricke and others (17) secured pretty good agreement with the actual response by dividing the exchanger lengthwise into six sections, with perfect mixing assumed in each tube and shell section. Other recent work is reviewed in Refs. 18 and 19.

11-3. CONTROL SCHEMES

The exit temperature of one fluid can be controlled by changing the temperature of the other fluid, changing the heat-transfer coefficient or the area, or passing part of the fluid around the exchanger. The usual method for steam heaters is to throttle the flow of steam, as shown in Fig. 11-5a.

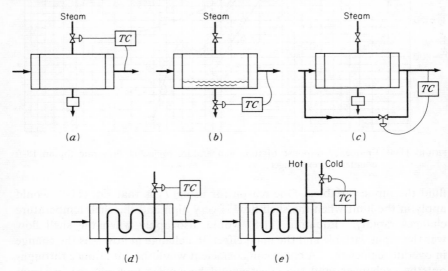

FIGURE 11-5   Control schemes for heat exchangers. (a) Throttle steam flow; (b) throttle condensate; (c) bypass method; (d) throttle liquid flow; (e) regulate inlet temperature.

Increasing the steam flow quickly leads to a higher pressure in the shell and a higher condensing temperature.

The shell pressure is sometimes controlled by a secondary controller that is reset by a temperature controller. Cascade control of this type is not much help for load changes on the tube side, and steam-pressure disturbances can be more easily corrected by using a pressure regulator ahead of the control valve.

A third scheme for steam heaters involves throttling the condensate, as shown in Fig. 11-5b. When the temperature is too high, the valve closes a little more, which causes condensate to cover a few more of the tubes. Because the holdup time for condensate is appreciable, the response is slower than with the other systems. The outlet temperature also shows significant fluctuations (4), perhaps because of splashing of the condensate.

If control of temperature is critical and the exchanger is oversized, the fast control scheme shown in Fig. 11-5c can be used. About 10 to 30 per cent of the process stream is bypassed and blended with hot fluid from the exchanger. If the exchanger heats the fluid almost to the steam temperature, the exit temperature varies only slightly with flow and the exchanger lags are almost insignificant in the control loop.

Countercurrent exchangers can be controlled by varying the flow of one stream, which changes both the overall coefficient and the driving force for heat transfer. Steady-state calculations should be made to make sure that control is possible over the desired range and to check the change in gain with flow rate. Equal-percentage valves are usually used (20), but they can only partially compensate for changes in process gain.

Control over a wider range and a more nearly constant gain can be obtained by blending hot and cold fluids to change the inlet temperature. A three-way valve can be used or cold fluid bled into a larger hot stream. Of course the bypass scheme of Fig. 11-5c can be used if very rapid response is needed.

Example 11-3

A shell-and-tube exchanger is used to heat 500,000 lb/hr of oil from about 100 to 200°F. The oil flows inside the tubes. There are four tube passes in a 3-ft shell, with a total of 560 steel tubes, 1-in. OD, 13 B.W.G., and 12 ft long. The shell fluid enters at 300°F, and the normal flow rate is 800,000 lb/hr. Both the oil and the shell fluid have a specific heat of 0.5 and a density of 55 lb/ft³.

The exit oil temperature is controlled by adjusting the flow of fluid to the shell. The time constant for the temperature bulb plus well is 12 sec, and there is a measurement delay of 1 sec because the well is 4 ft downstream from the exchanger. The total

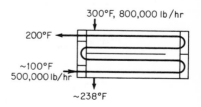

FIGURE E11-3   Diagram for Example 11-3.

holdup time in the tube headers is 6 sec, which can be treated as a time delay. Estimate the critical frequency of the control loop and the possible benefit of a twofold reduction in measurement time constant.

$$\Delta_1 = 138°F$$
$$\Delta_2 = 100°F$$
$$\Delta_{1.m.} = 119°F$$

Correction factor for reversed flow $= 0.98$ (Ref. 24, p. 194).

$$UA = \frac{Q}{\Delta} = \frac{500,000(0.5)(100°F)}{(119)(0.98)} = 214,000 \text{ Btu/(hr)}(°F)$$

$$= \frac{214,000}{12} = 17,800 \text{ Btu/(hr)}(°F)(\text{ft of exchanger})$$

The approximate time constants for a 2-lag model would be $1.0T_1'$ and $0.2T_2'$. The effective time constants will also be calculated using Mozley's model II (12). The equations for T_1' and T_2' are modified to include the capacities of the tube and shell walls, which are appreciable for this example.

Tube side: $M_f c_f = \dfrac{0.515 \text{ in.}^2}{144} (55)(0.5) = 0.099 \text{ Btu/(°F)(ft of tube)}$

$$M_w c_w = (1.00 \text{ lb/ft})(0.11) = 0.11$$
$$C_1 = M_f c_f + \tfrac{1}{2}M_w c_w = 0.154 \text{ Btu/(°F)(ft of tube)}$$
$$T_1' = \frac{C_1}{UA} = \frac{0.154 \times 560 \text{ tubes}}{17,800} \times 3,600 = 17.4 \text{ sec}$$

With 140 tubes per pass, $S = \dfrac{0.515}{144} \text{ in.}^2 \times 140 = 0.50 \text{ ft}^2$

$$v = \frac{500,000}{55 \times 0.5} \times \frac{1}{3,600} = 5.05 \text{ fps}$$

$$L_1 = \frac{12 \times 4}{5.05} = 9.5 \text{ sec}$$

Shell side: OD $= 36$ in., ID $= 35$ in.

$$\text{Free area} = \frac{\pi}{(4)(144)} [35^2 - (560)(1)^2] = 3.64 \text{ ft}^2$$

$$M_f' c_f = 3.64(55)(0.5) = 100 \text{ Btu/(°F)(ft of exchanger)}$$

$$\tfrac{1}{2}M_w c_w = \frac{0.11}{2} (560) = 31$$

$$M_{sh} c_{sh} = \frac{3\pi}{24} (500)(0.11) = 22$$

$$C_2 = 153$$

$$T_2' = \frac{C_2}{UA} = \frac{153}{17,800} \times 3,600 = 31 \text{ sec}$$

$$L_2 = \frac{3.64 \text{ ft}^2 \times 12 \text{ ft} \times 55 \text{ lb/ft}^3}{800,000} \times 3,600 = 10.8 \text{ sec}$$

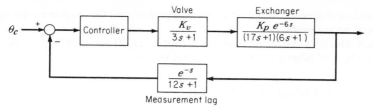

FIGURE 11-6   Block diagram for Example 11-3.

The recommended values are

$$T_a = T_1' = 17 \text{ sec}$$
$$T_b = 0.2T_2' = 6 \text{ sec}$$

From Eq. (11-19)

$$\frac{\theta}{\Delta F_{sh}} = \frac{K}{(T_1's + 2T_1'/L_1 + 1)(T_2's + 2T_2'/L_2 + 1) - 1}$$

$$= \frac{K}{(17.4s + 4.67)(31s + 6.75) - 1}$$

$$= \frac{K_P}{(5.3s + 1)(3.3s + 1)}$$

The block diagram for the control system is shown in Fig. 11-6. The time constant of the valve and transmission line is assumed to be 3 sec.

ω, rad/sec	Phase lag, deg					Amplitude ratio
	Valve	Exchanger		Bulb	Total	
		Lags	Delay			
Original case						
0.09	15	85	31	47	178	0.32
Measurement lag reduced to 6 sec						
0.11	18	92	38	33	181	0.31

Original case: $\omega_c = 0.09$ rad/sec, $K_{max} = 3.1$
Faster bulb: $\omega_c = 0.11$ rad/sec, $K_{max} = 3.2$

Using $T_a = 5.3$, $T_b = 3.3$ sec,

Original case: $\omega_c = 0.12$ rad/sec, $K_{max} = 2.5$
Faster bulb: $\omega_c = 0.14$ rad/sec, $K_{max} = 2.2$

The measurement time constant could probably be reduced to 6 sec by increasing the velocity past the well four- to fivefold. With the lower measurement lag, the critical frequency would be 20 per cent greater, and the maximum controller gain would be about the same.

Using Mozley's model gives critical frequencies 30 per cent higher, but the predicted effect of using a faster bulb is about the same.

11-4. MEASUREMENT LAG

Most temperature control systems use a filled bulb, a resistance thermometer, or a thermocouple as the measuring element. The measurement lag comes from the resistance and capacity of the element and the resistance (plus a little capacity) of the surrounding layer of nearly stagnant fluid. Usually the main resistance is in the fluid layer and the main capacity in the element itself, and so the response is approximately first-order.

When the external and internal resistance are of the same magnitude, as for a bulb in a well or a bulb in a high-velocity liquid stream, it takes two or more time constants accurately to describe the response, since the bulb wall and the bulb fluid are not at the same temperature. However, the frequency response for phase lags up to 45° can still be fairly well represented by using a single lumped time constant, as recommended by several workers (21, 22, 23). The lumped time constant is the sum of the effective time constants, $T_a + T_b$, when there are two interacting lags in series. The first-order approximation is considerably in error as the phase lag approaches 90°, but this is unimportant as long as the error is small at the critical frequency of the process. If the measuring element does contribute more than 45° phase lag at the critical frequency, it is probably worthwhile to redesign the measuring system to reduce the lag.

The effective time constant of a temperature bulb can be obtained from the transient response to a step input. Usually the time for 63 per cent response is taken, since this is always within 10 per cent of $T_a + T_b$ for two time constants in series. If the response curve is definitely second-order, a somewhat better procedure is to use the time for 73 per cent response, where $t/(T_a + T_b) = 1.3$ for all cases (see Fig. 3-21). The effective time constant $T = T_a + T_b$ is also equal to the steady time lag when a ramp input is applied. However, it is not easy to measure the bulb response at the proper fluid velocity and temperature, and the time constant must often be calculated from the properties of the bulb and the fluid.

The external resistance to heat transfer is predicted from published correlations for flow normal to cylinders, for flow parallel to cylinders, or for natural convection. For flow normal to cylinders, the following equation (24) can be used for both liquids and gases at moderate temperatures up to Reynolds numbers of 10,000:

$$\left(\frac{hD}{k}\right)\left(\frac{c_p\mu}{k}\right)^{-0.3} = 0.35 + 0.56\left(\frac{Dv\rho}{\mu}\right)^{0.52} \tag{11-20}$$

For small pipes, the bulb or well is placed in an elbow or T, and both parallel flow and normal flow are present.   If the bulb faces upstream, the coefficient at the tip is about twice that given by Eq. (11-20), but the coefficient for the parallel flow section is less than the average value for cross flow.   For a thermocouple in a well, the response at the tip is critical, and there is a definite advantage in having the well face upstream.   The average resistance for a long, filled bulb mounted axially in a pipe can be estimated by using a modification of the Colburn equation (25), where D_e is the equivalent diameter of the annulus.

$$\frac{h}{c_p G} \left(\frac{c_p \mu}{k}\right)^{2/3} = \frac{0.023 X}{(D_e G/\mu)^{0.2}} \tag{11-21}$$

where
$$X = 0.87 \left(\frac{\text{outer diameter}}{\text{inner diameter}}\right)^{0.53}$$

The resistance of the bulb wall per unit length is given by

$$R_w = \frac{\ln (D_o/D_i)}{2\pi k}$$

or
$$R_w \cong \frac{D_o - D_i}{2\pi k D_{\text{av}}} \tag{11-22}$$

If the fluid in the bulb is stagnant, the internal resistance can be approximated from the solution for the average temperature of a cylinder subjected to a step input (26).   The average response is 63 per cent complete when

$$\frac{k}{\rho c_p} \frac{t}{r^2} = 0.105$$

which corresponds to an internal resistance

$$R_i \cong \frac{1}{10\pi k} \quad \text{per unit length} \tag{11-23}$$

RESPONSE OF FILLED BULBS

The filled thermal systems are traditionally divided into four classes according to whether the bulb contains mercury, any other liquid, a gas, or a mixture of a liquid and its vapors (a vapor-pressure bulb).   For all types, the bulb is connected by capillary tubing to a Bourdon spiral or a bellows which responds to the increase in bulb pressure with temperature.   A mercury-filled bulb was chosen for Example 11-4 because it is the most common type, but the other classes have about the same response time, since much of the capacity is in the metal wall.

Example 11-4

a. Calculate the effective time constants for a mercury-filled bulb in an air stream at 100°F, 1 atm, and 20 fps.   The bulb is made of stainless-steel tubing 0.50 in. OD and 0.40 in. ID.

b. Repeat the calculation for water at 100°F and 1 fps.

External resistance for air

$$k = 0.0157 \text{ Btu/(hr)(ft}^2)(°F/ft)$$

$$D = \tfrac{1}{24} \text{ ft}$$

$$v = 20 \text{ fps}$$

$$\rho = 0.071 \text{ lb/ft}^3$$

$$\mu = 0.019 \text{ cp} \times 0.000672 = \frac{1.28}{10^5} \text{ lb/(ft)(sec)}$$

$$N_{Re} = \frac{Dv\rho}{\mu} = 4,630 \qquad N_{Re}^{0.52} = 81$$

$$N_{Pr} = c_p \frac{\mu}{k} = 0.71 \qquad N_{Pr}^{0.3} = 0.90$$

From Eq. (11-20), $N_{Nu} = \dfrac{hD}{k} = 0.9[0.35 + 0.56(81)] = 41$

$$h_o = 41(0.0157)(24) = 15.5 \text{ Btu/(hr)(ft}^2)(°F)$$

The resistance of the wall and the mercury are negligible compared with the external resistance, and so the time constant is the product of the external resistance and the capacity of the wall plus the mercury.   A basis of 1 ft of bulb is used for convenience, and conduction through the end of the bulb is neglected.

$$A = \frac{\pi}{24} = 0.131 \text{ ft}^2$$

$$R = \frac{1}{hA} = \frac{1}{(15.5)(0.131)} = 0.492 \text{ hr/(Btu)(°F)}$$

Capacity

Wall:

$$c_p = 0.11 \text{ Btu/(lb)(°F)}$$

$$\rho = 500 \text{ lb/ft}^3$$

$$c_w = \frac{\pi}{4} \frac{0.5^2 - 0.4^2}{144} (500)(0.11) = 0.027 \text{ Btu/(°F)(ft)}$$

Mercury:

$$c_p = 0.033$$

$$\rho = 840$$

$$c_{Hg} = \frac{\pi}{4} \frac{0.4^2}{144} (840)(0.033) = 0.024$$

$$C_{total} = 0.051 \text{ Btu/(°F)(ft)}$$

$$T = RC = (0.492)(0.051)(3,600) = 90 \text{ sec}$$

External resistance for water

$$\mu = 0.68 \text{ cp} \qquad k = 0.36$$

$$N_{Re} = \left(\frac{1}{24}\right)\frac{(1)(62)}{(0.68)(0.000672)} = 5,650$$

$$N_{Pr} = 4.6$$

From Eq. (11-20), $N_{Nu} = 79$

$$h_o = 79(0.36)(24) = 680 \text{ Btu/(hr)(ft}^2)(°F)$$

$$R_o = \frac{1}{(680)(0.131)} = 0.0112$$

The other resistances are not negligible in this case.

For stainless steel, $k = 9$

$$R_w = \frac{\ln (0.5/0.4)}{2\pi(9)} = 0.0032$$

For mercury, $k = 4.83$

$$R_i = \frac{1}{10\pi(4.83)} = 0.0066$$

The equivalent diagram for the system is given in Fig. E11-4.   This is a system with interacting time constants, and the transfer function is given by Eq. (3-76).

$$\frac{\theta}{\theta_{in}} = \frac{1}{T_1 T_2 s^2 + (T_1 + T_2 + R_1 C_2)s + 1} = \frac{1}{(T_a s + 1)(T_b s + 1)}$$

For low frequencies the effective time constant is

$$T = T_a + T_b = T_1 + T_2 + R_1 C_2$$

$$T_1 = R_1 C_1 = (0.0112 + 0.0016)(0.027)(3,600) = 1.2 \text{ sec}$$

$$T_2 = R_2 C_2 = (0.0016 + 0.0066)(0.024)(3,600) = 0.7 \text{ sec}$$

$$R_1 C_2 = (0.0128)(0.024)(3,600) = 1.1 \text{ sec}$$

$$T = 3.0 \text{ sec } (T_a = 2.7, \ T_b = 0.3)$$

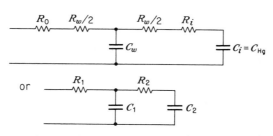

FIGURE E11-4   Equivalent diagram for Example 11-4.

As shown by Example 11-4, the response of a bulb in air is very close to first-order because the external resistance is 50 times the other resistances. The response in water can be approximated by a first-order lag only up to moderate phase angles, since the internal and external resistances are about the same.   However, even the simple procedure of lumping all the resistances and capacities $R_o + R_w + R_i$ times $C_w + C_{Hg}$ gives a value of 3.9 sec, which may be close enough to the more exact value of 3.0 sec, especially considering the 10 to 20 per cent uncertainty in the heat-transfer correlations.

A really thorough analysis of the bulb response would have to consider the distributed nature of the resistances of the mercury, the wall, and the external boundary layer, as well as the thermal expansion of the steel bulb and the resistance to flow in the capillary leading to the Bourdon tube (27, 28).   The last factor is perhaps the most significant according to the results of Eykman and Verhagen (28).   They found that a capillary 10 ft long had a time constant of 0.55 sec, and some filled systems have capillaries over 100 ft long.

Bulbs with short capillaries connected to temperature transmitters are often recommended for remote control.   This simplifies installation and maintenance and avoids the problem of correcting for changes in temperature of a long capillary.   Short capillaries permit the use of small helium-filled bulbs, which are about twice as fast as mercury bulbs.   The low conductivity of the gas does not lead to a large time constant, because the volumetric heat capacity is small.   (The thermal diffusivity $k/\rho c_p$ for helium at 10 atm is four times that for mercury.)

Example 11-5

Calculate the effective time constant for a stainless-steel bulb filled with helium at 10 atm.   The bulb is 0.50 in. OD and 0.40 in. ID, and the external coefficient is 680 (water at 1 fps).

$$h_o = 680 \qquad R_o = \frac{1}{(680)(0.131)} = 0.0113$$

From Eqs. (11-22), $R_w = 0.0032$

For He at 100°F, 10 atm,

$$k = 0.090$$

$$\rho = 0.098$$

$$c_p = 1.25$$

$$R_i \cong \frac{1}{10\pi(0.090)} = 0.35$$

$$C_{He} = \frac{\pi}{4}\frac{0.4^2}{144}(0.098)(1.25) = 1.07 \times 10^{-4} \text{ Btu/(°F)(ft)}$$

A two-capacity model of the system is given in Fig. E11-5. The effective time constant is $T_1 + T_2 + R_1C_2$,

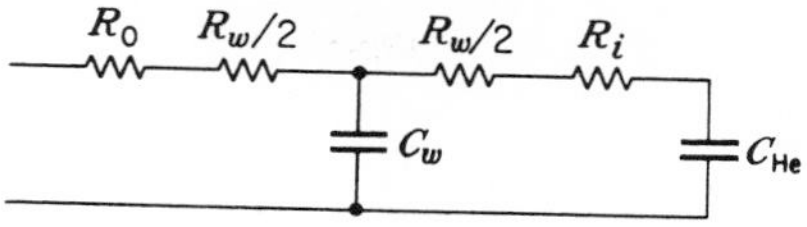

FIGURE E11-5 Equivalent diagram for Example 11-5.

$$R_1 = \left(R_o + \frac{R_w}{2}\right)$$

$$= 0.0113 + 0.0016 = 0.0129$$

$$C_1 = C_w = 0.027$$

$$T_1 = R_1C_1 = (0.0129)(0.027)(3,600) = 1.26 \text{ sec}$$

$$R_2 = \frac{R_w}{2} + R_{He} \cong R_{He} = 0.35$$

$$T_2 = R_2C_2 = (0.35)(1.07 \times 10^{-4})(3,600) = 0.14 \text{ sec}$$

$$R_1C_2 = (0.0129)(1.07 \times 10^{-4})(3,600) = 0.005 \text{ sec}$$

The interaction term R_1C_2 can be neglected, and the approximate response can be obtained by using an effective time constant equal to $T_1 + T_2$.

$$T_{\text{effective}} = 1.42 + 0.14 = 1.6 \text{ sec}$$

BULBS IN WELLS

The response of a bulb in a thermowell depends to a great extent on the clearance between the bulb and the well. If the bulb is placed in a dry well with an average clearance of 0.01 in., the heat-transfer coefficient from well to bulb is only 19 Btu/(hr)(ft²)(°F) ($0.0157 \times 12/0.01$). The time constant of the bulb and well in liquids would be 1 to 2 min, as large as that for a bare bulb in air. To improve the heat transfer, water or other liquids can be placed in the well if the well is upright. A film of water 0.01 in. thick has a coefficient of 430 Btu/(hr)(ft²)(°F) at 100°F; the water resistance could be neglected for a well in a gas stream, but it is significant for a well placed in a liquid stream.

One way of obtaining good contact between the well and bulb without having a force fit is to wrap the bulb with crinkled aluminum foil. This provides metal-to-metal contact at many points, makes it easy to remove the bulb, and avoids the problems of evaporation and spillage of liquids. The capacity of the well is usually one to two times that of the bulb; so if good contact is provided, the time constant of well plus bulb should be two to three times that of the bare bulb. With a ceramic well or a very thick metal well, the extra resistance may make the time constant several times that of the bare bulb. Using a thicker well than necessary gives poorer dynamic performance and also a larger static error because of conduction along the well.

THERMOCOUPLE RESPONSE

The speed of response of a bare thermocouple depends on the external resistance and the heat capacity of the wires. The time constant for a

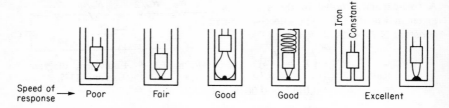

Speed of response → Poor   Fair   Good   Good   Excellent

FIGURE 11-7   Methods of installing thermocouples in wells.

butt-welded couple can be estimated by using the correlation for heat transfer to cylinders.   For 20-gauge wire (0.032 in.), the time constant is about 2 sec in air at 100°F, 1 atm, and 20 fps.   Data showing the effects of junction shape and bead size are presented by Moffat (29).   These predicted and published time constants serve mainly as a guide in designing temperature-measuring systems for experimental work.   When the temperature changes are very rapid, as in combustion or engine tests, the response of the thermo-couples should be measured by additional experiments to permit accurate calculation of the true temperatures.

In plant installations, thermocouples are nearly always enclosed in a well.   In the ideal case, a thermocouple in a well would respond as rapidly as the well itself, since the thermocouple has a very small mass and there should be good contact between the junction and the well.   A ½-in. stainless-steel well would have an effective time constant of 1 to 2 min in air and a few seconds in water.   The data of Aikman, McMillan, and Morrison (30) show that it is not easy to achieve this ideal performance.   For a thermo-couple and well in water, the time constants (based on 63· per cent response) were 1.5 min with the thermocouple in the center, 0.5 min with the thermo-couple touching the wall, and 0.1 min with the thermocouple brazed to the well.

The contact resistance can be eliminated by making the thermowell one of the elements of the couple, as, for example, an iron well with a constantan wire welded to the bottom.   The recommended method of securing good contact with a removable couple is to form a loop so that the wires contact the well at three places (see Fig. 11-7) or to use a spring that presses the junction against the bottom of the well.

RESISTANCE THERMOMETERS

It is difficult to predict the response of a resistance thermometer, because of the complex internal construction.   For a cylindrical thermometer in air, the capacity and time constant would be about the same as for a liquid-filled bulb of the same size.   The response in water is generally slower than for filled systems, since the greater internal resistance becomes important.   Some data are given in Refs. 31 and 32.

11-5. REDUCING THE MEASUREMENT LAG

When the measurement lag is critical, it may pay to use a small-diameter bulb or thermowell. The external heat-transfer coefficient varies with $D^{-0.5}$, and the ratio of surface to volume varies with D^{-1}; so a twofold reduction in bulb diameter decreases the time constant almost threefold. The lag is also decreased by increasing the velocity, which can sometimes be done by using a smaller-sized pipe or a flow restriction at the bulb. When the bulb can be placed either in the liquid or in the vapor, as in a distillation column, it should be placed in the liquid to gain the benefit of faster heat transfer.

Some compensation for a large measuring lag can be obtained by using derivative action in the transmitter. For a bulb that responds as a first-order system, the transfer function is

$$\frac{\theta_m}{\theta} = \frac{1}{T_m s + 1}$$

A commercial transmitter that has adjustable derivative action is shown in Fig. 11-8. The transfer function is

$$\frac{\theta_o}{\theta_i} = \frac{T_a s + 1}{T_b s + 1}$$

where T_a/T_b ranges from 4 to 10 (33). By setting $T_a = T_m$, the effective measurement lag is considerably reduced. The effect is similar to that obtained with derivative action in the controller, but a more accurate record of the temperature is obtained by using derivative action in the transmitter.

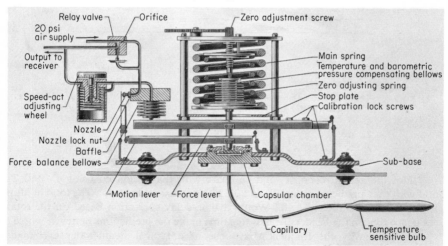

FIGURE 11-8 A temperature transmitter with derivative action. (*Courtesy of Taylor Instrument Companies.*)

PROBLEMS

1. Show how the transient response of a steam-water exchanger to a step change in steam temperature is influenced by both the time constant T_1 and the time delay L. Neglecting the capacity of the wall and the condensate film and assuming that the coefficient is constant along the length of the tubes, plot the response for an 8-ft and a 16-ft exchanger. Use the conditions of Example 11-1. Can the major time constant be obtained from the time for 63 per cent response or for 86.5 per cent response? ($y = 0.865$ at $t = 2T$ for a first-order system.)

2. The frequency response of a four-pass stainless-steel exchanger was fitted by a transfer function with two time constants (18, 34).

$$\frac{\text{Change in tube outlet temperature}}{\text{Change in shell flow}} = \frac{K}{(T_1 s + 1)(T_2 s + 1)}$$

where $T_1 = \dfrac{M_1 c_{p1}}{\pi D_o h_1}$ (π omitted in Ref. 18)

$T_2 = \dfrac{M_2 c_{p2}}{\pi D_o h_2}$

$Mc_p = $ capacity/ft for tube or shell fluid

Data for a typical run are given below:

ω, rad/sec	ϕ, deg	A.R.
0.01	30	
0.02	45	0.85
0.05	90	0.5
0.1	140	0.2

	Tube	Shell
Fluid	Water	Water
Inlet temperature, °F	171	79
Flow, gpm	21	50
h, Btu/(hr)(ft²)(°F)	540	357
U	...	194
T_1, sec	27.5	
T_2 sec	...	6.6

Plot the actual response, and compare with that predicted using T_1 and T_2 and with that predicted using T_1' and $0.2T_2'$, as in Example 11-3. The total holdup in the headers is about 7 gal. Would a time-delay correction based on the total residence time in the headers or half this time make the predictions better or worse?

3. Show why the frequency response of a typical steam-water exchanger (exit temperature/steam flow) is relatively insensitive to the value of the shell time constant. Use a lumped-capacity model for the fluid side, neglect the tube capacity,

and assume critical flow through the steam valve. Show that the transfer function has the form $\dfrac{\theta}{\Delta F} = \dfrac{K(T_c s + 1)}{(T_a s + 1)(T_b s + 1)}$, and calculate T_a, T_b, and T_c for several assumed values of the shell film resistance.

4. A 500-gal kettle has an internal heating coil made of 150 ft of 1-in.-ID tubing. Derive the transfer function relating kettle temperature and coil inlet temperature, and plot the frequency response for the following conditions:

$$F_2 = 50 \text{ gpm}$$
$$V_2 = 500 \text{ gal}$$
$$\rho_2 = 8.5 \text{ lb/gal}$$
$$c_{p_1} = 1 \text{ Btu/(lb)(°F)}$$
$$v_1 = 3 \text{ fps}$$
$$U_1 = 300 \text{ Btu/(hr)(ft}^2)(°F)$$

How well can the response be approximated by using a single time constant and a time delay equal to half the fluid residence time in the coil?

5a. Predict the effective time constants for thermocouples in brass, steel, and stainless-steel wells $\frac{9}{16}$ in. OD and $\frac{13}{64}$ in. ID with crossflow of water at 1.5 fps and 100°F. Assume perfect contact between the thermocouple and well.

b. Based on the frequency at 45° phase lag, the data of Linahan (21) correspond to time constants of 1.2, 2.7, and 4.8 sec for longitudinal flow of water past brass, steel, and stainless-steel wells of this size. Comment on the differences between these values and the predictions in part a.

6. Predict the effective time constant for a thermistor in water at 1 fps and 150°F. The thermistor is encased in a glass bead $\frac{1}{16}$ in. in diameter.

For spheres, $\qquad \dfrac{hD}{k} \cong 2 + 0.6 \left(\dfrac{Dv\rho}{\mu}\right)^{\frac{1}{2}} \left(\dfrac{c_p\mu}{k}\right)^{\frac{1}{3}}$

7. A mercury-in-steel thermometer sometimes shows an inverse response. The reading drops momentarily when the bulb is immersed in hot water, because the shell expands more rapidly than the mercury. How complex a model is needed to explain this behavior? Can inverse response be obtained if the wall and internal capacities are lumped? Show why inverse response is less likely with a gas-filled bulb.

REFERENCES

1. Gould, L. A.: The Dynamic Behaviour and Control of Heat Transfer Processes, Sc.D. thesis, Massachusetts Institute of Technology, 1953.
2. Cohen, W. C., and E. F. Johnson: Dynamic Characteristics of Double-pipe Heat Exchangers, *Ind. Eng. Chem.*, **48**:1031 (1956).
3. Campbell, D. P.: "Process Dynamics," p. 192, John Wiley & Sons, Inc., New York, 1958.
4. Catheron, A. R., S. H. Goodhue, and P. D. Hansen: Control of Shell-and-Tube Heat Exchangers, *Trans. ASME Paper* 59-IRD-14.

5. DeBolt, R. R.: Dynamic Characteristics of a Steam-Water Heat Exchanger, M.S. thesis, University of California, 1954.
6. Lees, S., and J. O. Hougen: Pulse Testing a Model Heat Exchange Process, *Ind. Eng. Chem.*, **48**:1064 (1956).
7. Thal-Larsen, H.: Dynamics of Heat Exchangers and Their Models, *Trans. ASME J. Basic Eng.*, **82**:489 (1960).
8. Koppel, L. B.: Dynamics of a Flow-forced Heat Exchanger, *I and EC Fundamentals*, **1**:131 (1962).
9. Morris, H. J.: Dynamic Response of Shell and Tube Heat Exchangers to Temperature Disturbances, in "Automatic and Remote Control," vol. 4, p. 354, Butterworth & Co. (Publishers), Ltd., London, 1961.
10. Takahashi, Y.: Transfer Function Analysis of Heat Exchange Processes, in A. Tustin (ed.), "Automatic and Manual Control," p. 235, Academic Press Inc., New York, 1952.
11. Hsu, J. P., and N. Gilbert: Transfer Functions of Heat Exchangers, *Am. Inst. Chem. Eng. J.*, **8**:593 (1962).
12. Mozley, J. M.: Predicting Dynamics of Concentric Pipe Heat Exchangers, *Ind. Eng. Chem.*, **48**:1035 (1956).
13. Paynter, H. M., and Y. Takahashi: New Method of Evaluating Dynamic Response of Counterflow and Parallel-flow Heat Exchangers, *Trans. ASME*, **78**:749 (1956).
14. Hansen, P. D.: The Dynamics of Heat Exchange Processes, Sc.D. thesis, Massachusetts Institute of Technology, 1960.
15. Iscol, L., and R. J. Altpeter: Frequency Response of Multipass Shell-and-Tube Exchangers, *Trans. ASME Paper* 59-IRD-4.
16. Masubuchi, M.: Dynamic Response and Control of Multipass Heat Exchangers, *Trans. ASME J. Basic Eng.*, **82D**:51 (1960).
17. Fricke, L. H., H. J. Morris, R. E. Otto, and T. J. Williams: Process Dynamics and Analog-computer Simulations of Shell-and-Tube Heat Exchangers, *Chem. Eng. Progr. Symposium Ser.*, **56**(31):80 (1960).
18. Hougen, J. O., and R. A. Walsh: Pulse Testing Method, *Chem. Eng. Progr.*, **57**(3):69 (March, 1961).
19. Williams, T. J., and H. J. Morris: A Survey of the Literature on Heat Exchanger Dynamics and Control, *Chem. Eng. Progr. Symposium Ser.*, **57**(36):20 (1961).
20. Sanders, C. W.: Better Control of Heat Exchangers, *Chem. Eng.*, **66**:145 (Sept. 21, 1959).
21. Linahan, T. C.: The Dynamic Response of Industrial Thermometers in Wells, *Trans. ASME*, **78**:759 (1956).
22. Coon, G. A.: Response of Temperature-sensing-element Analogs, *Trans. ASME*, **79**:1857 (1957).
23. Looney, R.: Method for Presenting the Response of Temperature-measuring Systems, *Trans. ASME*, **79**:1851 (1957).
24. McAdams, W. H.: "Heat Transmission," 3d ed., p. 267, McGraw-Hill Book Company, New York, 1954.
25. McAdams, W. H.: "Heat Transmission," 3d ed., p. 242, McGraw-Hill Book Company, New York, 1954.
26. Jakob, M.: "Heat Transfer," vol. I, p. 278, John Wiley & Sons, Inc., New York, 1949.

27. Muller-Girard, O.: The Dynamic Behavior of Mercury-filled Tube Systems, *Trans. ASME*, **77**:591 (1955).
28. Eykman, E. G. J., and C. J. D. M. Verhagen: Response and Phase Lag of Thermometers, in R. Oldenburger (ed.), "Frequency Response," p. 158, The Macmillan Company, New York, 1954.
29. Moffat, R. J.: How to Specify Thermocouple Response, *ISA J.*, **4**:219 (1957).
30. Aikman, A. R., J. McMillan, and A. W. Morrison: Static and Dynamic Performance of Sheathed Industrial Thermometers, *Trans. Soc. Instr. Technol.*, **5**:138 (1953).
31. Hornfeck, A. J.: Response Characteristics of Thermometer Elements, *Trans. ASME*, **71**:121 (1949).
32. Hainsworth, B. D., V. V. Tivy, and H. M. Paynter: Dynamic Analysis of Heat Exchanger Control, *ISA J.*, **4**:230 (1957).
33. Caldwell, W. I., G. A. Coon, and L. M. Zoss: Frequency Response for Process Control, p. 307, McGraw-Hill Book Company, New York, 1959.
34. Hearn, J. S.: The Dynamic Response of a Heat Exchanger to Shell-side Flow Disturbances, M.S. thesis, St. Louis University, 1959.

12 LEVEL CONTROL

Level control systems can be placed in two categories—systems where level is an important process variable and those where the flow from the tank is the important variable. The control of pressure is similar to the control of level, in that both exact pressure control and averaging control are encountered. In the first case, the system is designed to hold the level almost constant in spite of load changes. In the second class of systems, the exact level is unimportant as long as the tank does not overflow or run dry, and surges in flow are damped by allowing the level to change.

12-1. LEVEL A MAJOR VARIABLE

A stirred-tank reactor is an example of the first type. The level control system is usually arranged as shown in Fig. 12-1, with the level controller regulating a valve in the product line. A decrease in level decreases the residence time, which lowers the conversion and may also decrease the heat-transfer area. An increase in level can lead to excessive entrainment. Possible load changes include changes in feed rate, reactor pressure, and viscosity of the fluid. A reasonable goal for the control system is to keep the level within 5 per cent of the normal depth. A 1 per cent change in level would have a negligible effect on most processes.

A level control system for a reboiler is also shown in Fig. 12-1. The exit flow is regulated by a flow controller or a composition controller, and the input is controlled to maintain a constant reboiler level. In some cases, the feed to the system is fixed and the level controller used to adjust the

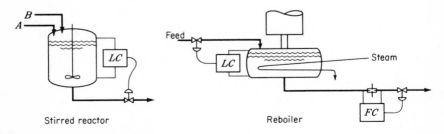

Stirred reactor   Reboiler

FIGURE 12-1   Level control systems.

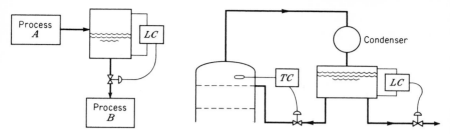

FIGURE 12-2   Averaging control systems.

exit flow.   The exact level is not critical as long as all the tubes are covered and there is sufficient disengaging space for the vapor.

12-2. AVERAGING CONTROL

An example where the level itself is not important is shown in Fig. 12-2. The output streams from several reactors are sent to a storage tank, and a level controller regulates the flow to the next step of the process.   When one of the reactors is shut down, there is a step change in flow to the storage tank and the level starts to drop.   A wide proportional band is used on the controller, so that a substantial change in level occurs before the outflow again equals the inflow.   Thus the change in outflow is spread out over several minutes, and the disturbance to process B is much less severe than a step change in flow.   If a high-gain controller were used, the level would stay almost constant, the outflow would drop very rapidly to the new value, and the storage capacity of the tank would be wasted as far as flow fluctuations are concerned.

Level control may be used on the reflux accumulator of a distillation column to average out fluctuations in product rate.   In the scheme shown in Fig. 12-2, the reflux flow is adjusted to maintain a constant composition at the top of the column, and the rest of the flow from the condenser eventually forms the overhead product.   If the product went to storage, fluctuations in flow would be unimportant, but in many cases the product from one column is the feed to the next column.   Control of the next column is much easier if sudden changes in flow are eliminated by an intermediate storage tank.

12-3. TANK DYNAMICS

The dynamics of level control are influenced by the lags in the tank, the measuring device, and the control valve.   For small changes in level, a tank with a control valve in the exit line behaves as a first-order system, as was shown in Chap. 3.   The time constant is twice the holdup time for an open tank discharging through a control valve to atmospheric pressure

[Eq. (3-45)].   Pressure above the liquid makes the time constant somewhat greater, and back pressure on the discharge line decreases the time constant. The time constant for the tank is usually several minutes, much greater than the other lags in the system.

When the level controller regulates the inflow to the tank and the outflow is fixed by a flow controller or a pump, the tank is a pure capacitance and has no self-regulation.   The transfer function relating level and flow is $1/As$, and the level lags the input by 90° at all frequencies.   The large phase lag is not really a disadvantage of this scheme, since a tank with a time constant of several minutes would contribute almost 90° lag at the critical frequency.

12-4. MEASUREMENT LAG

Many of the devices used to measure liquid level really measure the pressure at the bottom of the tank or, if the tank is under pressure or vacuum, the differential pressure from top to bottom.   If only an indication of level is needed, a pressure gauge, a diaphragm box, or an air bubbler may be satisfactory but for remote control a level transmitter is needed.

Any of the differential pressure transmitters used for flow control can be used for level control, and a typical installation is shown in Fig. 12-3. The high-pressure leg is filled with the tank fluid to avoid the problem caused by vapor condensing and partially filling the line.   When the level drops, the differential pressure rises, but this reversal in sign is unimportant, since the controller action can be either positive or negative for a given error signal.   A force-balance transmitter usually has a time constant of less than 1 sec.   If a conventional mercury meter is used, the effective time constant is 1 to 10 sec depending on the degree of damping (see Chap. 13).

DYNAMICS OF FLOAT CAGES

There are many level instruments which use a buoyant float connected to a cable, lever arm, or other device to record the float position.   Alternatively, the float movement can be restrained and the force on the float used to determine the fraction submerged and thus the level.   Floats used to measure the level in stirred tanks are usually enclosed in a chamber or cage made of pipe, as shown in Fig. 12-4.   The lag of the float can usually be neglected, but the response of the liquid in the cage is important in the system analysis.

The internal chamber shown in Fig. 12-4a is equivalent to a manometer with

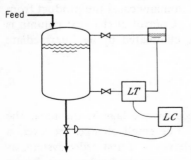

FIGURE 12-3   Level control based on pressure differential.

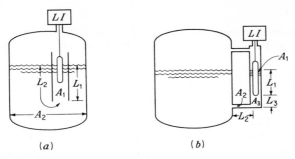

(a)   (b)

FIGURE 12-4   Float cages for level measurement.

arms of unequal diameter.   A force balance similar to Eq. (3-58) shows that the natural frequency is

$$\omega_n = \sqrt{\frac{g(1 + A_1/A_2)}{L_1 + L_2 A_1/A_2}}$$

(12-1)

or

$$\omega_n \cong \sqrt{\frac{g}{L_1}} \quad \text{if } A_1/A_2 \ll 1$$

For a chamber length of 2 ft, the natural frequency is 4 rad/sec.   The damping coefficient cannot be predicted accurately, because of entrance losses and uncertain flow patterns, but it would be quite small.   For laminar flow of water in a 1-in. manometer, the damping coefficient is only 0.006 (see Table 3-1).

For the external cage shown in Fig. 12-4b, the natural frequency depends on the length and size of connecting pipe as well as on the dimensions of the chamber.   If the float motion is negligible and if L_3 is small,

$$\omega_n \cong \sqrt{\frac{g}{L_1 + L_2(A_1/A_2)}}$$

(12-2)

Damping ratios obtained experimentally for a number of installations of this type were 0.05 to 0.06 for water or other low-viscosity fluids and 0.33 for a viscous oil (1).

A second-order system with a damping coefficient of 0.05 shows a peak amplitude ratio of 10.   If the lags of the controller and valve are small, the critical frequency of the level control system will be close to the natural frequency of the float cage and the large amplitude ratio will make it hard to obtain stable control.   With external cages, a valve in the connecting line can be used to provide damping.   For an internal cage, a restriction at the bottom would have to be added or a perforated cage used to damp the oscillations.

Example 12-1

What controller settings should be used for the following level control system? Can the level be kept at 8 ± 0.2 ft following 25 per cent changes in input flow?

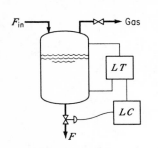

Tank area:	$A = 24$ ft^2
Normal depth:	$\bar{h} = 8$ ft
Liquid density:	$\rho = 60$ lb/ft^3
Tank pressure:	$P_0 = 2$ psig
Tank pressure:	$h_0 = 4.8$ ft $(2 \times 144/60)$
Normal flow:	$\bar{F} = 40$ ft^3/min

Transmitter range:

3–15 psi for 2-ft level change

FIGURE E12-1a Diagram for Example 12-1.

Measurement lag:	$T_m = 2$ sec
Valve lag:	$T_v = 20$ sec

Valve type:

Equal percentage with 5% change in flow/1% change in input

The block diagram for proportional control of the system is shown below. The output of the control valve is considered to be the valve position x, which means that the valve characteristic is included in the process gain K_1. The tank pressure is assumed to remain constant is spite of fluctuations in level. This is a reasonable assumption if the holdup time for gas is very small.

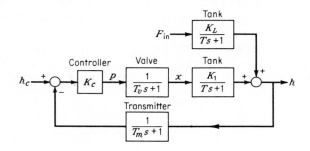

FIGURE E12-1b Block diagram for Example 12-1.

The time constant for the tank and the steady-state gains K_1 and K_L come from the material-balance equation, in which F_{in}, h, and x represent deviations from the normal values.

$$A\left(\frac{dh}{dt}\right) = F_{in} - \left(\frac{\partial F}{\partial h}\right)h - \left(\frac{\partial F}{\partial x}\right)x$$

Rearranging and applying the Laplace transform,

$$\frac{Ash}{\partial F/\partial h} + h = \frac{F_{in}}{\partial F/\partial h} - \frac{\partial F/\partial x}{\partial F/\partial h}x$$

or

$$(Ts + 1)h = K_L F_{in} + K_1 x$$

From Eq. (3-46),

$$\left(\frac{\partial F}{\partial h}\right)_{av} = \frac{\bar{F}}{2(\bar{h} + h_0)} = \frac{40}{2(8 + 4.8)} = 1.56 \text{ cfm/ft}$$

$$T = \frac{A}{(\partial F/\partial h)_{av}} = \frac{24}{1.56} = 15.4 \text{ min}$$

The gain K_L is made dimensionless by dividing the level change by 2 ft, the transmitter range, and dividing the flow change by 40 cfm, the normal flow.

$$K_L = \frac{1}{1.56} \text{ ft/cfm} \times \frac{40 \text{ cfm}}{2 \text{ ft}} = 12.8 \frac{\% \text{ change in level}}{\% \text{ change in flow}}$$

The process gain K_1 includes the factor 5, since the flow is increased 5 per cent above the normal value for a 1 per cent change in valve position.

$$K_1 = -\left(\frac{\partial F}{\partial x}\right)\left(\frac{1}{\partial F/\partial h}\right) = -5(12.8) = -64 \frac{\% \text{ change in level}}{\% \text{ change in valve lift}}$$

For frequency-response analysis, $T = 15.4$, $T_v = 0.33$, $T_m = 0.033$ min.
At $\omega = 4$ rad/min

$$\omega T = 61.6 \qquad \phi = 89° \qquad \text{A.R.} = 1/61.6$$

$$\omega T = 1.33 \qquad \phi = 53° \qquad \text{A.R.} = 0.6$$

$$\omega T_m = 0.13 \qquad \phi = \frac{7°}{149°} \qquad \frac{\text{A.R.} = 1}{\text{A.R.}_{\text{overall}} = 1/103}$$

Use $K_{\text{overall}} = 103$ for 30° phase margin.

$$K_c = \frac{K}{K_1} = \frac{103}{64} = 1.6$$

The steady-state error following a load change is

$$h_{ss} = \frac{K_L}{K + 1} F_{in} = \frac{12.8}{104} F_{in} = 0.12 F_{in}$$

For $F_{in} = 10$ cfm or 25 per cent of 40

$$h_{ss} = 0.12(0.25)(2 \text{ ft}) = 0.06 \text{ ft}$$

Therefore reset action is not needed.

The peak error following a step change in input flow could be estimated by using the method presented in Chap. 7. Since the tank time constant is so much larger than the others, the response to load changes is similar to that of a second-order system and the peak error would probably be about 1.5 times the final error, which is within the design limits.

It was easy to achieve control of level for the conditions of Example 12-1, where the tank time constant was 15 min and the valve lag 20 sec. If the ratio of the two largest time constants is this large or even larger, as is often the case, a complete dynamic analysis is hardly even needed. Difficult problems can arise in controlling the level in small standpipes, where the holdup time is less than 1 min (see Ref. 12 of Chap. 8).

Close control of level in a reboiler is sometimes a problem because of inverse response to changes in vapor rate (2, 3).   If the pressure is suddenly reduced to increase the vapor rate, the expansion of vapor bubbles causes the level to rise even though the total holdup has decreased.   A simple solution is to use a differential pressure transmitter to measure the total head in the vessel rather than a narrow-range device which measures the interface level.

12-5. PERFORMANCE OF AVERAGING CONTROLLERS

A tank with a level control system can be designed to give any desired degree of damping of flow surges.   Since the controller gain will be much smaller than the maximum, the closed-loop system will be overdamped and the response can be approximated by using only the large time constant of the tank.   For proportional control, the block diagram is shown in Fig. 12-5.   This system shows a first-order response to a step change in load, as was given by Eq. (4-10).

$$h = \frac{F_{in} K_L}{1 + K_c K_1} (1 - e^{-t/T'})$$

where

$$T' = \frac{T}{1 + K_c K_1} = \frac{T}{1 + K}$$

Following a step change in inflow, the important variable is the rate of change of outflow, dF/dt.   Since

$$A \left(\frac{dh}{dt}\right) = F_{in} - F$$

$$A \left(\frac{d^2h}{dt^2}\right) = -\frac{dF}{dt}$$

From the transient response of h

$$\frac{dh}{dt} = \frac{F_{in} K_L}{1 + K} \frac{e^{-t/T'}}{T'}$$

$$\frac{d^2h}{dt^2} = -\frac{F_{in} K_L}{1 + K} \left(\frac{1}{T'}\right)^2 e^{-t/T'}$$

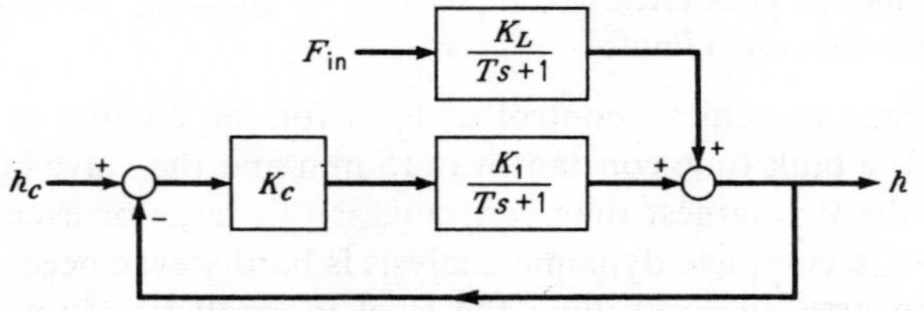

FIGURE 12-5   Block diagram for proportional control of tank level.

The maximum rate of change is at $t = 0$,

$$\left(\frac{dF}{dt}\right)_{max} = \frac{AF_{in}K_L}{1 + K}\left(\frac{1 + K}{T}\right)^2$$

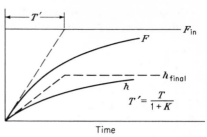

As was shown in Example 12-1,

$$T = RA \qquad \text{and} \qquad K_L = R$$

where $$R = \frac{1}{\partial F/\partial h}$$

FIGURE 12-6 Transient response of averaging-type level control system.

Therefore $$\left(\frac{dF}{dt}\right)_{max} = \frac{ARF_{in}(1 + K)}{T^2} = \frac{F_{in}}{T}(1 + K) \qquad (12\text{-}3)$$

This result could have been reached more quickly by realizing that the outflow will show an exponential response with the same time constant as for the level response, as shown by Fig. 12-6.

To determine what size tank is needed, the overall gain must be fixed. If the maximum and minimum flows are known, the gain can be chosen so that the tank will be full at the maximum flow and almost empty at the minimum flow. Then the time constant and size of the tank can be calculated from Eq. (12-3).

Example 12-2

The normal flow to a process is 300 gpm, and the extremes are 200 gpm and 400 gpm. The flow is subject to sudden changes of up to 50 gpm. What sized storage tank is needed to limit the rate of change of flow to 5 gpm/min? The incoming stream is at 40 psig, and the downstream process operates at 20 psig.

Solution

Choose a vertical cylindrical tank with a normal depth of 5 ft and a maximum depth of 10 ft, and solve for the required area. Use a level transmitter with a range of 10 ft and a controller gain such that the tank would be empty at 200 gpm and full at 400 gpm.

Using the same nomenclature as for Example 12-1,

$$K_L = R = \frac{\partial h}{\partial F} = \frac{2(\bar{h} + h_0)}{\bar{F}}$$

$$h_0 = 20 \text{ psig} \times {}^{144}\!/_{60} = 48 \text{ ft}$$

$$K_L = R = \frac{(2)(5 + 48)}{300} = 0.353 \text{ ft/gpm}$$

or $$K_L = 0.353 \times {}^{300}\!/_{10} = 10.6 \frac{\% \text{ change in level}}{\% \text{ change in flow}}$$

$$K_1 = \frac{\partial h}{\partial x} = \left(\frac{\partial h}{\partial F}\right)\left(\frac{\partial F}{\partial x}\right) = R\frac{\partial F}{\partial x}$$

Since the range of flows is small and almost all the pressure drop can be across the valve, use a poppet valve with a linear characteristic.   Assume

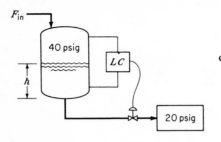

$$\left(\frac{\partial F}{\partial x}\right) = \frac{6 \text{ gpm}}{1\% \text{ lift}}$$

or

$$\frac{6}{300} \times 100 = 2 \frac{\% \text{ change in flow}}{\% \text{ change in lift}}$$

$$K_1 = (2)(10.6) = 21.2 \frac{\% \text{ change in level}}{\% \text{ change in lift}}$$

FIGURE E12-2 Diagram for Example 12-2.

$$h_{ss} = \frac{K_L}{1 + K} F_{in}$$

For $F_{in} = 100$ gpm and $h_{ss} = 5$ ft,

$$1 + K = 100 \text{ gpm} \frac{0.353 \text{ ft}}{\text{gpm}} \frac{1}{5 \text{ ft}} = 7.06$$

$$K = 6.06 = K_1 K_c$$

$$K_c = \frac{6.06}{21.2} = 0.29 = \text{controller gain}$$

To get tank area,

$$\left(\frac{dF}{dt}\right)_{max} = \frac{F_{in}(1 + K)}{T}$$

$$5 \text{ gpm/min} = \frac{(50)(7.06)}{T}$$

$$T = 70.6 \text{ min} = RA$$

$$R = 0.353 \text{ ft/gpm} \times 7.48 \text{ gal/ft}^3 = 2.64 \text{ ft/cfm}$$

$$A = \frac{70.6}{2.64} = 26.8 \text{ ft}^2$$

The total tank volume is 268 ft^3 or 2,000 gal.   Since the tank is half full at normal flow, the normal holdup time is $1,000/300 = 3.3$ min.

It may seem strange that the tank time constant for Example 12-2 is 21 times the holdup time.   This occurs because the depth of liquid contributes only a small fraction of the available pressure for flow and the process has little self-regulation.   If the pressure above the tank were much smaller, the time constant would be closer to the holdup time but the tank size needed for a given damping would not be changed.   A lower controller gain must be used when most of the pressure comes from the liquid head, as is shown in the following example.

Example 12-3

What sized tank is needed to satisfy the requirements of Example 12-2 if the static pressure difference is only 5 psi instead of 20 psi?

$$h_0 = 12 \text{ ft}$$

$$K_L = \frac{2(\bar{h} + h_0)}{\bar{F}} = 2\frac{5 + 12}{300} = 0.113 \text{ ft/gpm}$$

$$K_L = 0.113 \times 300/10 = 3.4 \frac{\% \text{ change in level}}{\% \text{ change in flow}}$$

$$K_1 = 2\frac{\% \text{ change in flow}}{\% \text{ change in lift}} \times K_L = 6.8 \frac{\% \text{ change in level}}{\% \text{ change in lift}}$$

For $F_{\text{in}} = 100$ gpm, $h_{ss} = 5$ ft,

$$1 + K = \frac{K_L F_{\text{in}}}{h_{ss}} = \frac{0.113 \text{ ft}}{\text{gpm}} \times \frac{100 \text{ gpm}}{5 \text{ ft}} = 2.26$$

$$K = 1.26 = K_1 K_c$$

$$K_c = \frac{1.26}{6.8} = 0.185$$

$$\text{Required time constant } T = \frac{50 \text{ gpm}}{5 \text{ gpm/min}} \times 2.26 = 22.6 \text{ min}$$

$$R = 0.113 \times 1.48 = 0.846 \text{ ft/cfm}$$

$$A = \frac{22.6}{0.846} = 26.8 \text{ ft}^2$$

The tank size is exactly the same as before, since the controller gain was picked to give the same change in level with flow rate.

PROBLEMS

1. For the conditions of Example 12-1, calculate the critical frequency and maximum gain of a control system that regulates the input flow. Assume that the output flow is fixed by a constant-displacement pump.

2. A small standpipe with a level controller is suggested for maintaining a liquid seal at the bottom of a flash chamber that operates at 20 psig. The liquid is discharged through a control valve and cooler to a storage tank at atmospheric pressure. An equal-percentage valve is suggested, and, at the normal flow of 20 gpm, three-fourths the pressure drop would be across the control valve. Because the liquid decomposes at high temperatures, a minimum holdup is desired. How much holdup time should be provided if sudden changes of ±30 per cent in liquid flow are likely? What diameter and normal depth should be used for the standpipe? Assume that the measurement lag is 1 sec and the valve lag 5 sec.

3. The equations presented for the response of tanks have assumed a constant area. How is the time constant related to the holdup time for a tank that is a

horizontal cylinder? Answer for a tank that is one-fourth full, one-half full, and three-fourths full.

4. Derive the expression for the natural frequency of the float cage in Fig. 12-4b. Include L_1, L_2, and L_3, but neglect the contribution of the fluid in the tank.

5. The level in a gasoline storage tank is measured by using a submerged diaphragm box which is connected to the bellows of a pressure recorder by 200 ft of $\frac{1}{16}$-in.-ID tubing. The diaphragm box is 6 in. in diameter and 2 in. deep, and the normal liquid depth is 40 ft. If the instrument bellows has a volume of 2 in.^{3}, how fast does the instrument respond to changes in level? What would be the response time if $\frac{1}{8}$-in.-ID tubing were used?

6. A reaction is carried out by using three parallel reactors, and the product is sent to a storage tank to permit more gradual changes in flow to the recovery column. Each reactor is shut down about every 10 days for an 8-hr period. While operating, the flow from each reactor is 2,000 gal/hr. What type of control system and what size of storage tank are needed to limit the change in column feed to 10 per cent per hour?

7. Repeat Example 12-3 for the case of zero pressure difference between the top of the storage tank and the downstream process. What is the significance of the negative value obtained for K_c?

REFERENCES

1. Schuder, C. B.: The Dynamics of Level and Pressure Control, *Bull.* TM-7, Fisher Governor Co.
2. Holzmann, E. G.: Dynamic Analysis of Chemical Processes, *Trans. ASME,* **78**:251 (1956).
3. Iinoya, K., and R. J. Altpeter: Inverse Response in Process Control, *Ind. Eng. Chem.,* **54** (7):39 (July, 1961).

13 FLOW CONTROL

Flow control differs in two respects from most other process control problems. The lag in the process itself is often negligible; after the control valve moves to a new position, the final flow rate is reached in a fraction of a second, or at most a few seconds. This means that the response of the system depends mainly on the size of the lags in the measuring device, the controller, the transmission lines, and the valve. Flow control systems generally show complete response to step changes in less than 1 min, and if close control is needed, the response time can be reduced to a few seconds by making all the above lags as small as possible. Of course, there is an appreciable flow lag in gas or oil piplines which are many miles long, because of compression of the fluid and expansion of the pipe. Only flow in short lines is considered here.

Another characteristic of flow systems is that the flow signal has a lot of noise, defined as fluctuations with frequencies of 1 cps or higher. The noise is not apparent in many cases, because the flowmeter is heavily damped, but it is noticeable when an undamped manometer is placed across an orifice, and sometimes the noise can be seen in fluctuations of a rotameter bob. Part of the noise represents actual changes in flow which are too fast to be corrected by the control system. Flow fluctuations may originate in the pump or compressor, and they may also come from random changes in the flow pattern at the valve, orifice, or other irregularities in the system (1, 2). When an orifice is used, the differential pressure signal contains additional spurious noise which comes from random pressure fluctuations at one or the other of the orifice taps. A record of the actual flow fluctuations can be obtained with a magnetic flowmeter, which measures the average velocity in the pipe and has a very fast response.

If the noisy signal reaches the controller, the chart record is ragged and hard to read. Furthermore, a low gain must be used to avoid amplifying the disturbances. As was shown in Chap. 7 (see Figs. 7-5 and 7-6), periodic disturbances at frequencies close to the critical frequency are made much worse by a feedback system operating at half the maximum gain. If the noise is completely damped out, significant changes in flow are not corrected so quickly. Damping the orifice signal also leads to an error in the measurement of pulsing flows; since the pressure drop varies with the square of the flow, the damped or average pressure drop does not correspond to the average flow.

13-1. PROCESS LAG

Although the lag of the flow process can often be neglected, it is worthwhile to use a simplified theory to show conditions where the lag might be appreciable. Consider a pipeline containing an orifice meter, a control valve, and other valves and fittings that offer localized resistance to flow. The line is fed from a constant-head tank and discharges at atmospheric pressure (Fig. 13-1). If the control valve opening is increased slightly, the pressure drop through the valve decreases and the difference between the available head and the total friction loss causes the fluid to accelerate.

$$\frac{AL\rho}{g_c}\left(\frac{dv}{dt}\right) = A(-\Delta P_f) \tag{13-1}$$

The change in friction loss, ΔP_f, is expressed as a linear function of the change in velocity and the change in valve position.

$$\Delta P_f = \left(\frac{d\,\Delta P_{\text{total}}}{dv}\right)\Delta v + \left(\frac{d\,\Delta P_v}{dx}\right)\Delta x \tag{13-2}$$

The total pressure drop is the drop in the valve and fixed resistances plus the drop in the line, and this is expressed in terms of the line drop and R, the ratio of localized pressure drop to line pressure drop. For turbulent flow, R is almost independent of velocity.

$$\Delta P_{\text{total}} = \Delta P_v + \Delta P_{1,2,3,4} + L\,\Delta P_L$$
$$= (1 + R)L\,\Delta P_L \tag{13-3}$$

where $R = \dfrac{\Delta P_v + \Delta P_{1,2,3,4}}{L\,\Delta P_L}$

ΔP_L = pressure drop per foot of pipe

Substituting in Eq. (13-1),

$$\frac{L\rho}{g_c}\left(\frac{dv}{dt}\right) + (1 + R)L\left(\frac{d\,\Delta P_L}{dv}\right)\Delta v = -\left(\frac{d\,\Delta P_v}{dx}\right)\Delta x \tag{13-4}$$

Thus the process responds as a first-order system to changes in valve position, and the time constant is

$$T = \frac{\rho}{g_c(1 + R)(d\,\Delta P_L/dv)} \tag{13-5}$$

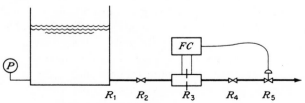

FIGURE 13-1   Typical flow system.

For turbulent flow, the pressure drop is given by the Fanning equation.

$$\Delta P_L = \frac{2fv^2\rho}{g_c D} \qquad (13\text{-}6)$$

$$\frac{d\,\Delta P_L}{dv} = \frac{4fv\rho}{g_c D} \qquad (13\text{-}7)$$

From Eq. (13-5)

$$T = \frac{D}{4fv(1+R)} \qquad (13\text{-}8)$$

For water flowing at 5 fps in a 2-in. pipe, the predicted time constant is 1.6 sec if all the pressure drop is due to pipe friction ($R = 0$). For this velocity and pipe size, Catheron and Hainsworth (1) measured a 63 per cent response time of 0.3 sec. The fivefold lower value indicates that 80 per cent of the pressure drop occurred in the valves and localized resistances, which seems reasonable for the pipe loop used in their studies. No data are available for industrial systems, but the time constants probably range from 0.1 to 5 sec. In measuring the response of flowmeters, the flow should not be assumed to follow the valve position; a magnetic flowmeter should be used to show the actual changes in flow.

13-2. MEASUREMENT LAG

The basis for most flow measurements is the pressure drop across an orifice, a venturi, or a flow nozzle. For many years it was standard practice to use a mercury manometer with a float chamber to measure the pressure drop across the orifice (see Fig. 13-2). The manometer is a second-order system, and the natural frequency is usually 0.5 to 1 cps. Adjustable damping is provided by a valve between the two legs of the manometer. Without this restriction the damping coefficient would be 0.1 or smaller (see Table 3-1). When adjusted for critical damping ($\zeta = 1$), the amplitude ratio is 0.5 and the phase lag 90° at the natural frequency. The low-frequency response and the transient response are about the same as that of a first-order system with a time constant of $2/\omega_n$ or 0.7 sec for $\omega_n = 3$ rad/sec. Commercial mercury flowmeters are reported to have response times of 2 to 20 sec (1, 3), the high values indicating much more damping than is necessary. Even with the adjustable resistance wide open, mercury meters are likely to be overdamped.

In new installations, the measuring device is often a force-balance differential pressure transmitter, a diaphragm device which converts the orifice differential to a standard pneumatic or electrical signal. A pneumatic transmitter is shown in Fig. 13-3, and the diaphragm portion of an electronic transmitter would be similar. The response is very rapid because the volume of fluid in the transmitter is small and the diaphragm motion is slight. An effective time constant of 0.2 sec can be obtained with a commercial electronic

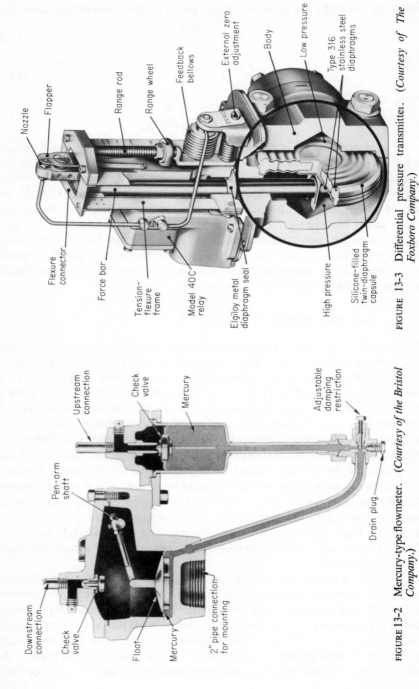

FIGURE 13-3 Differential pressure transmitter. (*Courtesy of The Foxboro Company.*)

Nozzle
Flopper
Range rod
Range wheel
Feedback bellows
External zero adjustment
Body
Low pressure
Type 316 stainless steel diaphragms
Flexure connector
Force bar
Tension-flexure frame
Model 40C relay
Elgiloy metal diaphragm seal
High pressure
Silicone-filled twin-diaphragm capsule

FIGURE 13-2 Mercury-type flowmeter. (*Courtesy of the Bristol Company.*)

Upstream connection
Check valve
Mercury
Adjustable damping restriction
Pen-arm shaft
Drain plug
Downstream connection
Check valve
Float
Mercury
2" pipe connection for mounting

transmitter (8), though somewhat larger values are ordinarily used to get more damping. The transmitter shown in Fig. 13-3, when dead-ended, has a natural frequency of about 3 cps.

The exact response of a pneumatic transmitter depends on the length of transmission line and the volume at the end of the line. With a short line an undamped transmitter shows the resonance of a second-order system. If the transmitter is damped or if the transmitter is connected to a long line, the response is approximately first-order. Frequency-response data for a Taylor differential pressure transmitter are shown in Fig. 13-4. With $\frac{1}{4}$-in. lines that are 100 ft or shorter, the phase lag of the transmitter is greater than that of the line. With longer lines, the transmitter response is almost independent of line length, and the line response becomes dominant.

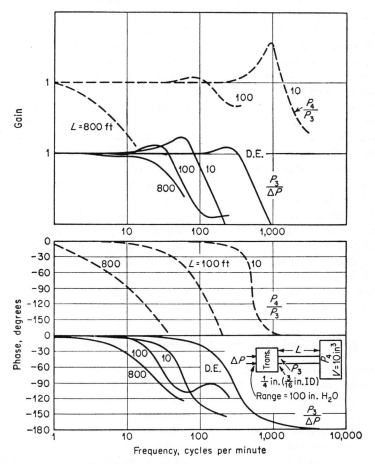

FIGURE 13-4   Frequency response of a differential pressure transmitter. (*Courtesy of Taylor Instrument Companies.*)

The fast response of a force-balance transmitter has been cited as an advantage for close control of flow. However, with long pneumatic transmission lines, the force-balance transmitter is only slightly faster than a properly damped mercury meter. For this case a much greater improvement in speed of response is possible by switching to an electronic transmitter and controller. The main advantages of the newer transmitters are easier installation, elimination of mercury, rugged construction, easy range change, and similar practical features, rather than the faster response.

13-3. EFFECT OF TRANSMISSION LAG ON FLOW CONTROL

The lags in pneumatic transmission lines are significant in flow control because the process and measurement lags are so small. The line from transmitter to controller has the smaller lag because it ends in a bellows. The effective time constant is 3 to 6 sec for a 500-ft line (see Fig. 10-11). The lag of a line plus a large valve motor may be several times that of the dead-ended line and is usually the largest lag in the flow control system.

Various methods could be used to increase the speed of response of flow control systems. The lag of the transmission line and valve can be reduced almost to that for a dead-ended line by using a valve positioner or a booster relay at the valve motor. However, since the line and valve form the largest time constant of the system, the increase in critical frequency is not very great. Using $\frac{3}{8}$-in. tubing instead of $\frac{1}{4}$-in. tubing would lower both the transmission lags two- to sixfold, depending on the terminal volumes. A similar reduction in the lag of both lines can be obtained by modifying the controller and transmitter to include an RC analog of the line (4). The biggest improvement comes from eliminating the transmission lag completely by using electronic control or by installing a pneumatic controller at the valve. One disadvantage of a valve-mounted controller is that four pneumatic lines are usually needed: the flow signal and valve signal are sent to a distant indicator or recorder, and two more lines are needed for the set-point signal and a seal for manual operation.

13-4. CONTROL WITH NOISY SIGNALS

The problem of noisy signals has been attacked in several ways (5). The differential pressure transmitter can be damped by restrictions in the lines to the orifice taps or a restriction in the outlet line. The danger in this procedure is that it is easy to apply too much damping, and the resulting smooth chart record is misleading. With a heavily overdamped transmitter, load fluctuations that could be corrected are never noticed by the controller. Some newer transmitters have calibrated damping adjustments that can give measurement time constants of 0.2 to 4 sec, and the amount of damping can be more carefully controlled.

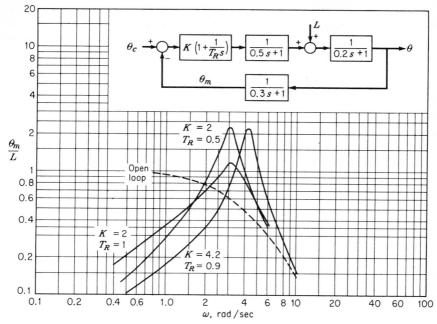

FIGURE 13-5   Calculated response of a flow control system to load fluctuations. (K_{max} = 9.4.)

A wide proportional band has also been used to minimize the effect of noisy signals. If the controller gain is only 0.1 to $0.3K_{c,max}$ instead of $0.5K_{c,max}$, the resonant peak in the closed-loop frequency response is much smaller. Figure 13-5 shows the measured response to load changes for a flow system with time constants of 0.5, 0.2, and 0.3 sec. The controller settings obtained from the usual recommendations [Eq. (9-2)] are $K = 4.2$ and $T_R = 0.9$ sec. With these settings the peak error is 2.1 at 0.7 cps, over four times the value for the open-loop system. Reducing the overall gain to 2 cuts the peak error in half, though it makes the transient response and the response at low frequencies somewhat poorer. Decreasing the reset time to 0.5 sec in an attempt to compensate for the lower gain makes the peak error too great. The optimum settings cannot be determined without knowing the nature of the load fluctuations, but values of $K = 2$ and $T_R = 1.0$ sec seem reasonable.

Some older controllers do not have reset times less than 6 sec and are not so suitable for fast processes as newer controllers with reset times adjustable to as low as 0.3 sec. Reset action is nearly always used with flow control because the maximum controller gain is fairly low and the offset after load changes would be significant. Derivative action is not recommended, since it amplifies the high-frequency noise. A three-mode controller should not be installed on the basis that the derivative action can be tuned out if necessary, since the minimum derivative time on some instruments is 0.1 min. A simple valve-mounted controller developed mainly for flow control (6) uses a fixed proportional band of 250 per cent ($K_c = 0.4$) and adjustable reset times of 1 to 50 sec.

An inverse derivative unit can be added to a conventional controller to stabilize processes with noisy signals (7). When the signal is changing rapidly, the change in controller output is smaller than for a steady error of the same size. Thus the proportional band is effectively widened for high-frequency signals but unchanged for low-frequency signals. In the example given by Moore (7), inverse derivative action permitted the controller gain to be increased sixfold, and the response time after an upset was only one-third as great.

13-5. NONLINEARITIES IN FLOW SYSTEMS

The nonlinear relationship between orifice differential and flow makes the stability of flow control systems change with flow rate. The increase in process gain with flow rate [Eq. (10-2)] can in theory be canceled if the effective valve gain varies inversely with flow rate. In practice no valve has just the right characteristic. If close control over more than a twofold range of flows is needed, a square-root converter can be used to give a signal proportional to the flow. Of course this nonlinearity does not arise if a turbine meter or magnetic flowmeter is used.

A different type of nonlinearity may be encountered if a valve positioner is used (8). A small change in the signal to the positioner calls for full air supply to the valve. A large signal to the positioner cannot produce any faster response. Thus the frequency response depends on the magnitude of the signals, and controller settings that seem suitable for large disturbances are not suitable for small ones.

Example 13-1

Predict the open-loop frequency response of a flow system with the following characteristics. What are the recommended controller settings based on the maximum gain and the ultimate period?

Process:	$T_1 = 1$ sec
Differential pressure transmitter damped to give an approximately first-order response:	$T_2 = 0.5$ sec
Line to controller:	500 ft $\frac{1}{4}$ in. OD
Line to valve:	500 ft $\frac{1}{4}$ in. OD
Valve motor:	No. 8 (see Fig. 10-11)
Valve trim:	Equal percentage
Valve pressure drop:	$\frac{1}{3} \Delta P_{total}$ at normal flow
Orifice differential:	30 in. water at normal flow
Transmitter range:	3–15 psi for 0–50 in. water

The effective time constants for the transmission lines are taken from Fig. 10-11, which gives conservative values because of the large upset used in the tests. The limiting values for very small signals would be about half as great.

For the line from transmitter to valve, $T_3 = 6$ sec.

For 500 ft of line plus a No. 8 motor, $T_4 = 27$ sec.

	Phase lag, deg			Amplitude ratio at $\omega = 0.35$ rad/sec
	$\omega = 0.2$ rad/sec	$\omega = 0.3$ rad/sec	$\omega = 0.35$ rad/sec	
$T_1 = 1$	11	17	19	0.94
$T_2 = 0.5$	6	8	10	0.98
$T_3 = 6$	51	61	64	0.42
$T_4 = 27$	80	83	84	0.105
	148	169	177	Product = 0.041

$$K_{max} \cong \frac{1}{0.041} = 24$$

$$P_u = \frac{2\pi}{0.35} = 18 \text{ sec}$$

The steady-state gain of the process is defined as a dimensionless ratio,

$$K_1 = \frac{d\Delta H/\Delta\bar{H}}{dF/\bar{F}} = \frac{\% \text{ change in differential}}{\% \text{ change in flow}}$$

$$\Delta H = kF^2$$

$$d\Delta H = 2kF\, dF$$

$$\frac{d\Delta H}{\Delta\bar{H}} = \frac{2\, dF}{\bar{F}} \quad \text{or} \quad K_1 = 2$$

The transmitter gain is 0.6, since a change of 30 in. of water, equal to the normal differential, changes the output only $^{30}\!/_{50} \times 12$ psi.

$$K_2 = 0.6 \frac{\% \text{ change in pressure}}{\% \text{ change in differential}}$$

The valve gain is estimated from Fig. 10-6, by using the curve for

$$\Delta P_v = 20 \qquad \Delta P_L = 40$$

$$K_4 = \frac{1.3\% \text{ change in flow}}{\% \text{ change in lift}}$$

The gain of the transmission line is 1.0; so $K_3 = 1.0$.

$$K = K_c K_1 K_2 K_3 K_4 = K_c(2)(0.6)(1.0)(1.3) = 1.56 K_c$$

$$K_{c,\,max} = \frac{24}{1.56} = 15.4$$

Based on the general Ziegler-Nichols equations [Eq. (9-2)], the recommended settings for a two-mode controller are

$$K_c = (0.45)(15.4) = 6.9$$

$$T_R = \frac{P_u}{1.2} = \frac{18}{1.2} = 15 \text{ sec}$$

Example 13-2

For the flow system of Example 13-1, how much could the response be improved by using a valve positioner? Assume that the lag of the positioner plus valve is negligible at frequencies below 1 rad/sec.

The 27-sec lag in Example 13-1 is reduced to 6 sec by terminating the line with a small bellows.

	Phase lag, deg at $\omega = 0.5$	A.R. at $\omega = 0.5$
$T_1 = 1$	27	0.9
$T_2 = 0.5$	14	0.96
$T_3 = 6$	71	0.31
$T_4 = 6$	71	0.31
	183	Product = 0.083

$$K_{max} \cong \frac{1}{0.083} = 12$$

$$P_u = \frac{2\pi}{0.5} = 13 \text{ sec}$$

While the period is 30 per cent lower, the maximum gain is only half as great. Following the method used in Example 7-4, the peak error after a step disturbance is estimated to be about 20 per cent greater than before when half the maximum gain is used for both cases. Thus there is hardly any improvement in controllability.

PROBLEMS

1. Show that a rotameter is a second-order system with a natural frequency of $\omega_n = \sqrt{\dfrac{2g(1 - \rho_{fluid}/\rho_{float})}{h}}$, where h is the distance above the bottom of the tube. (Assume that the annular area is proportional to the distance, and linearize the pressure-drop relationship.)

2. A gas flow fluctuates continuously, with an amplitude of 5 per cent of the mean flow. If the manometer across the orifice is damped to produce a straight-line record, what is the error in the measured flow?

3. Construct or find in the literature a piping diagram for a typical flow control installation. What fraction of the pressure drop occurs in straight runs of pipe? What is the predicted time constant for the process?

4. A flow control system uses an electronic controller and an electronic differential pressure transmitter. The major time constants are 0.4 sec for the process, 0.3 sec for the transmitter, and 1.5 sec for the valve. If the transmitter signal is not noisy, what would be the recommended controller settings for a two-mode controller? Would derivative action be helpful?

5. Ceaglske (9) calculated the optimum controller settings for a flow control system that had a negligible process lag, and he concluded that control would be better if a greater process lag existed. Can you show why adding a process time constant improves the control even though the critical frequency is lowered? Suppose that a flow system had four equal time constants: the process, transmitter, controller, and valve. Would increasing the process time constant improve the control?

6. A flow control system has a valve lag of 0.5 sec, a process lag of 0.2 sec, and a transmitter lag of 0.3 sec (see Fig. 13-5). For the settings $K = 2$, $T_R = 1$, and $K = 4.2$, $T_R = 0.9$, calculate the actual response to sinusoidal load changes which enter the loop just after the valve. Compare the peak values of θ/L with the values of θ_m/L given in Fig. 13-5.

7. A valve-mounted controller developed for flow control (6) has a fixed proportional band of 250 per cent and adjustable reset times of 1 to 50 sec. Would there be any flow systems for which this controller could not give stable operation? If so, give a numerical example.

8. The inner loop of a cascade control system regulates the flow of oil to a furnace. The process time constant is estimated to be 0.5 sec. The flow is measured with an orifice and pneumatic differential pressure transmitter and controlled by a small diaphragm valve. There is 100 ft of $\frac{1}{4}$-in. tubing from transmitter to controller and from controller to valve. The valve has an effective gain of 3 per cent change in flow per per cent change in pressure. The transmitter has a range of 12 in. of mercury, and the normal pressure drop is 8 in. of mercury.

What are the recommended controller settings? Use the data in Fig. 13-4 for the line response, even though the bellows in the controller has less than 10 in.[3] and the control valve has more than 10 in.[3] Assume that the dynamic response of the controller is the same as that of the transmitter.

REFERENCES

1. Catheron, A. R., and B. D. Hainsworth: Dynamics of Flow Control, *Ind. Eng. Chem.*, **48**:1042 (1956).
2. Catheron, A. R.: Factors in Precise Control of Liquid Flow, *Instruments*, **24**:705 (1951).
3. Caldwell, W. I., G. A. Coon, and L. M. Zoss: "Frequency Response for Process Control," p. 343, McGraw-Hill Book Company, New York, 1959.
4. Vannah, W. E., and A. R. Catheron: Improved Flow Control with Long Lines, *Instruments*, **25**:1733 (1952).
5. Balls, B. W.: Liquid Flow Control, *Trans. Soc. Instr. Technol.*, **8**(2):43 (1956).
6. Bowditch, H. L.: Development of a Universal Controller for Liquid Flow, *ISA Preprint* 52-15-3 (1952).

7. Moore, C. B.: The Inverse Derivative—A New Mode of Automatic Control, *Instruments*, **22**:216 (1949).

8. Ziegler, J. G., and N. B. Nichols: Electronic Flow Control, *ISA J.*, **6**(1):58 (1959).

9. Ceaglske, N. H.: Automatic Control of Fluid Flow, *Am. Inst. Chem. Eng. J.*, **5**:524 (1959).

14 THE DYNAMICS AND CONTROL OF DISTILLATION COLUMNS

Part I SURVEY OF CONTROL SCHEMES

The control of continuous distillation columns to give products of constant composition is one of the more difficult problems in process control. To begin with, it is very difficult to obtain accurate, continuous measurements of the composition of nearly pure streams. The usual practice has been to control the temperature several plates from one end of the column, where the temperature changes from upsets are large enough to be easily measured. However, the boiling temperature is not an accurate measure of the separation achieved, because it depends on the pressure and the amounts of minor components. Also, a constant composition at some intermediate plate does not assure a constant composition at the top or bottom plate. Instruments such as infrared analyzers, refractometers, or chromatographs have been used for direct control of product purity in a few recent installations. At present, the analytical instrument systems generally have a greater measurement lag (a minute or more) and are less reliable than temperature systems, but they will certainly be more widely used in the future.

The large holdup of liquid in the column, reboiler, and reflux drum tends to make distillation control systems sluggish. Open-loop tests may show a major time constant of 1 to 3 hr for the response of the product compositions to changes in feed rate or composition. Because the column holdup is distributed among many plates, there is also an effective time delay of up to several minutes before a change in the feed or reflux stream is noticed at the intermediate plates. As will be shown later, the period of the control system depends mainly on the time delay in the control loop, and the period can be made quite small by locating the control point at or close to the end of the column.

The dynamic analysis of distillation columns is difficult because of the counterflow of vapor and liquid, which introduces a special type of composition interaction between the stages, and because of the nonlinear equilibrium relationship, which makes the gain of each stage different. The equations for a series of thermal capacities or interacting pneumatic stages do not apply, though the solutions for counterflow heat exchangers may be of some help in predicting the response of distillation columns. In addition to composition lags, there are also lags in the flow of vapor and liquid through a column, and the lags in liquid flow have a large effect on the control-system performance.

There have been several theoretical studies of column dynamics employing analog or digital computers, but these are not yet broad enough in scope to be of direct use for design calculations. In the first part of this chapter the general features of various control schemes are discussed, with emphasis on the steady-state performance and the qualitative aspects of the dynamic response. The second part of the chapter is an introduction to the quantitative aspects of column dynamics. The factors that influence the various types of lags are discussed, and approximate methods of predicting the lags are presented. Some published data from actual columns are included in the discussion and in the problems.

14-1. BASIC FEATURES OF COMPOSITION CONTROL SCHEMES

Consider a typical distillation column which is fed F moles/hr of a feed stream containing two major components. The column separates the feed into D moles/hr of overhead or distillate product, which contains most of the more volatile component, and B moles/hr of bottom product, which contains most of the less volatile component. The main job of the control system is to adjust the split between distillate and bottoms by controlling the flow of one or the other of these streams. The ratio D/F (or B/F) is the most important adjustable parameter in the control system. The separation achieved also depends on the operating line slope L/V (liquid flow/vapor flow) in each section of the column, and the ratio V/F is a second adjustable parameter. In some control schemes, D is changed indirectly by changing V, and D/F and V/F are not independent.

Just from material-balance considerations, there is a very narrow range of overhead product flows (D/F) which will permit high-purity products to be obtained from a given feed. For example, suppose that the feed contains 50 per cent benzene and 50 per cent toluene and that 98 per cent purity is desired for both products. The normal values of D and B are 50 moles/hr if $F = 100$ moles/hr. If D is increased to 52, the overhead purity must drop to less than 96 per cent, since the total benzene feed rate is only 50 moles/hr. Since there will still be a little benzene in the bottoms, the overhead purity would probably be about 95 per cent for $D = 52$. A high overhead purity could easily be obtained by making D small, say 40 to 45, but the bottoms would be very impure.

If the feed rate and compositions are almost constant, satisfactory control can be achieved by using a flow controller to regulate either D or B.

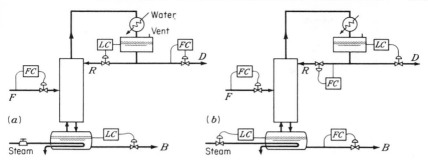

FIGURE 14-1   Material-balance control schemes.

With flow control of the distillate, the bottoms flow would be regulated to keep a constant level in the reboiler, as shown in Fig. 14-1a. With direct control of the bottoms flow, the distillate flow is regulated by the level in the reflux drum as in Fig. 14-1b. Direct flow control of both D and B would make the system overdetermined and would lead to unstable operation, since the sum of the product flows would not exactly match the feed rate. The material-balance equations at each end of the column must also be kept in mind to avoid an overdetermined system. The liquid flow to the reboiler equals the bottoms flow plus the vapor flow to the column. Therefore the steam rate, which determines the vapor flow, cannot be directly controlled if flow control of the bottoms is used.

In most distillation systems the split between overhead and bottoms is automatically adjusted to maintain a constant composition of one product or a constant temperature at some point in the column. (The term "composition control" is used here in the general sense and includes control of boiling temperature or other measures of composition.) If the overhead product is more important than the bottoms, the sensing element is placed in the top section of the column and decreases in purity are corrected by decreasing the overhead product rate. If the main function of the column is to produce a pure bottoms product, the sensing element is placed near the bottom of the column and the bottoms flow is reduced when the measured purity decreases. For either case the product flow may be controlled directly by using a control valve in the product line, or it may be controlled indirectly by changing the reflux rate or the steam flow. Usually no attempt is made to control both product compositions, though, as will be shown later, two-point control is theoretically possible by regulating both D/F and V/F.

14-2. CONTROL OF OVERHEAD COMPOSITION

DIRECT CONTROL OF D

Three methods of controlling the overhead composition x_D are shown in Fig. 14-2. The feed rate in each case could be controlled by a flow

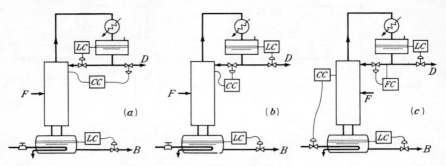

FIGURE 14-2   Control of overhead composition.

controller, if the feed comes from a storage tank, or allowed to vary, as when the feed comes directly from a previous process. The most direct scheme for controlling x_D is to use the signal from the composition controller to adjust the distillate flow D. The reflux flow is adjusted by an averaging-type level controller, and a decrease in D will eventually produce an equal increase in R.

The reflux drum or accumulator serves mainly to keep a liquid seal on the reflux and product lines and to provide disengaging space for non-condensables. Large accumulators can be a help in damping changes in flow and in providing a reservoir of high-purity material for reflux, but they have sometimes made control worse because of inadequate mixing in the tank (1).

A nearly constant vapor flow is maintained by regulating the steam pressure or by controlling the steam flow. An upstream pressure regulator is necessary for accurate flow control, because the flow for a given orifice pressure drop depends on the square root of the upstream density. Another way of controlling the vapor flow is to control the pressure drop across the column. The pressure drop across a plate depends on the gas velocity and the liquid depth, but since the liquid depth changes only slightly with liquid rate, the pressure drop is usually a satisfactory measure of vapor-flow rate.

INDIRECT CONTROL OF D

Instead of controlling D directly, the composition controller can adjust R and the level controller can adjust D. This has been more popular than the first scheme, though it has no apparent advantages. For very high reflux ratios R/D, this second scheme seems particularly sensitive to upsets. If $V = 550$, $R = 500$, and $D = 50$, a slight change in V to 560 would lead to $D = 60$, a 20 per cent higher flow than desired, until the composition controller sensed the change and increased the reflux. With the first scheme, the extra vapor flow would be returned as reflux without disturbing the product flow. The second system would also be more sensitive to changes in reflux temperature.

The third method of controlling the overhead composition is to decrease the vapor rate when the overhead purity drops.   The reflux flow is controlled at a fixed value, and so the lower value of V actually means a lower value for D.   The wide separation of the sensing element and the controlled flow is bad from a dynamic standpoint, and the system is slower than the first two schemes if the sensing element is almost at the top of the column. However, if the sensing element is several plates from the top, the third scheme is faster than the others because a change in vapor flow is transmitted through the column much faster than a change in liquid flow.   A cascade system with the composition controller adjusting the set point of a steam-flow controller is sometimes used as a modification of the third scheme.

14-3. CONTROL OF BOTTOM COMPOSITION

Methods of controlling the composition of the bottom product are shown in Fig. 14-3.   When direct control of the bottoms flow rate is used, the flow of steam is adjusted by the reboiler level controller.   The second scheme has the reverse arrangement; the composition controller regulates the flow of steam, and the level controller adjusts the bottoms flow.   For both cases, the reflux is controlled at a constant rate, and the overhead product rate is adjusted by the reflux-drum level controller.   The method analogous to that of Fig. 14-2c is not used, since there is a large delay before changes in reflux rate are noticed at the bottom of the column.

14-4. CONTROL OF BOTH PRODUCT COMPOSITIONS

By controlling the distillate-flow rate, the overhead composition x_D can generally be kept constant (in the steady state) in spite of changes in the feed composition.   However, the concentration of the bottom product x_B does change as the feed composition changes.   If the feed plate is correctly located for a normal feed, a shift in feed composition means that the feed is now introduced too high or too low in the column, which tends to decrease the separation that is achieved (the difference between x_D and x_B).   The

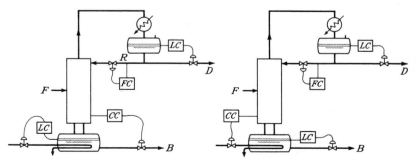

FIGURE 14-3   Control of bottom composition.

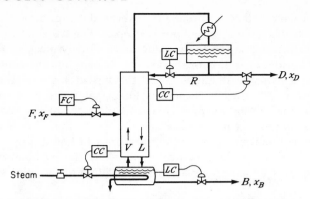

FIGURE 14-4   Control of both product compositions.

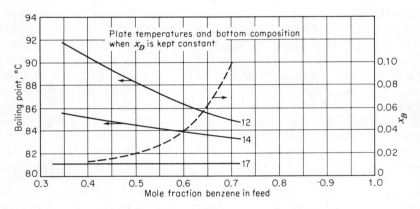

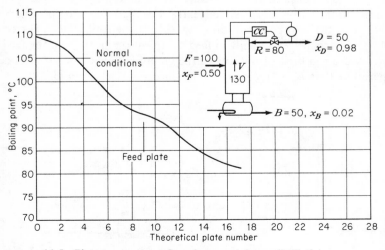

FIGURE 14-5   Plate temperatures for a benzene-toluene distillation.

separation can be increased by using a higher vapor rate and a higher reflux rate. Therefore, from a steady-state viewpoint, both x_D and x_B can be controlled by varying D/F and V/F independently, as shown in Fig. 14-4.

The conditions needed to keep both product compositions constant can be calculated by using either a McCabe-Thiele diagram or plate-to-plate calculations, with care being taken to switch operating lines at the actual feed plate and not at the optimum plate.

Example 14-1

A plate column is operated to produce 98 per cent benzene and 98 per cent toluene from a feed containing 50 per cent benzene. The feed is introduced as saturated liquid to the middle plate, and the normal boilup rate is 1.3 times the feed rate. If the reflux rate is changed to maintain a constant boiling point on the top plate, what happens to the other compositions when x_F changes? How could both x_D and x_B be kept constant? What happens to x_B and x_D if an intermediate plate is chosen as the control point?

A conventional McCabe-Thiele diagram for normal conditions shows that 17 theoretical plates are required, and the feed plate is number 9. The plate temperatures are plotted in Fig. 14-5. For a new value of x_F, say, 0.60, a new value of D is assumed, say, 0.59F. This fixes R, B, and x_B; so operating lines for both sections are drawn. If fewer than 17 plates are required for the assumed conditions, a higher value of D is chosen and the procedure repeated until exactly 17 stages are used. The flows and compositions for this case are the steady-state values that would result from controlling x_D at 0.98.

As shown by Fig. 14-5, an increase in x_F leads to more benzene on plates in the top section of the column if x_D is kept constant. The benzene content of the bottom product is slightly less than normal when x_F is low but up to fivefold higher with high values of x_F.

The vapor flow required to keep both x_D and x_B constant can be obtained by a double trial-and-error calculation, i.e., by assuming both D and V. If x_F is increased from 0.5 to 0.6, a 20 per cent increase in D and a 4 per cent increase in V gives steady-state values of $x_D = 0.98$ and $x_B = 0.02$.

Figure 14-5 also suggests that, if the temperature on plate 12 were kept constant, the top temperature would decrease, which is confirmed by further calculations. For $x_F = 0.6$, the product compositions would be $x_D = 0.965$ and $x_B = 0.021$. For $x_F = 0.7$, the values would be $x_D = 0.943$, $x_B = 0.011$.

The principle of two-point control can be applied to multicomponent distillation as well, though the problem of getting a reliable measure of composition is more difficult. Using temperature measurements near the top and bottom of the column could lead to trouble because of accumulation of medium boiling impurities in the center of the column. With high-speed chromatographs, a column could be controlled to keep both the heavy key in the overhead and the light key in the bottoms at constant values, subject of course to the fractionating capacity of the column.

Although control of both product compositions has a potential advantage over single product control, it has received little attention in the literature. The scheme is recommended by Robinson and Gilliland (2), and some columns are operating without trouble on two-point control. However, others have implied that such a scheme is unconventional or even unworkable (3, 4). The trouble comes from the interaction between control loops and the lags between changes in flow at one end of the column and effects at the other end. The problem is discussed further in the section on column dynamics.

14-5. LOCATION OF THE SENSING ELEMENT

Both the dynamic and the steady-state performance of the control system should be considered in locating the sensing element for composition control. Since control of column temperature is the standard approach, the location of temperature bulbs is considered first.

TEMPERATURE BULBS

From a dynamic standpoint, the temperature bulb should be placed so as to minimize the number and magnitude of the lags in the control loop. If the controller adjusts the reflux flow, the bulb should be placed on the top plate, where changes in reflux flow are felt immediately. Putting the bulb in the overhead vapor line introduces a small time delay and also makes the time constant of the bulb much greater. Putting the bulb several plates from the top introduces a hydraulic lag for each plate, and the control system has a much slower response. It is true that a bulb near the middle of the column senses changes in feed composition sooner than a bulb at the top, but this advantage does not offset the much larger period of oscillation.

In practice, very few columns are controlled by the temperature on the top (or bottom) plate. The temperature bulb is placed several plates below the top in a region where the temperature change from plate to plate is fairly large. This makes the measurement signal for a given change in product composition greater than if the bulb were on the top plate. However, even from a steady-state viewpoint, it would be unsound to use a maximum temperature gradient or a maximum measurement signal as the criterion for locating the temperature bulb. A constant temperature (or even a constant composition) several plates from the top does not mean a constant distillate composition, as was shown in Example 14-1. The scheme of controlling the temperature difference between two intermediate plates (5) eliminates any error due to pressure changes but still does not ensure constant product composition. For the example of Fig. 14-5, the temperature difference $\theta_{12} - \theta_{14}$ changes if x_D is constant; so keeping $\theta_{12} - \theta_{14}$ constant would make x_D change with changes in x_F.

Why is the top temperature not controlled to ensure a constant composition of the overhead product? The small changes in boiling point need not be a limitation. A standard filled bulb and transmitter respond to changes of 0.1°C, which corresponds to 98.5 ± 0.5 per cent benzene. For higher purities or more rigid specifications, a resistance bulb could be used. In an example reported by Boyd (6), the top temperature of a commercial column was held to within 0.02°F of the set point by using a resistance thermometer and an electronic controller. One problem is that the boiling point depends on the pressure, and pressure fluctuations may have more influence on the top temperature than composition changes. Compensation for pressure changes could be achieved by using an absolute pressure transmitter to adjust the set point of the temperature controller. A simpler method using just one instrument is to measure the difference between column pressure and the vapor pressure in a sealed bulb at column temperature. Vapor-pressure transmitters of this type have been available for some time (7), but they have not been widely used. A real limitation of top temperature control is that low-boiling impurities become concentrated in the top of the column, and they have a relatively large effect on the top temperature. Often the concentration of these non-key components in the liquid phase is relatively constant in the center of the column and increases rapidly over the last few plates. An intermediate temperature may therefore be better than the top temperature as a measure of the separation achieved. Plate-to-plate calculations for various feeds or a lot of experience is needed to choose the best location under these conditions.

The following summary applies to temperature bulbs or other non-specific types of sensing elements: The sensing element should be placed on the top or bottom plate if the sensitivity of the device is great enough and if minor components do not interfere. If the element is placed several plates from the end to minimize the effects of minor components, the ratio of key components in the product will probably change with changes in this ratio for the feed. Some information about the probable changes in feed composition is needed to strike a balance between these effects. If the only satisfactory location from a steady-state viewpoint is several plates from the end of the column, the use of specific analytical instruments should be considered.

Some recent articles show how steady-state performance calculations can be used to help select a control scheme (1, 8, 9). Berger and Short (10) present the example of a de-isopentanizing column controlled by measuring the refractive index 10 plates from the top in a 50-plate column. This location was chosen after plate-to-plate calculations showed that adding 2 per cent butane to the feed would result in less than ½ per cent butane in the liquid on plate 10 or lower plates but up to 5 per cent butane in the liquid on the top plates.

ANALYTICAL INSTRUMENTS

The use of chromatographs to control distillation columns avoids the uncertainties of physical-property measurements and also avoids the need to sample at points several plates from the ends of the column.   A minor component can be determined with the same relative accuracy as a major component; so control should always be based on product composition or, to get slightly faster response, on the compositions on the top or bottom plates.   The chromatograph is a batch instrument, but automatic devices are available for injecting samples at regular intervals and for reading the peak heights.   A converter can be used to give a continuous signal proportional to the peak height of a selected component, say, the heavy key in the overhead product (11).   When the concentration is changing, the signal is a series of short, connected steps, and this signal is suitable as the input to a conventional proportional-reset controller, provided that the steps are of short duration compared with the period of the controlled system (12, 13). A chromatograph with a sample time of 1 min has been used to control a depropanizer where the period of oscillation at the sample point was about 8 min (14).   Even more frequent sampling will be possible with future instruments.

Chromatographs, refractometers, infrared analyzers, and other delicate instruments are usually mounted on the ground for ease of servicing, and there is a time delay in getting samples from the column to the analyzer. This time delay may be one of the major lags in the control loop and should be minimized by careful design of the sampling system.   It is best to use a large flow in the sample line and recycle most of this flow, rather than using just the low flow required by the measuring instrument.

14-6. CONTROL OF COLUMNS WITH VARYING FEED RATES

The feed rate to a distillation column is often determined by the flow from upstream processing equipment such as reactors, extraction columns, or other distillation towers.   A storage tank with an averaging-type level controller (see Example 12-2) can be used to damp out fluctuations in flow or composition, but it is generally uneconomic to provide a very large tank just to permit constant feed to a distillation column over a long interval. When the feed to one column is the overhead or bottoms product from another column, a change in composition of the first feed means a change in flow to the subsequent columns.   These flow changes can be damped somewhat by letting the level in the reboiler or reflux drum change, but it is not worthwhile to make these vessels very large just for flow damping.

One way of coping with feed-rate changes is to use a ratio controller to make the steam flow proportional to the flow rate and a composition controller to adjust the product flow.   The ratio control can be accomplished with just one instrument, which measures and records the feed rate and sends to the steam control valve an air pressure proportional to the feed rate.   If the reboiler lag is large and fluctuations in pressure of the heating fluid are severe, a cascade system can be used, with the ratio controller adjusting the set point of a boilup controller. Of course the use of ratio control is not essential; it just starts corrective action before the change in feed rate is noticed at the end of the column.

14-7. PRESSURE CONTROL

For distillations at other than atmospheric pressure, the pressure must be closely controlled if column temperature is used as a measure of product composition. In many columns using temperature control, the temperature fluctuations resulting from pressure fluctuations (of perhaps 1 per cent in magnitude) are greater than those resulting from actual changes in composition. The steady-state performance of the column is not strongly dependent on the pressure; probably a 10 per cent change in pressure would be required to cause significant changes in the relative volatility or plate efficiency. If close control of boiling point is needed, it is better to use a vapor-pressure transmitter and an ordinary pressure control system than to try to design a system that will keep pressure fluctuations to a fraction of 1 per cent. The pressure is generally measured at the top of the column or at the reflux drum; since changes in vent rate or condenser operation affect the top pressure first, the control loop has smaller lags than if the mid-tower pressure or bottom pressure were controlled.

For vacuum distillation the top pressure is often controlled by adjusting a bleed of air or inert gas into the line from the reflux drum to the steam jet (Fig. 14-6a). This seems wasteful but it gives much better control than trying to regulate the steam supply to the jet. Most steam jets operate effectively only over a very narrow range of steam supply pressures. A control valve in the line between the accumulator and the jet could be used, but the pressure drop across the valve would be a disadvantage when very low column pressures were desired.

LARGE AMOUNTS OF NONCONDENSABLES

The type of control system chosen for pressure operation depends primarily on the amount of noncondensables in the overhead vapor. With a large amount of inerts, a throttling valve in the line venting the reflux drum gives satisfactory control (Fig. 14-6b). A separate vent condenser can be used to reduce the loss of product vapors. The major time constant in the control system depends on the holdup time of inerts in the top of the column, condenser, and reflux drum, which all operate at almost the same pressure. A proportional-reset controller is generally used, but a proportional controller would be satisfactory if the composition-sensing element is not affected by pressure changes.

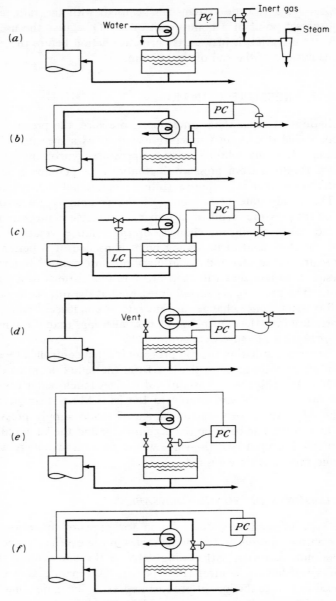

FIGURE 14-6   Methods of controlling the pressure in a distillation
column. (a) Vacuum tower; (b) noncondensables
in vapor; (c) vapor product; (d) cooling-water regulation; (e) flooded condenser; (f) hot-gas bypass.

If the overhead product is a vapor and all the liquid formed in the partial condenser is refluxed, the control scheme shown in Fig. 14-6c can be used. The pressure controller regulates the valve in the product line, and the supply of cooling water is regulated to maintain a level in the reflux drum. The flow of reflux is controlled by the temperature or composition at the top of the column. The major time constant is quite small, and good control is easily obtained.

SMALL AMOUNT OF NONCONDENSABLES

Several methods can be used to control the column pressure if the noncondensables are negligible (15, 16, 17). They depend on regulating the area or driving force for heat transfer in the condenser or on bypassing some gas to the reflux drum. The simplest method is to control the flow of cooling water, as shown in Fig. 14-6d. If the condensate temperature and pressure start to drop (because of a drop in inlet-water temperature, for example), the controller decreases the flow of water until a new heat balance is reached with a slightly higher exit-water temperature. This system has a fairly rapid response (as do most heat exchangers) and uses the minimum amount of cooling water, but the exit water may be almost as hot as the condensate, because condensers are often sized with large safety factors. Since fouling of the tubes is more rapid at high temperatures, this method is not recommended when the water temperature would be greater than 120°F, unless the water has been chemically treated to retard fouling. Some provision must be made for venting the small amount of inerts which may accumulate in the condenser and reflux drum. An intermittent or continuous purge can be used, and this can be adjusted manually by using the condensate temperature as a guide.

Even if there is a continuous flow of inerts into the system, satisfactory control of the pressure cannot be achieved just by throttling this gas stream if the flow is very small. Consider the example of a 10-ft column operating with a vapor velocity of 2 fps and 0.1 per cent inerts in the vapor. With perhaps 500 ft³ of vapor holdup in the condenser and reflux drum, the time constant based on the flow of inerts would be about 1 hr. A drop in water temperature would lead to a drop in condenser pressure which could be restored only very slowly even with the vent valve closed. Note that a large time constant in the control loop is unsatisfactory here because the controller saturates, whereas systems with one very large lag are generally very easy to control. The explanation is that there are two time constants to consider. The time constant relating pressure to changes in water flow or water temperature is about 1 min or less (see the discussion on heat exchangers in Chap. 11) because it is the partial pressure of condensables that is changing. The time constant for changes in the partial pressure of inerts is much larger. After a change in cooling rate, the control system must change the partial pressure of inerts to keep a constant total pressure, and, since this is more like servo-operation than regulator operation, the large time constant is detrimental.

In some columns the pressure is controlled by throttling the liquid flow from the condenser to keep some of the tubes submerged. Thus the heat-transfer area rather than the temperature drop is varied to keep a constant condensing temperature and pressure. This type of control is more sluggish than the other systems, because the level cannot be changed very rapidly. In the system shown in Fig. 14-6e, the throttling valve is placed between the condenser and the reflux drum. In some installations, the pressure control valve is in the product line, and the reflux drum is kept flooded; this system is even slower, because the product flow is only a fraction of the total liquid flow, and the reflux drum no longer serves to average out fluctuations in product flow. More rapid response can be achieved with a partially flooded condenser when the condenser is mounted several feet below the reflux drum. The pressure controller regulates a small bypass flow to the reflux drum. With a drop in column pressure, more gas is bypassed to the drum, and the increase in drum pressure forces liquid back into the condenser to cover more of the tubes.

BYPASS CONTROL SCHEMES

A common method of pressure control for petroleum distillations involves bypassing a significant fraction of vapor around the condenser, as shown in Fig. 14-6f. For a typical case the liquid leaving the condenser might be subcooled 30°F and represent 90 per cent of the total vapor. The remaining 10 per cent of the vapor condenses in the reflux drum and heats the subcooled liquid. As the column pressure drops, the controller sends more gas through the bypass line, which raises the pressure in the reflux drum, condenser, and column. If a large amount of gas must be bypassed, it may be necessary to partially close a valve in the main vapor line to provide more pressure drop between the column and the reflux drum. Of course putting the control valve in the main vapor line would give satisfactory control, but the valve would have to be much larger than a bypass control valve.

The maximum amount of gas to be bypassed can be calculated from a heat balance by using the minimum water temperature, as shown by Hollander (16). A possible disadvantage of bypass control is that the reflux temperature may vary because of incomplete mixing in the reflux drum.

Of the many systems mentioned for pressure control, only regulation of the cooling water or bypassing hot gas to the reflux drum can give a constant reflux temperature for different water temperatures. With the other methods, particularly the use of a flooded condenser, the degree of subcooling varies with the water temperature. Changes in reflux temperature change the internal reflux ratio if the reflux flow is constant and thus disturb column operation, but the changes in temperature are usually not rapid enough to make this a vital point in choosing a pressure control system.

14-8. CONTROL OF FEED TEMPERATURE

Automatic control of the feed temperature is a refinement that makes operation of some columns a little steadier. A cold feed condenses some vapor at the feed plate, which leads to different operating line slopes and a slightly different fractionation capacity. A more important effect of a drop in feed temperature is the decrease in overhead product rate which occurs if the reflux flow is fixed. For example, a 10°F drop in the feed temperature for the previous benzene-toluene example would lower the overhead vapor rate from 130 to 127 and the overhead product rate from 50 to 47, if the reflux flow were kept at 80 by a composition controller. The overhead purity would be as great as or greater than before, but if these flows were maintained, the bottoms would have to have twice as much benzene just from material-balance considerations. The composition controller or controllers would tend to restore the proper product split, but not until after deviations had occurred. Feed temperature control acts to keep the deviations from entering the system.

For the usual case where the feed temperature does not change very much or very rapidly, simple schemes can be used. If steam is used to preheat the feed, a pressure regulator on the steam line would give satisfactory control. The feed temperature would change if the feed rate changed, but the resulting upset in column conditions would come mainly from the change in flow rate and would be about the same with or without close control of the temperature. If feedback temperature control is used, a proportional controller should suffice. A steady-state offset is not harmful, and the period of the temperature-control loop is not critical, since the exchanger response is much faster than the column response. There is rarely justification for elaborate control schemes such as a temperature controller adjusting the set point of a steam-flow controller.

In some cases, the feed is heated in a feed-bottoms exchanger followed by a steam heater. Because the bottoms flow rate is not constant, automatic feedback control is required in order to keep a constant feed temperature. If automatic control is used, the temperature set point should be slightly below the boiling point of the feed. If the controller is set to give feed at the boiling point or partially vaporized feed, small changes in feed composition or column pressure can result in significant changes in feed quality, which upset the material balance at the top of the column.

14-9. INTERNAL REFLUX CONTROL

Cold reflux has the same effect as cold feed—it generates a small additional quantity of reflux at the point where it is introduced. Though the internal reflux is only a few per cent greater than the external reflux, the difference becomes important when the reflux ratio is high and the product flow small.

Consider a case where the vapor flow in the column is 550 moles/hr and the external reflux is kept at 500 moles/hr. If the reflux is at the boiling point, there is no extra condensation and the product flow is 50 moles/hr. If the reflux is subcooled 10°F, about an additional 20 moles/hr is condensed in the column, the flow of vapor to the condenser is reduced to 530 moles/hr, and the product flow drops to 30 moles/hr. The reduced product flow would affect the product quality and would eventually be corrected by the composition controller, but meanwhile some off-specification product would have been produced. The section on pressure control showed how changes in reflux temperature could arise from changes in the amount of inerts or from changes in cooling-water temperature, but these are likely to be relatively slow changes. With air-cooled condensers, sudden changes in reflux temperature do occur because of rapidly changing ambient conditions, and it was to cope with this problem that engineers at the Phillips Petroleum Company developed a pneumatic analog computer and controller to keep the internal reflux constant. A brief description of the system is presented here as an example of how control can be improved with a relatively simple combination of standard components.

Control of reflux temperature was formerly accomplished by manipulating the fan pitch, as shown in Fig. 14-7 (18). The temperature in the reflux accumulator cycled with a period of about 40 min and an amplitude of 5°F. The system is not self-regulating since a decrease in reflux temperature leads to a decrease in vapor flow and thus to a further decrease in reflux temperature. In this particular case, control is made worse by the large holdup and incomplete mixing in the reflux drum. Measuring the temperature at the condenser exit would have given better control.

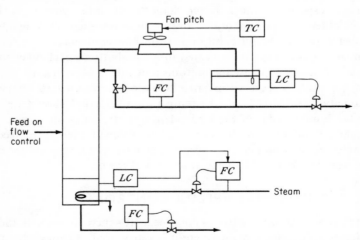

FIGURE 14-7   Reflux temperature and steam flow oscillate with this control scheme. [*After Lupfer and Berger* (18), *courtesy of ISA Journal.*]

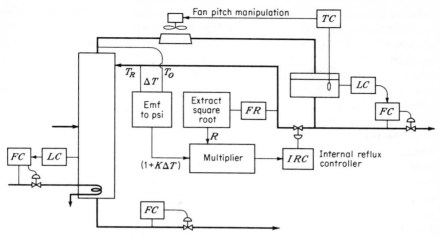

FIGURE 14-8   Internal reflux control gives stable operation.   [*After Lupfer and Berger* (18), *courtesy of ISA Journal.*]

In the system shown in Fig. 14-8 the measured degree of subcooling and the external flow are used to calculate the internal reflux and keep it constant.   The internal reflux is a linear function of the degree of subcooling.

$$R_{\text{int}} = R_{\text{ext}}(1 + K\,\Delta T) \qquad (14\text{-}1)$$

Thermocouples measure the difference between the top temperature and the reflux temperature, and the difference is converted to a pneumatic signal $(1 + K\,\Delta T)$.   The square-root extractor gives a pressure signal that is proportional to the external flow.   Multiplying these signals gives a pressure proportional to the internal reflux, which is used as the input to a conventional controller.   The reflux temperature is measured just before the reflux enters the column, to avoid a time delay in the control loop.   The controller adjusting the pitch of the fan was eliminated in other installations and satisfactory control of product flows obtained in spite of large changes in reflux temperature (18).

Part II   DYNAMIC BEHAVIOR OF DISTILLATION COLUMNS

14-10. TYPES OF LAGS

Each of the stages of a distillation column has three major capacities associated with it, and thus there are three types of lags or effective time constants which determine the dynamic behavior of a column.   The largest lags are the concentration lags arising from the capacity in the liquid held on the plates.   Next largest are the lags in liquid-flow rate resulting from the change in holdup with flow rate (the capacity of the plates for increased holdup).   Still smaller and often negligible are the lags in vapor-flow rate which arise because the vapor holdup increases with pressure and thus with flow rate.   A fourth type of lag, the concentration lag associated with the vapor holdup, can be either neglected or lumped with the much larger liquid-concentration lag.

The time required for 63 per cent response after a change in external conditions depends primarily on the concentration lags.   It is shown later that the largest lag for all plates and all upsets is approximately equal to the total column holdup divided by the feed rate, which may be several minutes to a few hours.   However, the initial response at any plate depends mainly on the fluid-flow lags and the distance from the source of the disturbance.   The concentration on an intermediate plate cannot change until the composition or flow rate of an entering stream changes, and changes in liquid rate, vapor rate, and stream composition spread at different rates through the column.   The initial response of the measured variable to the controlled flow or to load changes has a great effect on the behavior of the control system, and therefore the flow lags, though only a few seconds to a fraction of a minute per plate, must be considered in an analysis of the control loop.   In the following sections, the factors influencing the flow lags and the concentration lags are discussed separately before dealing with actual control systems, since others (3) have shown that interactions between the two types of lags can probably be neglected.

14-11. LAG IN LIQUID FLOW

The hydraulic lag for liquid flow can be treated as a liquid-level problem which is complicated somewhat by the change in level across the plate.   At the center of a plate, the equivalent depth of clear liquid is usually less than at either end (19) and may even be less than the weir height, as shown in

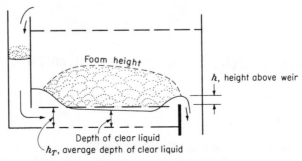

FIGURE 14-9   Variation of liquid depth across a plate [*After Robinson and Gilliland (19).*]

Fig. 14-9. The height above the exit weir, h, and the average depth h_T must increase with flow rate. If the effect of vapor-flow changes is neglected, the material-balance equation is

$$\frac{A\, dh_T}{dt} = \Delta L_{\text{in}} - \left(\frac{dL_{\text{out}}}{dh_T}\right) h_T \qquad (14\text{-}2)$$

or

$$A\left(\frac{dh_T}{dL_{\text{out}}}\right)\frac{dh_T}{dt} + h_T = \left(\frac{dh_T}{dL_{\text{out}}}\right)\Delta L_{\text{in}}$$

The time constant is the area times the rate of change of average depth with liquid rate or just the rate of change of holdup with flow rate.

$$T_L = A\left(\frac{dh_T}{dL_{\text{out}}}\right) = \frac{dH}{dL} \qquad (14\text{-}3)$$

where T_L = hydraulic lag for 1 plate, min
$\quad\quad H$ = holding of liquid per plate, ft^3
$\quad\quad L$ = liquid rate, cfm

To get some typical values of T_L, published holdup data for a bubble-cap plate operating with air and water (20) are plotted in Fig. 14-10. The clear liquid height is obtained from the static pressure at the center of the plate and is slightly less than the average depth h_T; but the difference is assumed to be small. The hydraulic gradient is also neglected in the following analysis.

For a 2-in. weir at a liquid rate of 10 to 20 gpm/ft and an F factor of 1.0 ($F = v\sqrt{\rho_G}$), the slope of Fig. 14-10 is 0.05 in./(gpm)(ft). The time

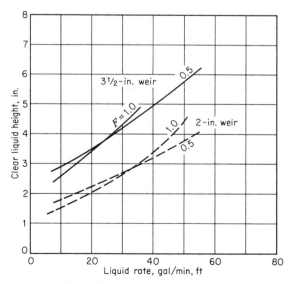

FIGURE 14-10   Holdup of water on a bubble-cap plate.
(*Data from Ref. 20.*)

constant does not depend on the width of the flow path; so a 1-ft-wide section of tray is used for a basis. For a tray with a liquid path 5 ft long the time constant is

$$T_L = 5 \text{ ft}^2 \times \frac{(0.05)(7.48)}{12} \text{ ft/cfm} = 0.156 \text{ min}$$

For a tray 10 ft long, the time constant would be twice as great, or 0.31 min. For a column 1 ft in diameter, the time constant would be only a few seconds unless the length of weir was much smaller than the width of the contact area. The time constant is roughly proportional to the column size D because the liquid rate varies with D^2 but the weir length and the flow per foot of weir are proportional to D. Because of the smaller flow lag, control schemes may appear better when tested on pilot columns than when tried in the plant.

An alternative method of predicting T_L is based on the Francis formula for a long rectangular weir.

$$L = kh^{3/2} \tag{14-4}$$

$$\frac{dL}{dh} = \tfrac{3}{2}kh^{1/2} = \frac{3}{2}\frac{L}{h} \tag{14-5}$$

Substituting Eq. (14-5) into Eq. (14-3) gives

$$T_L = A\frac{dh}{dL}\left(\frac{dh_T}{dh}\right) = \tfrac{2}{3}A\frac{h}{L}\left(\frac{dh_T}{dh}\right) \tag{14-6}$$

If dh_T/dh is assumed equal to $\bar{h}_T/\bar{h}$, which is nearly true at high values of L and F, the time constant is two-thirds the holdup time.

$$T_L = \tfrac{2}{3}A\frac{h_T}{L} = \tfrac{2}{3}T_H \tag{14-7}$$

At low gas and liquid rates, the change in h_T is not much greater than the change in h, and the time constant is approximately $\tfrac{2}{3}T_H(\bar{h}/\bar{h}_T)$, or perhaps 0.1 to $0.5T_H$.

As far as liquid flow is concerned, the plates in the column form a series of noninteracting first-order lags. A step increase in reflux rate leads to a first-order response in top-plate level and in liquid flow to the second plate, and an S-shaped response in liquid flow to the third or lower plates. The transient response several plates from the top can be approximated by a time delay plus an exponential response.

The effective time delay (based on the time for 5 per cent response) would be about $0.5nT_L$ for $n = 5$, $0.65nT_L$ for $n = 20$, and $0.80nT_L$ for $n = 50$ (see Fig. 3-22). If a composition controller on the tenth plate were used to adjust the reflux rate, there would be a time delay of about $5T_L$, or perhaps 2 min, before a change in reflux had a noticeable effect at plate 10. It is this time delay that makes control at intermediate plates much more sluggish than control at the top plate.

For large columns, the hydraulic gradient should be considered in a rigorous analysis of the flow lag. A sudden increase in inlet flow would result in a wave traveling across the plate, and this would mean a time delay before the exit flow started to increase. However, since several noninteracting lags in series give a large effective time delay, a small additional true delay on each plate would not be critical. Another factor to consider is the change in holdup with vapor flow. As shown by Fig. 4-10, at low liquid rates an increase in vapor flow *decreases* the liquid holdup and causes a momentary increase in liquid flow.

The variable holdup in the downcomer might also be given special treatment, but lumping this capacity with the plate holdup is probably satisfactory for present studies.

14-12. LAG IN VAPOR FLOW

The vapor spaces between the plates and the resistances at the plates form a series of interacting lags as far as vapor flows are concerned. The response is similar to that of a series of pressure tanks where the flow out of any tank depends on the pressure downstream as well as the pressure in the tank. For one tank with equal resistances upstream and downstream, the time constant is $RC/2$. With two tanks and three resistances in series, the transfer function relating pressure in the second tank and input pressure for the first tank has effective time constants of RC and $RC/3$. Adding more tanks increases the spread between the largest and the smallest time constant. With several tanks, the response approaches that of a distributed system, and the solutions for heat transfer or diffusion in a slab apply. The response at the outlet to a step change at the inlet can be approximated by a time delay $L = 0.05(\Sigma R)(\Sigma C)$ and a time constant $T = 0.45(\Sigma R)(\Sigma C)$, as shown by Fig. 3-27. The initial response is somewhat faster than if the elements were noninteracting.

The lag in vapor flow is usually an order of magnitude smaller than the lag in liquid flow because of the small capacity for vapor holdup on the plates. If this capacity were just the volume of vapor space, the vapor lag would indeed be negligible but the increase in pressure that goes with an increase in vapor flow means that vapor must be condensed to heat the liquid to a higher temperature. The holdup of liquid therefore acts as a capacity for vapor which is generally much higher than the volumetric capacity in the vapor space. The material balance for vapor flow for a single plate is (see Fig. 14-11)

$$\text{Accumulation} \qquad = \text{inflow} \; - \; \text{outflow}$$

$$H_V\left(\frac{dP_2}{dt}\right) + H_L\left(\frac{d\theta}{dP}\right)\left(\frac{c_L}{\lambda}\right)\left(\frac{dP_2}{dt}\right) = \frac{P_1 - P_2}{R} - \frac{(P_2 - P_3)}{R} \qquad (14\text{-}8)$$

where H_V = holdup in vapor space, scf/atm

H_L = liquid holdup, lb

$\dfrac{d\theta}{dP}$ = change in boiling point with pressure

c_L = heat capacity of liquid, Btu/(lb)(°F)

λ = heat of condensation, Btu/scf

R = flow resistance, atm/scfm

The following example shows values of the capacity for a typical plate based on the benzene-toluene system at 1 atm.

Example 14-2   Vapor Lag for Benzene-toluene System

Capacities are based on 1 ft² of column cross section.

$$H_V = 1.6 \text{ scf/atm} \qquad (2\text{-ft plate spacing} - 0.4 \text{ ft liquid})$$

$$C = 1.6 + 21 \text{ lb liquid}\left(\frac{40°\text{F}}{\text{atm}}\right)\left[\frac{0.5 \text{ Btu/(lb)(°F)}}{30 \text{ Btu/ft}^3}\right]$$

$$C = 1.6 + 14 = 15.6 \text{ scf/atm}$$

Assume $\Delta P = 0.01$ atm/plate and flow = 1 scf/sec. Estimate transient response 20 plates from reboiler.

$$\Sigma\,\Delta P = 0.2 \text{ atm}$$

$$\Sigma R = 2\,\frac{\Delta P}{q} = 0.4 \text{ atm/(scf)(sec)} \qquad \text{if } \Delta P = kF_v{}^2$$

$$\Sigma C = 20 \times 15.6 = 310 \text{ scf/atm}$$

$$(\Sigma R)(\Sigma C) = 124 \text{ sec}$$

$$\text{Effective time delay} = (0.05)(124) = 6 \text{ sec}$$

$$\text{Effective time constant} = (0.45)(124) = 56 \text{ sec}$$

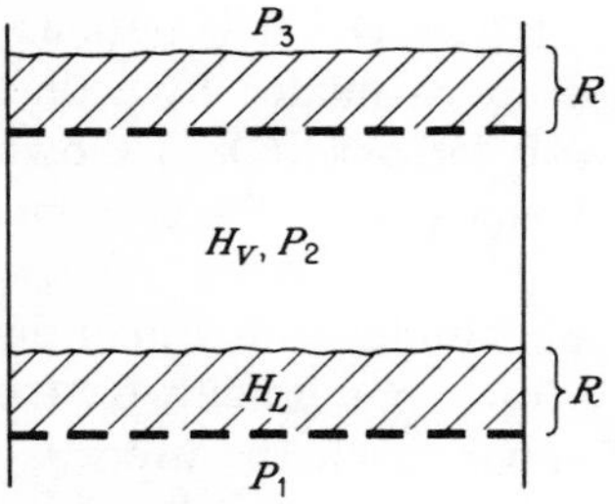

FIGURE 14-11 Capacities and resistances for vapor flow through a plate column.

The vapor lags would be greater than the above values if there were more liquid on the plates or a greater pressure drop across the plates, but they would still be smaller than other lags in the control loop in most cases.

14-13. EXPERIMENTAL VALUES OF FLOW LAGS

Values of the flow lags or effective time delays are given in some of the published studies of column dynamics.   Though none of the papers includes enough data for a complete prediction of the behavior, the values given afford a rough check of the theory outlined previously.   Armstrong and Wood (22) used a 4-in. column and reported a time constant for liquid flow of 0.12 min per plate, based on the measured change in holdup with flow rate.   This seems a little high for such a small column (the holdup time per plate was 0.6 min), but details of the weir arrangement were not given. After a step change in reflux rate, it was about 2 min before a significant change in composition was noticed 20 plates away, which agrees with the predicted delay of about 1.6 min ($0.65 \times 20 \times 0.12$).   Tests by Berger and Short (10) on an 8.5-ft 50-plate column showed that 5 min was required before a change in reflux rate affected the composition 10 trays from the top.   This indicates a hydraulic lag of about 1 min for a plate, since the flow to tray 10 would have started to change after a time equal to five time constants.   Similar studies of a 13.5-ft extractive distillation column showed a delay of 15 min before a change in furfural rate was noticed 94 plates away (23).   In this case the flow lag was about 0.2 min per plate, only slightly less than the average holdup time.   In extractive distillation, the liquid flows are much greater than the vapor flows, and the holdup times are less than for normal distillation.

Rademaker reported that a large turbogrid column, which has a low holdup, had a lag of about 2 sec per plate (24).

There is relatively little information on vapor-flow lags, because these lags have generally been too small to attract attention.   Tests by Aikman (25) on a 20-plate column indicated a dead time of 20 sec before changes in the heat supply were noticed by a sensitive temperature bulb in the overhead vapor line.   Since the dead time included a few seconds' delay in the vapor line and at least a few seconds' delay in the reboiler, the effective delay for vapor flow in the column was 10 to 15 sec.   For the 50-plate pentane fractionator, a step change in heat input resulted in an "immediate" change in overhead product rate (10), which probably means that the delay was less than one minute.   A large vapor lag was reported for the 100-plate extractive distillation column (23); a step change in heat input was noticed after 3 min 24 trays up and after 12 min at the reflux drum.   Since the vapor velocity was low (about 0.5 fps), the plate spacing high (3 ft), and the liquid holdup large, a larger than normal vapor lag would be expected for this column but a time delay of 12 min still seems too high.   It is hoped that more studies of flow lags will be reported and that these will include enough data on plate characteristics to permit prediction of the lags.

14-14. CONCENTRATION LAGS

The concentration lags describe the relatively slow changes in plate composition that follow changes in feed composition or that occur after changes in liquid or vapor flow have reached all the plates. The liquid on an isolated, well-mixed plate would show a first-order response to changes in the composition or flow of either the entering liquid or the entering vapor stream. With several plates in a countercurrent series, the time constants interact, and exact analysis, while straightforward, is rather tedious. The values of the effective time constants depend on the slope of the equilibrium line, the plate holdup, the flow rates in the column, and the feed rate. To show how these variables influence the lags and to provide some basis for generalizations, exact solutions are presented for the simple examples of one-, two-, and four-stage columns. In these examples, perfect mixing is assumed in each stage, and the lags in liquid and vapor flows are considered negligible.

EFFECT OF EQUILIBRIUM LINE SLOPE

Consider a one-stage separation, and assume that the vapor composition is a linear function of the liquid composition over the range of compositions covered.

$$y = a + bx \qquad (14\text{-}9)$$

The dynamic response for changes in feed composition is derived in the usual way, the symbols for concentration of the more volatile component denoting the changes from the normal values. For the system of Fig. 14-12,

$$H\frac{dx_1}{dt} = Fx_F - Bx_1 - Dx_D \qquad (14\text{-}10)$$

From Eq. (14-9)

$$x_D = bx_1 \qquad (14\text{-}11)$$

$$H\frac{dx_1}{dt} + (B + bD)x_1 = Fx_F \qquad (14\text{-}12)$$

Transforming and rearranging,

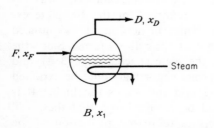

FIGURE 14-12   A single-stage distillation.

$$\frac{X_1}{X_F} = \frac{K}{Ts + 1} \qquad (14\text{-}13)$$

where $K = F/(B + bD)$ = steady-state gain

$T = H/(B + bD)$ = time constant

$X = \mathcal{L}(x)$

If $b = 1$, the time constant is $H/(B + D)$ or H/F, the holdup time based on the entering liquid rate. At low concentrations where b is greater than 1, the time constant is less than the nominal holdup time, but for the same system at high concentrations, the reverse is true. The time constant depends, not on the relative volatility or on the difference between y and x, but on the rate of change of y with x. Note that a low time constant is associated with a low value of steady-state gain. If $K = 0.5$, a change in x_F of 0.10 means an eventual change of only 0.05 in x_1 and

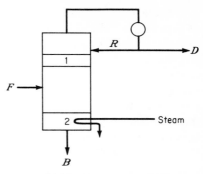

FIGURE 14-13   A two-stage distillation.

it is logical that the fractional response is more rapid than if $K = 1$ and $\Delta x_1 = \Delta x_F$.

EFFECT OF FEED RATE AND TOTAL HOLDUP

The importance of these variables in determining the major concentration lag can be demonstrated with a two-stage column. The equations for the column of Fig. 14-13 are conveniently handled in transformed notation. Following Eq. (14-9), the change in vapor composition is taken as b times the change in liquid composition.

For the top plate
$$H_1 \frac{dx_1}{dt} = Vy_2 - Rx_1 - Dx_D$$
(14-14)

or
$$(H_1 s + R + bD)X_1 = bVX_2$$

For the bottom plate
$$H_2 \frac{dx_2}{dt} = Fx_F - Bx_2 + Rx_1 - Vy_2$$
(14-15)

or
$$(H_2 s + B + bV)X_2 = FX_F + RX_1$$

Combining and solving for X_2 or X_1 gives a quadratic expression whose factors represent effective time constants.

$$\frac{X_1}{X_F} = \frac{bVF}{H_1 H_2 s^2 + [H_1(B + bV) + H_2(R + bD)]s + (BR + BbD + b^2 VD)}$$
or
(14-16)

$$\frac{X_1}{X_F} = \frac{K_1}{(T_a s + 1)(T_b s + 1)}$$

$$\frac{X_2}{X_F} = \frac{X_1}{X_F} \frac{H_1 s + R + bD}{bV}$$
or
(14-17)

$$\frac{X_2}{X_F} = \frac{K_2(T_c s + 1)}{(T_a s + 1)(T_b s + 1)}$$

Table 14-1 gives the effective time constants for several combinations of parameters. As with one stage, an increase in the slope of the equilibrium line means smaller time constants and smaller eventual changes in plate composition after changes in feed composition. With b greater than 1, the change in x_D is greater than the change in x_F, but the vapor holdup is negligible, and it is the change in x_1 and x_2 that determines the speed of response. With b close to 1, the largest time constant is approximately the total capacity of the column divided by the feed rate. Increasing the reflux ratio lowers the nominal holdup time per plate, but this affects mainly the smaller and not the larger time constants. With a very high reflux rate,

Table 14-1. Effective Time Constants for a Two-plate Column

Holdup, moles		Flows, moles/min					dy/dx	Time constants, min			Gain	
H_1	H_2	F	D	B	V	R	b	T_a	T_b	T_c	K_1	K_2
1	1	1	0.5	0.5	1	0.5	1	2.0	0.50	1.0	1.0	1.0
1	1	1	0.5	0.5	1	0.5	2	1.13	0.32	0.67	0.73	0.55
1	1	1	0.5	0.5	2	1.5	1	2.0	0.25	0.25	1.0	1.0
1	1	1	0.5	0.5	4	3.5	1	2.0	0.12	0.25	1.0	1.0
1	4	1	0.5	0.5	2	1.5	1	4.84	0.41	0.50	1.0	1.0
1	4	1	0.2	0.8	2	1.5	1	5.0	0.40	0.50	1.0	1.0

the smaller lags, T_b and T_c, would approach zero, and plates 1 and 2 would show almost identical first-order responses to changes in feed composition (and also to changes in feed rate or product rate). A high reflux rate tends to keep both plates at equilibrium with respect to each other when both are changing slowly because they are not in equilibrium with the feed. The feed rate controls the response, since any increase in the amount of volatile component held on the plates has to come from the feed. An overall material balance may help to emphasize this.

$$\text{(Total holdup)}\frac{dx_{\text{av}}}{dt} = Fx_F - Dx_D - Bx_B \qquad (14\text{-}18)$$

The general conclusions about the effects of F and H and Eq. (14-18) can be applied to columns with a large number of plates. With a very high reflux ratio, the plates will all respond at about the same rate (± 50 per cent), a rate determined primarily by the total holdup and the feed rate. With a low reflux ratio, differences in the initial response will be more noticeable, with plates far from the disturbance showing some time delay before responding. Columns for systems with a low relative volatility will tend to have a very large time constant since they will have many plates and also a low feed rate.   This is a possible explanation for the reported correlation between the major time constant and the square of the number of plates (3).

REBOILER HOLDUP

The effect of holdup in the reboiler, condenser, and reflux drum deserves further comment, since in some columns this extra holdup is as great as the holdup on 20 to 30 plates.   This extra holdup would have as much influence as the plate holdup in determining the major lag if the reboiler composition showed as much change as a typical plate after changes in operating conditions.   Usually the change in reboiler or distillate composition is smaller than the change in composition on an intermediate plate, and so the extra holdup does not count so much as plate holdup.   If the purity of both products is high and the change in concentration per plate is only 1 or 2 per cent at the ends of the column, a 5 per cent increase in feed concentration may result in 5 per cent increase in concentration for intermediate plates but only a 1 per cent increase at the ends of the column. What really counts is the average change in concentration for all the holdup in the system, and this could readily be calculated if necessary.   For rough estimation of column dynamics, a weighting factor of 0.5 could be applied to the reboiler holdup to get the major concentration lag.

TRANSIENT-RESPONSE STUDIES

There have been many experimental and theoretical studies of concentration lags in plate columns.   Digital computers were used for several theses at the Massachusetts Institute of Technology, and empirical correlations for the effective time constants are given by Gilliland and Mohr (26). Workers at the University of Delaware used an analog computer to simulate 5- and 16-plate columns (27), and experimental tests were performed on a 5-plate column (28).   Other work is cited in the review by Archer and Rothfus (29).   There is too much material in these studies to summarize here, and only a few experimental results are cited below in support of the general conclusions presented earlier.

The transient response of a 21-plate column was studied by Wood and Armstrong (21).   After a change in feed composition, plates near the feed plate started to change composition almost immediately and showed an approximately first-order response.   The top and bottom compositions started to change after a delay of about 10 times the holdup time per plate and then responded about as fast as the feed plate.   Based on a 63 per cent response, the major time constant for all plates was about $120H/R$, almost equal to the total holdup over the feed rate.   (The reflux rate was $4.73F$, and the total holdup was $24.9H$, including $3.9H$ in the reboiler.)   The 63 per cent response times for changes in reflux rate (22) were about the same as for changes in feed composition, but the initial delay for plates below the disturbance was somewhat smaller.   The composition on a plate starts to change when either the flow rate or the composition of the stream to the plate changes, and changes in liquid flow travel down the column faster than changes in liquid composition.

14-15. FREQUENCY RESPONSE OF COLUMNS

Exact solutions for the frequency response of a column are just as difficult to obtain as the transient response, even after linearizing all the equations and treating the concentration and flow lags separately.   The problem is to obtain the effective concentration lags which are the roots of the nth-order equations.   For particular cases, the linearized equations can be solved by standard procedures, or a digital computer can be used to calculate transients for either open- or closed-loop behavior.   The approach taken here is to present the general features of the frequency response deduced from solutions for short columns and from transient-response studies.   Examples presented later show that a simple approach to the frequency response can be used to predict the difference between the performance of various control schemes.

Consider the most common control scheme, which involves varying the reflux to keep a constant composition on the top plate or a plate near the top.   If the control point is n plates from the top, there will be $n - 1$ liquid-flow lags in the control loop, each of which can contribute up to 90° phase lag.   Following the suggestion of Armstrong and Wood (3), these lags can be considered to act in series with the concentration lags that would exist if there were no flow lags.   It seems logical to expect the major concentration lag to be the same as that for the transient response: column capacity divided by the flow rate.   As further proof of this, transfer functions for a four-plate column are given in Example 14-3.

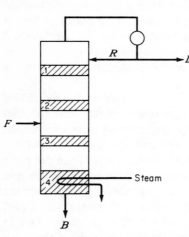

FIGURE E14-3   Diagram for Example 14-3.

Example 14-3   Transfer Functions for a Four-plate Column

Basis: liquid feed at the boiling point.

$$\frac{dy}{dx} = 1 \text{ for all plates}$$

Negligible flow lags

$F = 2$ $D = 1$ $B = 1$ moles/min

$V = 3$ $R = 2$ moles/min

$H_1 = H_2 = H_3 = H_4 = 1$ mole

If both the reflux rate and the feed composition are considered as variables, the four basic equations are

$$(H_1s + R + D)X_1 = VX_2 + \Delta R(x_D - x_1)$$

$$(H_2s + R + D)X_2 = VX_3 + RX_1 + \Delta R(x_1 - x_2)$$

$$(H_3s + R + F + V)X_3 = VX_4 + RX_2 + FX_F + \Delta R(x_2 - x_3)$$

$$(H_4s + B + V)X_4 = (R + F)X_3 + \Delta R(x_3 - x_4)$$

The terms $X_1, X_2, \ldots$ represent deviations from the normal values, which are x_1, $x_2, \ldots$. For the specific flows chosen, the transfer functions for changes in feed composition are

$$\frac{X_1}{X_F} = \frac{0.25s + 1}{(1.82s + 1)(0.46s + 1)(0.17s + 1)(0.097s + 1)}$$

$$\frac{X_2}{X_F} = \frac{(0.25s + 1)(0.33s + 1)}{(1.82s + 1)(0.46s + 1)(0.17s + 1)(0.097s + 1)}$$

The transfer function for changes in reflux rate are

$$\frac{X_1}{\Delta R} = \frac{0.236(0.41s + 1)(0.18s + 1)(0.095s + 1)}{(1.82s + 1)(0.46s + 1)(0.17s + 1)(0.097s + 1)}$$

$$\frac{X_2}{\Delta R} \cong \frac{0.20(0.03s + 1)}{1.82s + 1} \quad \text{based on} \quad \frac{X_1}{\Delta R} \cong \frac{0.24}{1.82s + 1}$$

The transfer functions in Example 14-3 are like those for systems with side capacities, such as the well of a thermowell and bulb system, a kettle with appreciable heat loss through the wall, or a series of pressure tanks with dead-end branches. The phase-lag curve may have maxima and minima, and the amplitude curve may have several inflection points before reaching a limiting slope. For changes in feed composition, which are one of the most important load changes, the top-plate response is somewhat slower than the response of the second plate, though both are dominated by the major time constant of 1.82 min, which is quite close to 2.0, the total holdup divided by the feed rate. For changes in reflux rate, the response of plate 1 has a maximum phase lag of 90°, and the response is fairly close to first-order, since the numerator terms almost cancel the smaller denominator terms. Plate 2 would show a similar response at low and moderate frequencies, but the phase lag approaches zero at high frequencies. Of course an actual column would have a large enough flow lag and time delay because of imperfect mixing to make the phase lag increase with frequency at high frequencies.

With one dominant concentration lag and perhaps several small flow lags (with a control point at an intermediate plate), the dominant lag should contribute almost 90° phase lag and a large amount of damping at the critical frequency, or at 180° lag. This would mean a high value of the maximum overall gain, which seems inconsistent with the fact that many columns have to be operated with a controller gain of less than 1 for stable control. The reason for low controller gain is that the process gain is often very large; i.e., a small change in reflux flow produces a large change in product composition, mainly because of material-balance considerations. For the benzene-toluene example, a decrease in reflux from 80 to 70 moles/hr means an increase in overhead product from 50 to 60, and x_D must drop from 0.98 to less than 0.83. Based on a maximum reflux flow of 100 and a concentration- (or temperature-) measuring device with a range from 80 to 100 per cent benzene, the steady-state process gain would be 75 per cent divided by 10 per cent, or 7.5.

14-16. EXPERIMENTAL FREQUENCY-RESPONSE STUDIES

One of the first reported studies was that of Endtz, Janssen, and Vermeulen (30), who measured the response of an 11-plate pilot column to sinusoidal changes in steam flow, reflux flow, and cooling-water rate. As for most other studies, not enough data are given to determine the zero-frequency gain or to predict the flow and concentration lags. For the response of the top temperature to changes in reflux rate, the largest time constant was at least 5 min, since the amplitude was still increasing with decreasing frequency at 0.03 cpm. The corresponding phase curve showed a minimum and then a maximum, all in the range of 50 to 100°, as the frequency was increased 100-fold. This is the type of behavior expected for a system with side capacities. The phase lag for the composition on plate 5 increased rapidly with frequency and exceeded 450°. A control system based on plate 5 sampling would have given a period at least 20 times that for plate 1 sampling. Surprisingly, the response to steam-flow changes 2 plates from the reboiler showed greater phase lag than the response to reflux changes 5 plates from the top. Perhaps the column was operated so that an increase in vapor flow meant an increase in reflux flow; this would introduce several liquid-flow lags into the system. Columns that are operating on boilup control normally have a fixed reflux flow, and an increase in vapor flow means an increase in overhead product rate.

The data reported by Aikman (25) for a 20-plate refinery column show the amplitude and phase of temperature fluctuations in the overhead vapor line for changes in reflux flow. The results were interpreted as equivalent to a dead time from the delay in the vapor line, a time constant of 0.083 min for the temperature bulb, and a time constant of 0.83 min corresponding to the holdup time on the top plate. However, the lower plates must have some effect on the frequency response, and a major time constant of 4 to 5 min would be expected, based on the total holdup and the feed rate. The zero-frequency gain used to normalize the amplitudes was only estimated by Aikman, and study of the data indicates that the reported gain and largest time constant may be low by a factor of 2 to 10. In an almost identical study by Boyd (6), a major time constant of 0.55 min was reported, but again the zero-frequency gain was not firmly established. In both these studies, the temperature fluctuations were very small and may have been influenced by pressure fluctuations.

The frequency response of a 30-plate depropanizer to changes in steam flow was investigated by Powell (31), and calculated values of zero-frequency gain were used to normalize the results. The response of the temperature 21 plates from the reboiler indicated effective time lags of 7.6 and 0.45 min and a time delay of 5 sec. The large time constant represents the total column capacity, and the small delay shows that increases in vapor flow pass quickly through the column. A more detailed analysis is not worthwhile without data on pressure fluctuations and plate characteristics.

Much work has been done at the Royal Dutch/Shell Laboratories, using both digital and analog computers to predict column performance (9, 32, 33). Frequency-response tests on an analog of a 32-plate column showed the amplitude ratio for plate concentration to be nearly the same at several points in the column for changes in either reflux rate or vapor rate (32). The slope of −1 indicates a large dominant concentration lag. When corrected for lags in flow rate, the phase curves were about the same and showed shallow maxima and minima of the type observed on actual columns (30). The response of the column to changes in position of the top vapor-line valve was also studied (9), and interactions between the pressure and temperature control systems were investigated. Good agreement with actual column tests was reported, though some of the parameters in the model had to be adjusted to get good results.

14-17. PREDICTING THE BEHAVIOR OF CONTROL SYSTEMS

The success of a feedback control system in compensating for load changes depends both on the elements in the closed loop and on the transfer function for load changes. Because the fluid-flow lags depend on the number of plates between the sample point and the flow change, the critical frequency of the system is decreased by putting the sample point closer to the end of the column. Sampling at the top plate for reflux control or the bottom plate for boilup control also permits a higher overall gain to be used, since the major concentration lag is about the same for any control point. The advantage of sampling at the end of the column would be less pronounced for boilup control than for reflux control, since the vapor-flow lags are small and may be negligible compared with other lags in the control loop (thermometer, valve, reboiler, etc.). A possible advantage of sampling near the feed plate is that changes in feed composition are detected sooner and corrective action can start before off-specification product results. The computer studies of Rose and Williams (34, 35) and of Brown (36) showed top-plate control to be much better than intermediate-plate control; however, the load change used in both studies was a simultaneous change in feed composition and feed quality, causing an instantaneous change in vapor flow to the top plate. With only a change in feed composition, the dynamic advantage of top-plate control might not be so great, but even if this were so, top-plate control would be recommended for most cases just for the steady-state characteristics.

The digital-computer study of a 15-plate column by Brown offers a chance to check a simple model for the column frequency response, which consists of $n - 1$ liquid-flow lags (n is the sample plate number) and a major time constant equal to the column capacity divided by the feed rate.

Example 14-4

Predict the critical frequency and maximum gain for control of the 15-plate column studied by Brown (36). The liquid-flow lag is 0.1 min per plate, and the sample cell has a time constant of 0.1 min. For distillate sampling, a delay of 0.5 min is introduced. Other column parameters are given in the following table, which compares the computed and predicted values of critical frequency and maximum controller gain.

Predicted and Computed Performance of Control Systems for a 15-plate Distillation Column

Basis: 15-plate column, including reboiler.

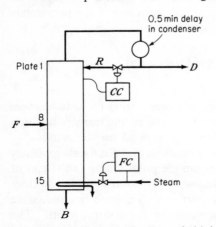

$H = $ holdup/plate $ = 2.33$ moles

$H_{15} = $ reboiler holdup $ = 23.3$ moles

$F = 3.33$ moles/min, at plate 8

$B = D = 1.67$ moles/min

$V_1 - V_8 = 8.33$ moles/min

$V_9 - V_{15} = 6.66$ moles/min

$R = 6.66$ moles/min

$q = 0.5$ (feed quality)

$\alpha = 2.0$ (relative volatility)

$T_L = 0.1$-min flow lag

$T_C = 0.1$-min sample cell lag

FIGURE E14-4   Diagram for Example 14-4.

Normal conditions: $x_F = 0.5$ $x_D = 0.97$ $x_B = 0.03$

Upset: Change in x_F to 0.65   and   change in q to 0.18

Sample point	Critical frequency, cpm		Maximum gain, moles/min, mole fraction	
	Computed	Predicted	Computed	Predicted
Plate 1	...	∞	>500	∞
Plate 2	3	1.6	400	240
Plate 4	1.2	0.7	80	60
Plate 6	0.7	0.4	60	20
Distillate	0.9	0.5	150	100

Sample Calculations, Plate 6 Sampling

$$G = \frac{K_p}{T_1 s + 1}\left(\frac{1}{T_L s + 1}\right)^5 \frac{1}{T_C s + 1}$$

$$T_1 = \frac{(14)(2.33) + (0.5)(23.3)}{3.33} = 13 \text{ min (only half of reboiler holdup included)}$$

$$T_L = T_C = 0.1 \text{ min}$$

At $\omega = 2.7$ rad/min

$$\omega T_1 = 35 \qquad \phi = 88° \qquad \text{A.R.} = \tfrac{1}{35}$$

$$\omega T_L = 0.27 \qquad \phi = 15.2° \qquad \text{A.R.} = 0.97$$

$$\phi_{total} = 88 + (6 \times 15.2) = 179°$$

$$\text{Max overall gain} = 35/(0.97)^6 = 42$$

$$K_p = \frac{\Delta x_6}{\Delta D} = 1.9 \frac{\text{mole fraction}}{\text{mole/min}} \qquad \left(\begin{array}{c}\text{trial and error by using}\\\text{McCabe-Thiele diagram}\end{array}\right)$$

$$K_{c,\max} = \frac{42}{1.9} = 22$$

For sampling on plate 1, there are only two time constants in the control loop, and the system is never unstable with just proportional control.   In a more realistic system, using a composition sampler would introduce a time delay, and the phase lag could exceed 180°.   For the other sample points, both the predicted frequencies and the predicted gains are about 0.6 times the computed values.   Apparently the side capacities act to make the phase lag for concentration changes appreciably less than that for a single lumped capacity, but without a correspondingly large change in the amplitude ratio.   Using the holdup time for one plate as the concentration lag would make the predicted frequencies closer to the computed values, but the predicted gains would then be in error by a factor of 20.   Actually the predicted results would be good enough for most purposes, particularly for just comparing the performance at different proposed sample locations.   The critical frequency and maximum overall gain are the two most important figures for characterizing the control performance, though suggested values of reset rate and derivative rate could also be obtained by using the frequency-response plot.   Derivative action might not be much help for intermediate-plate sampling, since the phase-lag curve is steep, but it has been used to advantage in control of top tower temperatures (6).

The problem of simultaneously controlling the compositions at both ends of the column is more difficult to analyze than the case of single-point control because of possible interactions between the control loops.   Suppose that the feed is introduced near the bottom of the column and that there is a sudden increase in feed concentration.   Eventually the vapor flow must be increased and the reflux flow decreased to keep the product composition constant.   The effect of a richer feed would be felt first at the bottom of the column, and the resulting increase in the flow of vapor would rapidly pass through the column.   The reflux rate would therefore increase at first, which is opposite to the final direction of change.   The dynamics of control under such circumstances can probably be predicted by using an analog computer with linearized equations for each plate and by making proper allowance for flow lags (27, 32), but to study a particular column in this way still takes much time.   Perhaps the approximate behavior can be predicted by using a model based on effective time delays and the major time constants; this would permit the response to be obtained with a simple analog computer or even by hand calculation.   A simple analog would not predict the steady-state behavior correctly, because of the nonlinearities of the system, but this may not be important, since the steady-state concentrations can be predicted by routine methods.   To include the effects of a curved equilibrium line, inefficient plates, and perhaps several components in models for dynamic studies makes the models very complex.   The recent paper by Rosenbrock (37) shows that interactions between control loops are predicted by using a relatively simple analog.   Lupfer and Parsons also used a simple analog controller to anticipate the effects of changes in feed rate and feed enthalpy (38).   It is hoped that more experimental results for actual columns will be made available to test the reliability of such analogs or models.

PROBLEMS

1. The following control scheme is proposed for a 36-stage isopentane superfractionator (39): The feed, containing 10 to 16 per cent i-C_5 is flow-controlled and fed to theoretical stage 14 (counting from the bottom).   A temperature controller

at stage 26 adjusts a valve in the steam line to the reboiler.   The reflux and bottoms flows are adjusted by level controllers at the reflux drum and reboiler.   The overhead product rate is fixed by a flow controller.

Criticize this scheme, and suggest alternative arrangements.

2. Draw a block diagram for controlling the column pressure by regulating the flow of water to the condenser.   Derive the transfer function relating condenser pressure and temperature of the condenser tubes.

3. A 100-plate distillation column has a feed rate $F = 100$ gpm.   The normal product and reflux rates are $D = 60$, $B = 40$, $R = 200$ gpm.   The liquid holdup is 250 gal per plate.   The time constant for each plate for liquid flow is about one-third the holdup time, and the vapor lag for one plate is estimated to be 3 sec. A two-point control scheme is considered, with the temperature 10 plates up used to adjust the steam flow and the temperature 80 plates up to adjust the distillate flow. Draw a block diagram, and fill in the approximate transfer functions, using equivalent time delays and no more than two first-order lags for each block.

4. The frequency response of a 20-plate distillation column was reported by Aikman (25).   The reflux flow was varied sinusoidally, and the temperature of the overhead vapor was measured with a resistance thermometer in a well ($T = 0.1$ min).   The time delay in the vapor line was 3 sec.   In a tentative control scheme for

Frequency, cpm	Input amplitude, % mean flow (peak–peak)	Output amplitude, °F (peak–peak)	Phase lag, deg
0.244	39	3.60	68
0.416	39	2.80	95
0.565	39	1.60	115
0.86	37	0.80	131
1.5	39	0.40	162
1.85	37	0.364	180
2.86	39	0.20	205
3.86	40	0.120	242
6.5	40	0.053	355
7.5	39	0.030	407
12.9	37	0.0294	530

a similar column, a sample stream from the top plate is pumped through a refractometer.   The measurement of refractive index is practically instantaneous, but there is a 30-sec time delay in the sample line.   How much improvement in controllability might be expected if the sampling delay were cut to 10 sec?

5. An ideal mixture containing 60 per cent A and 40 per cent B is fractionated at atmospheric pressure to give an overhead product with 95 per cent A and a bottoms product with 5 per cent A.   The relative volatility is 2.0.   For a reflux ratio 1.5 times the minimum value, what is the process gain expressed as per cent change in overhead concentration of A divided by per cent change in reflux flow? What would be a typical corresponding value of the gain in degrees Fahrenheit divided by fractional change in reflux flow?

6. The following data were reported for the response of a distillation column to a step change in reflux.

Time	% incomplete
0	100
3	100
4	90
4.5	83
5	71
5.5	57
6	37
6.5	20
7	14
7.5	7

a. Estimate the effective time constants and time delays, and compare the actual transient response with that calculated for the effective lags.   Also, compare your values with the effective lags reported by Woods (40).

b. Note that the effective time delay is apparently larger than the major time constant, whereas theory predicts that it should be considerably smaller.   Study the original article, and suggest a reason for this unusual behavior.

7. The frequency response of a depropanizer column is reported by Powell (31).

a. For this column, how much could control be improved by using derivative action in the controller?

b. How much could control action be improved by using derivative action in the temperature transmitter?

8. The pressure at the top of a 6-ft distillation column is controlled at 2.0 atm by throttling the flow of inerts from the 1,000-gal reflux drum, which is normally two-thirds full.   The superficial velocity in the column is 1.5 fps, and the vapor to the condenser contains 0.4 per cent inert gas.   Draw a block diagram for the control system, and calculate the time constant for pressure changes in the reflux drum, assuming 50 per cent inert gas in the vent stream.   What would be the next largest lag in the control loop?

9. If a vent condenser is added to a reflux drum to change the vent-gas composition from 20 per cent inerts to 90 per cent inerts, by what factor would the time constant for pressure changes be altered?

10. The relative volatility of A in an A-B mixture is 4.3 $\{y_A^* = 4.3x_A/[1 + (4.3 - 1)x_A]\}$.   Derive the transfer function relating x_A and x_F for a one-stage distillation unit which has a liquid holdup of 50 gal, a feed rate of 20 gpm, and a liquid-product composition $x_A = 0.2$.

a. What is the transfer function if the stage efficiency is only 80 per cent?

b. Derive the transfer function for x_B, using the relative volatility of B, and compare with that for x_A.

11. The following scheme is used to control a superfractionator that separates isobutane from *n*-butane (41): A chromatograph receives a vapor sample from a plate between the feed plate and the bottom plate and measures the i-C_4 content of the sample.   The signal from the composition controller is used to adjust the valve in the distillate line, which really regulates the column pressure, since the condenser is partly flooded.   The reflux flow is flow-controlled, and the bottoms

flow is controlled by the reboiler level.   The temperature seven plates below the control point is used to adjust the heat input to the reboiler.

Give the sequence of events following:

a. An increase in feed rate to the column

b. An increase in per cent i-C_4 in the feed

Will this system perform better for changes in feed composition or changes in feed rate?

REFERENCES

1. Wherry, T. C., and D. E. Berger: What's New in Fractionator Control, *Petroleum Refiner*, **37**(5):219 (May, 1958).
2. Robinson, C. S., and E. R. Gilliland: "Elements of Fractional Distillation," 4th ed., pp. 472–478, McGraw-Hill Book Company, New York, 1950.
3. Armstrong, W. D., and R. M. Wood: An Introduction to the Theoretical Evaluation of the Frequency Response of a Distillation Column to a Change in Reflux Flow Rate, *Trans. Inst. Chem. Engrs.*, **39**:80 (1961) (see also discussion by Rosenbrock and Wood, pp. 86–89).
4. Hengstebeck, R. J.: "Distillation Principles and Design Procedures," p. 281, Reinhold Publishing Corporation, New York, 1961.
5. Webber, W. O.: Control by Temperature Difference, *Petroleum Refiner*, **38**(5):187 (May, 1959).
6. Boyd, D. M., Jr.: The Effects of Hysteresis on Derivative Control, *ISA J.*, **4**:136 (1957).
7. Tivy, V. V. St. L.: Automatic Control of Fractionating Columns, *Petroleum Refiner*, **27**(11):123 (November, 1948).
8. Anisimov, I. R.: "Automatic Control of Rectification Processes," Consultants Bureau, Inc., New York, 1959.
9. Rademaker, O., and J. E. Rijnsdorp: Dynamics and Control of Continuous Distillation Columns, "Proceedings of Fifth World Petroleum Congress," sec. 7, p. 59, Butterworth Scientific Publications, London, 1959.
10. Berger, D. E., and G. R. Short: Sampling and Control Characteristics of Analysis-controlled Pentane Fractionator, *Ind. Eng. Chem.*, **48**:1027 (1956).
11. Burk, M. C., and F. W. Karasek: Data Converter Adapts Chromatograph to Process Control, *ISA J.*, **5**(10):28 (October, 1958).
12. Williams, T. J., and R. J. Harnett: Automatic Control in Continuous Distillation, *Chem. Eng. Progr.*, **53**:220 (1957).
13. Tou, J. T.: "Digital and Sampled-data Control Systems," pp. 80–137, McGraw-Hill Book Company, New York, 1959.
14. Fourroux, M. M., F. W. Karasek, and R. E. Wightman: High Speed Chromatography in Closed-loop Fractionator Control, *ISA J.*, **7**(5):76 (May, 1960).
15. Boyd, D. M., Jr.: Fractionation Instrumentation and Control, *Petroleum Refiner*, **27**(10, 11):533, 594 (October, November, 1948).
16. Hollander, L.: Pressure Control of Light-ends Fractionators, *ISA J.*, **4**:185 (1957).
17. Reidel, T. C.: How to Instrument the Fractionation Section in Field Processing Plants, *Oil and Gas J.*, **52**(51):136 (Apr. 26, 1954).

18. Lupfer, D. E., and D. E. Berger: Computer Control of Distillation Reflux, *ISA J.*, **6**(6):34 (June, 1959).
19. Robinson, C. S., and E. R. Gilliland: "Elements of Fractional Distillation," 4th ed., p. 415, McGraw-Hill Book Company, New York, 1950.
20. Williams, B., J. W. Begley, and C. Wu: "Tray Efficiencies in Distillation Columns," final report of American Institute of Chemical Engineers Research Committee, p. 13, New York, 1960.
21. Wood, R. M., and W. D. Armstrong: The Transient Response of a Distillation Column to Changes in Feed Composition, *Chem. Eng. Sci.*, **12**:272 (1960).
22. Armstrong, W. D., and R. M. Wood: The Dynamic Response of a Distillation Column to Changes in the Reflux and Vapor Flow Rates, *Trans. Inst. Chem. Engrs.*, **39**:66 (1961).
23. Berger, D. E., and G. G. Campbell: Experience in Controlling a Large Separations Column with the Continuous Infrared Analyzer, *Chem. Eng. Progr.*, **51**:348 (1955).
24. Rademaker, O.: Dynamic Measurements on a Tall Turbogrid Tray Column, "International Symposium on Distillation," p. 190, Institute of Chemical Engineers, London, 1960.
25. Aikman, A. R.: Frequency Response Analysis of a Fractionating Column, *Proc. Inst. Soc. Am.*, **11**(1) (1956).
26. Gilliland, E. R., and C. M. Mohr: Transient Behaviour in Plate-Tower Distillation of a Binary Mixture, *Chem. Eng. Progr.*, **58**(9):59 (1962).
27. Lamb, D. E., R. L. Pigford, and D. W. T. Rippins: Dynamic Characteristics and Analog Simulation of Distillation Columns, *Chem. Eng. Progr. Symposium Ser. No.* 36, **57**:132 (1961).
28. Baber, M. F., L. L. Edwards, Jr., W. T. Harper, Jr., M. D. Witte, and J. A. Gerster: Experimental Transient Response of a Pilot Plant Distillation Column, *Chem. Eng. Progr. Symposium Ser. No.* 36, **57**:148 (1961).
29. Archer, D. H., and R. R. Rothfus: The Dynamics and Control of Distillation Units and Other Mass Transfer Equipment, *Chem. Eng. Progr. Symposium Ser. No.* 36, **57**:2 (1961).
30. Endtz, J., J. M. L. Janssen, and J. C. Vermeulen: Measuring Dynamic Responses of Plant Units, in "Plant and Process Dynamic Characteristics," Butterworth Scientific Publications, London, 1957.
31. Powell, B. E.: How to Design Control Loops from Frequency Response Data, *ISA J.*, **5**(4):32 (April, 1958).
32. Rijnsdorp, J. E., and A. Maarleveld: Use of Electrical Analogues in the Study of the Dynamic Behaviour and Control of Distillation Columns, in "Instrumentation and Computation in Process Development and Plant Design," p. 135, Institute of Chemical Engineers, London, 1959.
33. Voetter, H.: Response of Concentrations in a Distillation Column to Disturbances in the Feed Composition, in "Plant and Process Dynamic Characteristics," Butterworth Scientific Publications, London, 1957.
34. Rose, A., and T. J. Williams: Automatic Control in Continuous Distillation, *Ind. Eng. Chem.*, **47**:2284 (1955).
35. Williams, T. J., R. T. Harnett, and A. Rose: Automatic Control in Continuous Distillation, *Ind. Eng. Chem.*, **48**:1008 (1956).

36. Brown, G. R., Jr.: Digital Computer Simulation of Distillation Tower Control, Ph.D. thesis, Case Institute of Technology, 1958.
37. Rosenbrock, H. H.: The Control of Distillation Columns, *Trans. Inst. Chem. Engrs. (London)*, **40**:35 (1962).
38. Lupfer, D. E., and J. R. Parsons: A Predictive Control System for Distillation Columns, *Chem. Eng. Progr.*, **58**(9):37 (1962).
39. Keating, J. M., and D. S. Townend: Superfractionator Controllability Data Obtained by the Use of a Digital Computer, in "Instrumentation and Computation in Process Development and Plant Design," p. 29, Institute of Chemical Engineers, London, 1959.
40. Woods, F. A.: Improving Fractionator Performance with Dynamic Analysis, *Control Eng.*, **5**(5):91 (May, 1958).
41. Tyler, C. M.: Process Analyzers for Control, *Chem. Eng. Progr.*, **58**(9):51 (1962).

15 THE STABILITY AND CONTROL OF CHEMICAL REACTORS

Chemical reactors are often the most difficult units to control in a chemical plant, particularly if the reactions are rapid and exothermic. An increase in temperature of 1°C may increase the reaction rate by 10 per cent, enough to cause significant changes in conversion and perhaps in yield. Furthermore, the increase in rate with increasing temperature tends to make the reactor unstable; if a 1° rise in temperature leads to a 10 per cent increase in rate of heat generation, the temperature will rise still further unless the heat-removal rate is quickly increased to restore the normal temperature.

The degree of self-regulation of a reactor depends on the changes in heat generation and heat removal with temperature. The reactor is quite stable if a 10 per cent increase in heat-generation rate is accompanied by a 100 per cent increase in heat-removal rate. If the heat-generation rate increases more rapidly than the heat-removal rate, the reactor is unstable by itself but can sometimes be automatically controlled. Commercial reactors are often operated at nearly unstable conditions, either for economic reasons or because the dynamics were overlooked, and there have been many cases of runaway reactions leading to plugged kettles, melted reactor tubes, or explosions.

The first step in the design of a reactor control system is to study the temperature stability. The basic features of the analysis are the same for all exothermic reactions, whether carried out continuously in a stirred tank, pipeline, or packed bed or carried out in a batch kettle. In some cases, particularly with packed-bed reactors, the reactor is designed to have inherent temperature stability without a feedback control system. For most stirred-tank reactors, an automatic control system is used to obtain faster response or to permit operation at an unstable point. Often an emergency control system is added to stop the reaction by dumping the charge or killing the catalyst in case the main control fails to halt a runaway temperature.

The general approach taken here is first to examine the temperature stability of reactors without any control system. The aim is to give a relatively simple and general method of predicting the approximate stability limits rather than all the complex equations or computer solutions which describe the exact behavior for various types of kinetics. The effect of

308

adding proportional control to a stirred-tank reactor is then considered, to learn how a reactor can sometimes be controlled in the unstable region. The emphasis is on reactors which are supposed to operate almost iso-thermally, but a later section shows that stability problems can arise in adiabatic reactors as well.

15-1. STIRRED-TANK REACTORS

Consider a continuous stirred reactor with a constant feed rate, feed con-centration, and holdup time, and a first-order, irreversible exothermic reaction. Some of the heat of reaction is removed as sensible heat of the product stream, and the rest is transferred to cooling water in the jacket. For a preliminary analysis the possible stable and unstable states are found by a graphical analysis of the type used by Heerden (1). The heat generated by chemical reaction and the heat removed by the jacket and the product stream are plotted against the reactor temperature, as shown in Fig. 15-1. The heat-generated graph shows the heat released for the steady-state conversion that corresponds to each reactor temperature. The heat gen-erated is calculated as follows:

$$Q = kVc_0(1 - x)(-\Delta H) = Fc_0(-\Delta H)x$$

$$\frac{x}{1 - x} = \frac{kV}{F} \tag{15-1}$$

$$k = ae^{-E/RT}$$

where Q = heat generated, Btu/hr
    c_0 = feed concentration
    x = fraction converted
    ΔH = heat of reaction
    k = first-order rate constant
    E = activation energy
    T = absolute temperature
    V = volume
    F = feed rate

    The slope of the graph increases at first because of the exponential effect of temperature on reaction rate. The slope eventually decreases because of the decreased reactant concentration, and the graph becomes asymptotic to the heat released at complete conversion.

    The heat-removed graphs are linear if the coefficient, jacket temperature, and specific heat of the product are constant.

$$Q_{\text{out}} = UA(T - T_j) + F\rho c_p(T - T_F) \tag{15-2}$$

Three cases are shown in Fig. 15-1 for one reaction and different com-binations of jacket temperature and overall coefficient. For case 1, there

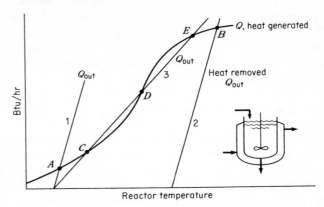

FIGURE 15-1   Possible operating states for a stirred-tank reactor.

is only one intersection, and point A gives the reactor temperature. This is a stable operating point, since a slight increase in reactor temperature leads to a much greater increase in heat transferred than heat generated, which tends to return the temperature to normal. For case 2, the heat-transfer coefficient is the same; but the jacket temperature is higher, and the only intersection is at B, a stable operating point at almost complete conversion. For case 3, that are three intersections, and D is clearly an unstable point, even though the heat-balance equation is still satisfied. A slight increase in temperature increases the rate of heat generation more than the rate of heat removal, which results in a rapid rise in temperature to point E. A decrease in temperature from D sends the temperature down to C.

The lines in Fig. 15-1 are the steady-state rates of heat generation and removal, and a necessary but not sufficient condition for stability is that the slope of the heat-removal line be greater than the slope of the heat-generation graph. A complete stability analysis must consider the dynamics of the reactor, in particular the rate of change of concentration with temperature. If the reactor is operating at point C, where the conversion is about 30 per cent, and the temperature is rapidly increased by 1°C, the increase in heat-generation rate will be more than that given by the steady-state graph, since the reactant concentration will not have dropped to the steady-state value at the new temperature. The maximum increase in heat-generation rate occurs if the reactant concentration is constant.

$$\left(\frac{\partial Q}{\partial T}\right)_x = \left(\frac{dk}{dT}\right) Vc_0(1-x)(-\Delta H) = a\frac{E}{RT^2}e^{-E/RT}Vc_0(1-x)(-\Delta H)$$

$$= \frac{kVc_0(1-x)(-\Delta H)}{RT^2/E} = \frac{\bar{Q}}{RT^2/E} \qquad \bar{Q} = Q_{av} \qquad (15\text{-}3)$$

For a first-order reaction, it can be shown that the slope of the steady-state graph and the maximum rate of change are related by the fraction of reactant unconverted.

$$\left(\frac{\partial Q}{\partial T}\right)_x = \left(\frac{dQ}{dT}\right)_{ss}\left(\frac{1}{1-x}\right) \tag{15-4}$$

Thus, for a normal conversion of 50 per cent, a sudden increase in temperature can increase Q up to twice as much as predicted from steady-state analysis. In practice, there will be some drop in concentration during a rise in temperature, and the change in Q will be between the value given by Eq. (15-3) and the value from steady-state analysis, the exact value depending on the order of the reaction and the nature of the upset. However, the reaction will be stable to small disturbances, no matter what the reaction order, if the rate of change of heat removal is made greater than the maximum rate of change of heat generation.

For stability $\qquad UA + F\rho c_p > \dfrac{\bar{Q}}{RT^2/E} = \left(\dfrac{\partial Q}{\partial T}\right)_x \tag{15-5}$

If most of the heat is removed through the jacket, the temperature difference must not exceed RT^2/E, which is called the "critical temperature difference."

$$\Delta T_c = \frac{RT^2}{E} \tag{15-6}$$

$$T - T_j < \Delta T_c \qquad \text{for stability, if } UA \gg F\rho c_p \tag{15-7}$$

The significance of the critical temperature difference for packed reactors was pointed out by Wilson (2), and the concept can be applied to stirred reactors, pipeline reactors, batch reactors, storage tanks, and single catalyst particles in a packed bed (3, 4, 5, 6). If the temperature rise above the surroundings is greater than RT^2/E, a runaway reaction is possible. For continuous-flow reactors, instability is most likely to occur when the conversion is low or moderate, say, less than 50 per cent. If the conversion is over 90 per cent, there is little danger of a runaway temperature, because not much more energy would be released at 100 per cent conversion.

The exact stability analysis for a stirred reactor and a first-order reaction is given in several references (7, 8, 9, 11). By linearizing the material-balance and heat-balance equations and finding the conditions for roots with negative real parts (stable system), two criteria are obtained.

$$1 + \alpha + 1 + \beta > \gamma \tag{15-8}$$

$$(1 + \alpha)(1 + \beta) > \gamma \tag{15-9}$$

where $\qquad \alpha = \dfrac{kV}{F} \qquad \beta = \dfrac{UA}{F\rho c_p} \qquad \gamma = \dfrac{\bar{Q}}{F\rho c_p}\dfrac{E}{RT^2}$

The criteria may be easier to understand in the following form: Equation (15-8) becomes

$$UA + F\rho c_p\left(2 + \frac{kV}{F}\right) > \left(\frac{\partial Q}{\partial T}\right)_x = \frac{\bar{Q}}{\Delta T_c}$$

$$(15\text{-}10)$$

or

$$UA + F_\rho c_p \frac{2 - x}{1 - x} > \left(\frac{\partial Q}{\partial T}\right)_x$$

Equation (15-9) becomes

$$UA + F\rho c_p > \left(\frac{\partial Q}{\partial T}\right)_x(1 - x) = \left(\frac{dQ}{dT}\right)_{ss} \qquad (15\text{-}11)$$

Equation (15-11) is just the steady-state criterion that the heat removed must increase more rapidly with temperature than the heat generated.

Equations (15-10) show that $UA + F\rho c_p$ may be less than $(\partial Q/\partial T)_x$ but that the general criterion given in Eq. (15-5) is only slightly conservative if $UA \gg F\rho c_p$. The difference between Eq. (15-5) and the exact criterion would be somewhat greater for a second-order reaction but would vanish as the reaction order approached zero.

The physical significance of Eqs. (15-10) is that the reactor can be stable when $(\partial Q/\partial T)_x > UA + F\rho c_p > (dQ/dT)_{ss}$ if the concentration changes relatively more rapidly than the temperature. Considering only temperature changes, the time constant of the reactor is

$$T_1 = \frac{V\rho c_p}{F\rho c_p + UA - (\partial Q/\partial T)_x}$$

The value is negative if $F\rho c_p + UA < (\partial Q/\partial T)_x$, but it is still a measure of the initial rate of change of temperature. The time constant for concentration changes at constant temperature is

$$T_2 = \frac{V}{F + kV}$$

Equating T_2 and $-T_1$ leads to Eqs. (15-10).

If a reactor is unstable according to Eqs. (15-10) yet the steady-state criterion [Eq. (15-11)] is satisfied, a limit cycle may develop in which the temperature and concentration exhibit constant amplitude but nonsinusoidal fluctuations. The amplitudes of the limit cycles are not easily predicted, since the nonlinearity of the chemical-reaction equations must be taken into account. Some examples of limit cycles are given by Aris and Amundson (8).

15-2. CONTROL OF STIRRED REACTORS

With an automatic control system, a reactor may be operated at what would otherwise be an unstable point. The effect of the control system is to make the heat-removal graph steeper than the heat-generated graph, either by decreasing the jacket temperature or by increasing the coolant-flow rate as the temperature rises.   Controlling the jacket temperature (for example, by changing the pressure of a boiling coolant) is generally a more effective scheme and is also easier to analyze.

In the following analysis a proportional controller is used, with control-system gain K_c' defined as the steady-state change in jacket temperature for a unit change in error (set-point–reactor temperature).   The gain K_c' includes the bulb, controller, valve, and jacket, and the definition is the same whether one controller or two (cascade system) are used.   The overall gain of the control loop is the control-system gain times the gain of the reactor.   The change in heat generation with temperature is taken as $(\partial Q/\partial T)_x$, to be somewhat conservative and to simplify the analysis.   (The change in Q could actually be greater than this for "large" temperature changes of 3 to 5°C.)

The heat balance for the reactor shown in Fig. 15-2 is written with θ, θ_F, and θ_j denoting changes from the normal values.   The symbol θ is used for temperature and T for time constant in the following equations:

$$V\rho c_p \left(\frac{d\theta}{dt}\right) = F\rho c_p(\theta_F - \theta) - UA(\theta - \theta_j) + \left(\frac{\partial Q}{\partial \theta}\right)_x \theta \qquad (15\text{-}12)$$

The transfer function for the reactor is

$$\left[\frac{V\rho c_p s}{UA + F\rho c_p - (\partial Q/\partial \theta)_x} + 1\right]\theta$$

$$= \frac{F\rho c_p}{UA + F\rho c_p - (\partial Q/\partial \theta)_x}\theta_F + \frac{UA}{UA + F\rho c_p - (\partial Q/\partial x)_x}\theta_j$$

or $\qquad\qquad (Ts + 1)\theta = K_F\theta_F + K_p\theta_j$ $\qquad\qquad\qquad (15\text{-}13)$

Note that T, K_F, and K_p are negative at the unstable point, since $UA + F\rho c_p$ is less than $(\partial Q/\partial \theta)_x$.   If the lags in the bulb, controller, valve, and jacket are negligible, the overall open-loop transfer function is

$$G = \frac{K_c' K_p}{Ts + 1}$$

For the closed loop

$$\frac{\theta}{\theta_c} = \frac{G}{1 + G} = \frac{K_c' K_p}{1 + K_c' K_p} \frac{1}{\dfrac{Ts}{1 + K_c' K_p} + 1} = \frac{K}{1 + K} \frac{1}{\dfrac{Ts}{1 + K} + 1} \qquad (15\text{-}14)$$

The overall gain K is negative, but if the *absolute value of K is greater* than 1, the terms $K/(1 + K)$ and $T/(1 + K)$ are positive and the system has

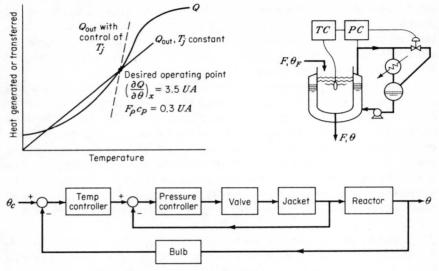

FIGURE 15-2   Reactor control in the unstable region.

the stable behavior of a conventional first-order system.   For the example in Fig. 15-2,

$$K_p = \frac{UA}{UA + 0.3UA - 3.5UA} = -0.455$$

Therefore, K_c' must be *greater than* 2.2 to get a stable system.

$$|K| > 1 \qquad \text{for operation at an unstable point}$$

$$K_c' > \frac{-1}{K_p} \tag{15-15}$$

The speed of response of the control system depends on the effective time constant $T/(1 + K)$.   If K_c' is only slightly larger than the minimum value, the reactor will be very slow to respond to set-point changes or to recover from load changes.   Using a large gain gives a rapid response.   Of course, there is still a maximum permissible controller gain, which depends on the amplitude ratio at $180°$ phase lag.   However, when the reactor has a fairly large time constant and the other lags are small, the maximum gain will be much higher than the minimum needed to permit operation in the unstable region.

Other control schemes have been considered to permit operation in the unstable region.   Theory shows that using proportional control of concentration, with the jacket temperature or flow rate as the controlled variable, does not work no matter what gain is used (8, 12); the reaction temperature gets out of control before there is an appreciable change in concentration.   Integral control of reactor temperature is also too sluggish

(8), though of course some integral action can be added to proportional control. A certain amount of derivative action used either in the temperature transmitter or in the controller would improve the speed of response and the stability of the system.

If a cooling coil or jacket is used, the reactor can be controlled by changing the flow of coolant, which changes both the overall coefficient and the driving force for heat transfer. Transfer functions for flow-rate changes for both lumped-parameter and distributed-parameter models are given in a thesis by Weber (11). The equations were checked by control studies with an 80-gal reactor, where a zero-order exothermic reaction was simulated by automatically injecting steam at a rate dependent on the temperature. The reactor proved fairly easy to control in the unstable region because the tank time constant was much larger than the other lags and delays. However, with a large measuring lag (bulb in a dry well), control was much poorer, and the difference between the maximum gain and the minimum gain was much smaller.

The reactor can be controlled by adjusting the feed temperature, and the fast response with split stream control (see Fig. 11-5) may make this method preferable to controlling the flow of cooling fluid to a jacket or coil. Most of the heat would still be removed by the jacket, and the small fluctuations in heat-generation rate would be compensated for by relatively large changes in feed temperature. To permit operation in the unstable region, the overall gain must be greater than 1.0, according to Eq. (15-15). Since the manipulated variable is feed temperature, the process gain K_p is small and K_c' must be large. In Eq. (15-13), K_F becomes K_p, and θ_j becomes a load variable.

$$K_p = \frac{Fpc_p}{UA + Fpc_p - (\partial Q/\partial \theta)_x}$$

For the example in Fig. 15-2,

$$K_p = \frac{0.3UA}{UA + 0.3UA - 3.5UA} = -0.136$$

Therefore,

$$K_c' \geq \frac{-1}{-0.136} = 7.35 = \frac{\text{change in feed temperature}}{-(\text{change in reactor temperature})}$$

15-3. EFFECT OF OTHER LAGS IN THE CONTROL LOOP

In the analysis so far, we have considered only the lag in the kettle and have neglected the dynamics of the jacket, temperature bulb, and control valve. The polar Nyquist diagram† is a good way to demonstrate the effects of other lags and controller gain on reactor stability. Consider a reactor with

† Appendix 1.

the transfer function $G = \theta/\theta_j = -0.455/(-13s + 1)$. The diagram for the reactor alone is obtained by substituting $j\omega$ for s in the transfer function.

$$G = \frac{K}{Ts + 1} \qquad \text{where } K = -0.455 \text{ and } T = -13$$

$$= \frac{(-1)K(Tj\omega - 1)}{(-1)(Tj\omega + 1)(Tj\omega - 1)} = \frac{K}{T^2\omega^2 + 1} - \frac{j\omega KT}{T^2\omega^2 + 1}$$

When $\omega = 0$, $\qquad G = K = -0.455 = 0.455\underline{/-180°}$

When $\omega = -1/T$, $\quad G = \dfrac{-0.455}{2} - \dfrac{j(-1)(-0.455)}{2} = \dfrac{0.455}{\sqrt{2}}\underline{/-135°}$

The polar plot thus starts at 180° lag and goes toward 90° lag, a mirror image of the plot for an ordinary first-order lag. The system is not stable with a

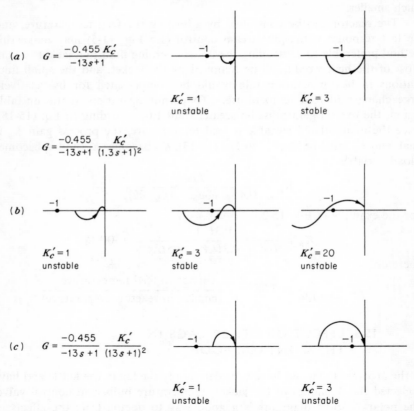

(a) $\quad G = \dfrac{-0.455\, K_c'}{-13s + 1}$

$K_c' = 1$ unstable $\qquad K_c' = 3$ stable

$G = \dfrac{-0.455}{-13s + 1}\dfrac{K_c'}{(1.3s + 1)^2}$

(b)

$K_c' = 1$ unstable $\qquad K_c' = 3$ stable $\qquad K_c' = 20$ unstable

(c) $\quad G = \dfrac{-0.455}{-13s + 1}\dfrac{K_c'}{(13s + 1)^2}$

$K_c' = 1$ unstable $\qquad K_c' = 3$ unstable

FIGURE 15-3   Nyquist diagrams for reactor systems (not drawn to scale). (a) Reactor alone; (b) reactor plus two small time constants; (c) reactor plus two large time constants.

low gain, since there is one positive denominator root ($P = 1$) and no counterclockwise encirclement of the -1 point ($N = 0, Z = P - N = 1$) (Fig. 15-3a).

Adding two small first-order lags to the system increases the maximum phase lag to 270°. The minimum controller gain is the same as before (that required for an overall steady-state gain of 1.0), but there is also a *maximum* controller gain which gives an overall gain of 1.0 at the higher frequency corresponding to 180° phase lag (Fig. 15-3b).

If two additional large lags are included, the phase lag is always greater than 180° and an overall gain greater than 1 leads to a clockwise encirclement of the -1 point (Fig. 15-3c). There is no controller gain which will make the system stable. The other lags would have to be reduced, perhaps by cascade control, or the time constant of the reactor increased to permit stable operation.

Nonlinear effects are expected with reactor control because of the rapid change in reaction rate with temperature. If the gain is adjusted to make the reactor stable by a 40 per cent margin at 100°C, an excursion to 105°C, where the reaction rate is, say, 1.5-fold higher, would bring the reactor to an unstable region and the temperature would rapidly rise to the upper stable point. Possible upsets to the system should be considered at the design stage, and the reactor should be made stable at the highest temperature that might be reached after a disturbance.

Example 15-1

The design conditions for a continuous stirred-tank reactor are given below. Would the reactor be stable with a constant jacket temperature? Would cooling the feed to 20°F to permit a higher jacket temperature improve the stability?

$$\text{Feed} = 2{,}000 \text{ lb/hr at } 70°F, \text{ containing 50 per cent } A$$

$$c_p = 0.75 \text{ Btu/(lb)(°F)}$$

$$W = 2{,}400 \text{ lb holdup} \qquad \frac{W}{F\rho} = 1.2 \text{ hr}$$

$$A = \text{jacket area} = 50 \text{ ft}^2$$

$$U = \text{overall coefficient} = 150 \text{ Btu/(hr)(ft}^2)(°F)$$

$$\Delta H = \text{heat of reaction} = -600 \text{ Btu/lb (exothermic)}$$

Kinetics: First-order reaction, $k = 1.3 \text{ hr}^{-1}$ at 150°F

$$E = 30{,}000 \text{ Btu/lb mole}$$

Desired reaction conditions:

$$50\% \text{ conversion at } 130\text{--}150°F$$

The temperature needed for 50 per cent conversion could be calculated directly and the stability checked by using Eqs. (15-10) and (15-11). However, to obtain a better grasp of the situation, conversions and heat-generation rates at several

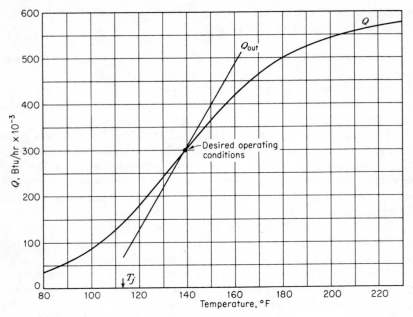

FIGURE 15-4   Stability analysis for Example 15-1.

temperatures should be calculated.   A sample calculation is given below, and the values of Q are plotted in Fig. 15-4.

At 140°F,

$$\ln \frac{k}{1.3} = \frac{30,000}{R} \frac{-10}{(600)(610)} \qquad k = 0.861$$

$$\frac{x}{1-x} = \frac{kV}{F} = (0.861)(1.2) = 1.03$$

$$x = 0.508 \qquad Q = 1{,}000\ \text{lb/hr} \times 600 \times 0.508 = 305{,}000\ \text{Btu/hr}$$

From Fig. 15-4,

$$x = 0.50 \qquad \text{at } 139°\text{F}$$

$$Q = 300{,}000$$

$$\text{Sensible heat out} = 2{,}000\,(0.75)(139 - 70) = 104{,}000\ \text{Btu/hr}$$

$$Q_j = 196{,}000\ \text{Btu/hr}$$

$$(T - T_j)\ \text{needed} = \frac{196{,}000}{(50)(150)} = 26.1°\text{F}$$

$$T_j = 139 - 26 = 113°\text{F}$$

As shown by Fig. 15-4, the heat-removed graph is steeper than the steady-state graph of heat generated at 139°F; so criterion (15-11) is satisfied.   However, the slight difference in slopes should be a warning that Eqs. (15-10) may not be satisfied.

$$\Delta T_c = \frac{RT^2}{E} = \frac{(1.987)(599)^2}{30,000} = 23.7°F$$

$$\left(\frac{\partial Q}{\partial T}\right)_x = \frac{300,000}{23.7} = 12,700 \qquad \text{Eq. (15-3)}$$

$$UA + Fpc_p\left(2 + \frac{kV}{F}\right) = 7,500 + (1,500)(3) = 12,000$$

Since $UA + Fpc_p(2 + kV/F) < (\partial Q/\partial T)_x$, the reactor would be unstable with a constant jacket temperature. A slight increase in temperature would lead to a still higher temperature. However, there is no upper stable point in Fig. 15-4, and so the temperature would eventually come down again after the reactant concentration had decreased somewhat. The temperature and concentration would oscillate in limit cycles of the kind shown in Ref. 8.

Lowering the feed temperature from 70 to 20°F would not make the process stable, since UA and Fpc_p are unchanged. With the feed at 20°F, all the heat could be removed with

$$T - T_j = \frac{196,000 - (1,500)(50)}{7,500} = 16.1°F$$

which is less than the critical temperature difference. The rule that a process is stable if $T - T_j < \Delta T_c$ is intended only as a general warning and must be checked by the other criteria when the sensible-heat term is appreciable.

The reactor could be operated at 139°F by using a feedback control system to regulate the jacket temperature. Since the process is unstable by only a small margin, the reactor temperature would change slowly even at constant jacket temperature and good control should be fairly easily obtained. An important factor in the success of the control system would be the minimum jacket temperature that could be obtained. It would be desirable to be able to increase the driving force for heat transfer to 1.5 to 2 times the normal value, to allow for changes in reaction rate or heat-transfer coefficient. The lower feed temperature (20°F instead of 70°F) would give a greater safety factor, since the normal jacket tempera- ture would be higher.

15-4. STABILITY OF TUBULAR REACTORS

If a high conversion is required, a tubular reactor (pipeline reactor) is often used instead of a stirred-tank reactor, because a much smaller reactor volume is needed. The problem considered here is how to carry out an exothermic reaction at nearly constant temperature. The reactor could be made from sections of jacketed pipe or from a long coil immersed in a cooling bath. A constant jacket temperature is assumed to simplify the analysis, and plug flow with no radical gradients of temperature or concentration is assumed inside the reactor.

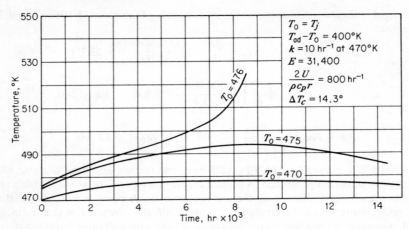

FIGURE 15-5   Temperature profiles for tubular reactor.   [*After Boynton, Nichols, and Spurlin* (3), *courtesy of Industrial and Engineering Chemistry.*]

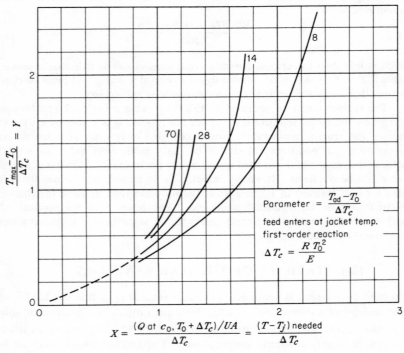

FIGURE 15-6   Peak temperatures in tubular reactors.   (*Data from Refs.* 3 *and* 15.)

If the feed enters at the jacket temperature, the reactor temperature usually rises to a maximum a short distance from the inlet. At this point the local rate of heat transfer to the jacket equals the rate of heat generation. Since the reaction rate usually decreases with increasing conversion, the reactor temperature then decreases and approaches the jacket temperature if the reactor operates near 100 per cent conversion. If the reaction were zero-order (rate independent of concentration), the reactor temperature would stay at the maximum value until the reaction was complete. Typical temperature profiles taken from the computer studies of Boynton, Nichols, and Spurlin (3) are shown in Fig. 15-5.

As shown by several workers (3, 14, 15), tubular reactors are quite sensitive to disturbances if the maximum temperature difference $T - T_j$ is close to the critical value ΔT_c. For the conditions of Fig. 15-5, increasing the feed temperature and the jacket temperature from 475 to 476°K raises the peak temperature from 494°K to over 530°K. Similar large changes can result from slight changes in feed concentration or feed temperature alone (14, 15). The reactors are not unstable in the usual sense, since there is a definite stable temperature profile for each combination of initial conditions. (There are no limit cycles of the kind found with stirred-tank reactors.) What makes the reactors potentially unstable in a practical sense is the very large change in maximum temperature that can result from almost unnoticed changes in jacket temperature, feed temperature, or heat-transfer coefficient.

The temperature profiles for any order reaction can be generalized by using dimensionless groups proportional to time (or reactor length) and to temperature. These groups contain the heat-transfer rate, the rate of heat generation at initial conditions, the adiabatic temperature rise $T_{ad} - T_0$, and the critical temperature difference. Results from Refs. 3 and 15 are shown in Fig. 15-6, where the peak temperature rise divided by ΔT_c is plotted as a function of a group X, which is a conservative measure of how close the reactor is to the critical temperature difference. The group X is defined by using the reaction rate that would exist at the initial concentration and a temperature ΔT_c above the jacket temperature.

$$X = \frac{(Q \text{ at } c_0, T_0 + \Delta T_c)/UA}{\Delta T_c} \simeq \frac{e(Q \text{ at } c_0, T_0)}{UA \, \Delta T_c} \qquad (15\text{-}16)$$

Since the reactant concentration would be less than the initial value by the time the temperature reached $T_0 + \Delta T_c$, the actual reaction rate and heat-generation rate would be somewhat less than the value used to calculate X.

The conservative criterion is that X must be less than 1 to avoid large changes in peak temperature from slight disturbances. This is the same as saying that the reactor temperature should not exceed the jacket temperature by more than the critical temperature drop. If the adiabatic rise is very large, this criterion is only slightly conservative. For $T_{ad} - T_0 = 28 \, \Delta T_c$, the reactor could be operated at $X = 1.2$, with a peak temperature rise of $1.0 \, \Delta T_c$. The actual rise is $1.0 \, \Delta T_c$, and not $1.2 \, \Delta T_c$, because the reactant concentration at the peak is $0.83 c_0$. However, a 5 per cent decrease in U would change X to 1.26, and the maximum temperature rise would be 20 per cent greater. An increase in X of 10 to 15 per cent would give a very large increase in peak temperature.

If the adiabatic temperature rise is small relative to the critical temperature drop, the reactor can be operated safely with a maximum temperature difference higher than ΔT_c. For $T_{ad} - T_0 = 8 \, \Delta T_c$ the reactor could be designed for $X = 1.9$ and $T_{max} - T_0 = 1.37 \, \Delta T_c$; a 10 per cent increase in X would change the temperature rise by 30 per cent, which is perhaps tolerable.

The curves in Fig. 15-6 are for a feed temperature equal to the jacket temperature. If the feed temperature is below the jacket temperature, there is hardly any reaction until $T = T_j$, and the maximum temperatures are almost as great as those shown. If the feed is preheated above the jacket temperature, the value of X must be closer to 1.0 to keep the same value of Y or the same sensitivity to disturbances.† The exact effect of feed temperature is shown by Barkelew (15), who calculated temperature profiles for several combinations of feed temperature, reaction order, and reaction rate. Barkelew's parameter S/N is equal to X/e, his S is $(T_{ad} - T_0)/\Delta T_c$, and his dimensionless temperature rise T_m is the same as Y or $(T_{max} - T_0)/\Delta T_c$.

For reaction orders greater than 1, the maximum temperature rise is slightly less than would be predicted from Fig. 15-6, since the stabilizing effect of decreasing reactant concentration is greater. For zero-order reactions there is no such stabilizing effect, and the value of X must be less than 1 for a stable reaction. Exact temperature profiles can be calculated with a computer for cases of parallel or consecutive reactions or changing physical properties. However, much can be learned about the feasibility of proposed designs from a simple calculation of X based on the maximum possible reaction rate at the desired temperature.

Example 15-2

A liquid-phase exothermic reaction $A + B \to C$ must be carried out in a pipeline reactor to achieve 99 per cent conversion of reactant A. There is a twofold excess of reactant B, and the rate constants below are the pseudo-first-order constants for A based on the initial concentration of B. The reaction should be carried out at 110 to 120°C to get reasonable rates without too much by-product formation. What jacket temperatures would be recommended for reactors made of 1-, 2-, or 3-in. standard pipe? How stable would reactors of these diameters be? A single

---

† The feed could be preheated without reaction if the reactants were heated separately and mixed at the reactor inlet. A feed temperature above the jacket temperature can also occur with packed-bed reactors, where no reaction occurs in the preheater.

pipe would do for a 2- or 3-in. reactor, but four parallel units would be used if made of 1-in. pipe. The estimated coefficients are given below.

	Possible reactor sizes		
	1-in.	2-in.	3-in.
A, ft²/ft³	45.8	23.3	16.8
v, fps	2.3	2.4	1.1
U, Btu/(hr)/(ft²)(°F)			
or PCU/(hr)(ft²)(°C)	400	400	300
UA	18,400	9,300	5,040

Temperature, °C	k, hr⁻¹
100	12
110	20
120	32

$$F = 200 \text{ ft}^3/\text{hr}$$
$$c_{A_0} = 0.3 \text{ lb mole/ft}^3$$
$$-\Delta H = 52,000 \text{ Btu/lb mole} = 28,900 \text{ PCU/lb mole}$$
$$\rho = 50 \text{ lb/ft}^3$$
$$c_p = 0.5 \text{ Btu/(lb)(°F)} = 0.5 \text{ PCU/(lb)(°C)}$$
$$\rho c_p = 25 \text{ PCU/(ft}^3)(°C)$$
$$T_{ad} - T_0 = \frac{(0.3)(28,900)}{25} = 347°C$$
$$\Delta T_c = \frac{RT^2}{E} = \frac{T_2 - T_1}{\ln (k_2/k_1)}$$

100–110°C data:
$$\Delta T_c = \frac{10}{0.511} = 19.6°C$$

110–120°C data:
$$\Delta T_c = \frac{10}{0.47} = 21.3°C$$

Use 20°C for ΔT_c.

$$\frac{T_{ad} - T_0}{\Delta T_c} = 347/20 = 17.3$$

Assume 100°C for a jacket temperature.

$$T_0 + \Delta T_c = 100 + 20 = 120°C$$
$$Q \text{ at } c_0, 120°C = 32(0.3)(28,900) = 277,000 \text{ PCU/(hr)(ft}^3)$$

For a 1-in. reactor

$$T - T_j \text{ needed at } c_0, 120°C = \frac{277,000}{18,400} = 15.1°C$$

$$X = \frac{15.1}{20} = 0.75$$

From Fig. 15-6 $Y \cong 0.4$

$$T_{max} \cong 100 + (0.4)(20) = 108°C \quad \text{if } T_j = 100$$

For a jacket temperature of 105°C,

$$k_{105} \cong 1.29 k_{100}$$

$$X \cong (0.75)(1.29) = 0.97 \qquad Y \cong 0.56$$

$$T_{max} \cong 105 + (0.56)(20) = 116°C$$

For the 2-in. reactor at $T_j = 100°C$

$$T - T_j \text{ needed} = \frac{277,000}{9,300} = 29.8$$

$$X = \frac{29.8}{20} = 1.49 \qquad Y = 1.3 - 1.5, \text{ in unstable region}$$

For $T_j = 95°C$, $X = 1.16$ $Y = 0.75$ $T_{max} = 110°C$

For $T_j = 97°C$, $X = 1.28$ $Y = 0.94$ $T_{max} = 116°C$

For the 3-in. reactor with $T_j = 95°C$

$$X = \frac{1}{1.29} \frac{277,000}{(5,040)(20)} = 2.13 \qquad Y > 2 \qquad \text{unsatisfactory}$$

If the jacket temperature were lowered to 85°C,

$$T_j + \Delta T = 105°C$$

$$X \cong 1\tfrac{2}{2}0(2.13) = 1.28 \qquad Y = 0.95$$

$$T_{max} = 85 + (0.95)(20) = 104°C$$

The 3-in. reactor could be operated at a jacket temperature of 85°C or lower, but the average reaction rate would be only half the controllable rate in the 2-in. reactor.

The 2-in. reactor seems best for reasonably stable operation at 110 to 120°C. Controlling the jacket temperature at $96 \pm 1°C$ would keep the peak temperature at 110 to 116°C.   There is no need to keep the jacket temperature at 96°C for the entire length of the reactor.   The reactor could be divided into two equal lengths to aim for 90 per cent conversion in the first section and 99 per cent conversion at the end of the second section.   The second section could have a jacket temperature of about 115°C and could even be made from larger pipe.   The final conversion could be controlled by adjusting the jacket temperature of the second section.

15-5. CONTROL OF TUBULAR REACTORS

Automatic temperature control of tubular reactors is a difficult problem, because the temperature is a function of length, and changes in reactor conditions shift the location of the peak temperature.   Furthermore, a

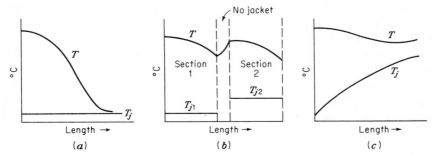

FIGURE 15-7   Temperature profiles for tubular reactors operating in the unstable region ($T - T_j \gg \Delta T_c$). (a) One section, constant T_j; (b) multiple sections; (c) parallel flow of coolant.

tubular reactor does not have the large lumped capacity of a stirred reactor, and the temperature changes much more rapidly after a disturbance. Generally, the best approach is to make the reactor stable at a constant jacket temperature and to design a jacket control system to keep T_j within 1 to 2°C of the set point.   Several thermocouples should be installed to measure the reactor temperature in the region of the temperature peak. The peak value can be used to reset the jacket temperature controller manually.

Suppose that it seems desirable to operate a tubular reactor in the unstable region where $T - T_j \gg \Delta T_c$.   One approach is to preheat the reactants separately to the desired reactor temperature and use a jacket temperature low enough just to balance the initial rate of heat generation. The jacket temperature could be automatically controlled to maintain a slight negative temperature gradient in the first few feet of the reactor. The reactor would have to be divided into sections with a higher jacket temperature for each succeeding section (see Fig. 15-7).   If parallel flow of coolant were used, the jacket temperature would rise with length and perhaps only one or two sections would be needed.

15-6. BATCH REACTORS

The changes in temperature and concentration with time for a perfectly mixed batch reactor are the same as for an ideal (plug flow, no axial diffusion) tubular reactor.   Therefore the curves in Figs. 15-5 and 15-6 can be used to predict safe operating conditions for batch reactors.

Batch reactors can be controlled at an otherwise unstable point by automatically adjusting the jacket temperature to make the rate of change of heat transfer with temperature greater than $(\partial Q / \partial T)_x$, as would be done for continuous reactors.   The reactor can be stabilized by using just proportional control, provided that the reactor is the major lag in the control system.   To put it another way, the reactor can be stabilized if the temperature changes fairly slowly compared with the rate at which the jacket temperature can be changed.

Extremely fast reactions are sometimes carried out in semibatch fashion to prevent runaway temperatures.   One reagent is added continuously to a cooled reactor containing the other reagent or the catalyst, and the addition rate is controlled to maintain a given batch temperature or a given heat-removal rate.   For safe operation, the temperature should be high enough to ensure a low concentration of the added reactant (3).

15-7. PACKED-BED REACTORS

In a packed-bed catalytic reactor all the heat of reaction is released in the pores of the catalyst and is conducted out to the pellet surface and transferred from the surface to the fluid; for a nearly isothermal reactor, most of the heat is then transferred radially through the packing, the wall, and the external resistance of the coolant.   Each step of the heat-transfer process requires a temperature driving force; and if the sum of these driving forces exceeds the critical temperature drop, the reactor is very sensitive to changes in the kinetic or heat-transfer variables.

The temperature drop from the center of a catalyst pellet to the pellet surface is usually quite small.   Other factors limit the reaction rate before the internal temperature drop becomes significant (16).   The temperature drop from pellet to gas is a more important factor, and there are probably many cases where $T_{solid} - T_{gas}$ is close to RT^2/E, the critical temperature drop.   At the critical temperature drop, a slight increase in rate leads to a rapid rise of 100 to 300°C in surface temperature, and the reaction becomes limited by diffusion.   Steady operation at intermediate solid temperatures is not possible.

Instability caused by heat transfer from the particles was observed in a study of ethylene hydrogenation (6).   An increase in feed temperature from 72 to 81°C increased the reaction rate 10- to 100-fold, because the temperature drop from solid to gas reached the critical value at about 80°C.   The temperature difference can be obtained from a fairly simple calculation (17) which should be made for all kinetic studies.   The temperature instability is also evident in many combustion reactions, including the regeneration of carbon-fouled catalysts.   For rapid regeneration, operation at the upper stable point may be desirable, but the sudden jump in temperature can fuse the catalyst unless the oxygen concentration is carefully controlled.   The existence of two stable states has been demonstrated even with the fine particles used in fluidized beds; some particles with a higher than average burning rate reached very much higher temperatures than the bulk of the catalyst (18).

Heat transfer from the center of a packed bed to the wall is most often the factor that causes trouble in packed-bed reactors.   Five mechanisms of radial heat transfer were described by Argo and Smith (19), but though the theory seems sound, it has proved difficult to separate the contributions and obtain an accurate correlation. Judging by the scatter in a recent summary (20), the effective thermal conductivity cannot be predicted to closer than 20 per cent. There is even more uncertainty in the value of h_w, the inside coefficient for transfer to the wall. Therefore packed-bed reactors should not be designed to operate close to the unstable point; slight differences in the packing density or gas-flow rate could make some tubes burn out while others were performing as expected.

A quick estimate of reactor stability can be obtained from a one-dimensional analysis. The radial heat-transfer rate is based on an overall coefficient which can be defined in two ways. The effective coefficient for the bed is $h = 4k_e/r$ if the average bed temperature is used to calculate the heat flux (21). If the heat flux is based on the center-line temperature, the coefficient is $h = 2k_e/r$. Barkelew (15) used the first definition, and so his temperature profiles and peak temperatures are the radial average values. Some of his peak values for first-order reactions were used in constructing Fig. 15-6, which of course can be used for heterogeneous as well as homogeneous reactions. The accuracy of the one-dimensional approach depends on the change in reaction rate from the center to the wall. As shown by Example 15-3, the method is fairly accurate even when there is a twofold change in rate. A computer study may still be needed to test various designs, but the value of a quick hand calculation in checking the computer or screening proposed systems should not be overlooked.

Because part of the resistance to heat transfer is distributed inside the bed, a packed reactor can be operated with a somewhat higher total driving force than would be possible for an open-pipe reactor. If all the resistance is in the bed, the critical temperature drop (center to wall) for a cylindrical tube is $1.39RT^2/E$, compared with $1.0RT^2/E$ if all the resistance is at the walls (3).

Automatic control of the peak temperature in a packed-bed reactor is not practical. A typical reactor has a few hundred tubes, and because of variation in packing density and gas-flow rate, the peak temperature is not at the same distance from the inlet in each tube. The main reason for using automatic control with other types of reactors is to permit higher temperature driving forces and lower transfer areas to be used. With a large-diameter packed bed, most of the temperature drop would be in the bed, and much of the catalyst would be at a lower temperature than desired. Automatic control is, of course, often used to maintain a constant temperature outside of the tubes.

Example 15-3

Estimate the maximum temperature and the temperature sensitivity of a packed-bed reactor for the following conditions. The data are taken from an example given by Beek (20).   The temperature drop from pellet to gas is negligible

$$A + B \rightarrow C + D$$
$$A + 2B \rightarrow E$$

Overall reaction $A + 1.068B \rightarrow 0.932C + 0.932D + 0.136E$

$$\Delta H = -84.3 \text{ kcal}$$

$$\text{Reaction rate} = \frac{ky_A}{1 + 10.1y_C}$$

$$y_{A_0} = 0.095 \qquad y_{B_0} = 0.887 \qquad y_{C_0} = 0$$

$$k = 77.7 \times 10^{-7} \text{ mole/(sec)(cm}^3) \qquad \text{at 320°C, the jacket temperature}$$

$$E/R = 11,580$$

$$c_p = 28.7 \text{ cal/(mole)(°C)}$$

$$T_0 = T_j = 320\text{°C} = 593\text{°K}$$

Reactor has cylindrical pellets in a 1.87-cm-radius tube.

For $N_{Re} = 166$

$$k_e = 2.5 \times 10^{-3} \text{ cal/(cm)(sec)(°C)}$$
$$h_w = 7.3 \times 10^{-3} \text{ cal/(cm}^2)(\text{sec})(\text{°C})$$
$$h_o = 28.5 \times 10^{-3} \qquad \text{(coefficient for boiling coolant)}$$

The overall coefficient is calculated from an equation based on the average bed temperature minus the jacket temperature (21).

$$\frac{1}{U} = \frac{1}{h_o} + \frac{1}{h_w} + \frac{r}{4k_e} = 35 + 137 + 187 = 359$$

$$U = 2.78 \times 10^{-3} \qquad A = \frac{2\pi r}{\pi r^2}$$

$$UA = 2.78 \times 10^{-3} \left(\frac{2}{1.87}\right) = 2.97 \times 10^{-3} \text{ cal/(cm}^3)(\text{sec})(\text{°C})$$

To use the solutions for first-order reactions, the rate for the initial concentration of A is based on the average concentration of C in the section up to the temperature peak.   Based on other studies (15, 16), the conversion at the peak temperature is about 30 per cent, or the average is 15 per cent.

$$\text{Reaction rate} \cong \frac{ky_A}{1 + 10.1(0.15)(0.095)} = \frac{ky_A}{1.145} \qquad \text{0–30\% conversion}$$

$$Q_0 = \frac{77.7 \times 10^{-7} \times 0.095 \times 84,300}{1.145} = 0.0542 \text{ cal/(sec)(cm}^3)$$

$$\Delta T_c = \frac{(593)^2}{11,580} = 30.5\text{°C}$$

$$X \cong \frac{eQ_0}{UA\,\Delta T_c} = \frac{(2.72)(0.0542)}{2.97 \times 10^{-3}} \left(\frac{1}{30.5}\right) = 1.63$$

$$T_{ad} - T_0 = \frac{(0.095)(84,300)}{28.7} = 279\text{°C}$$

$$\frac{T_{ad} - T_0}{\Delta T_c} = 9.2$$

From Fig. 15-6, $Y \cong 1.05$.

$$\text{Average temperature rise} = (1.05)(30.5) = 32°C$$

The *average* bed temperature at the temperature peak is about $320 + 32 = 352°C$. If the average temperature is halfway between the center temperature and the temperature near the wall,

$$T_{av} - T_{near\ wall} = 32(^{187}\!/_{359}) = 17°C$$

$$T_{center} \cong 352 + 17 = 369°C$$

This compares quite well with the value of $372°C$ obtained by a computer solution that allowed for radial temperature gradients.

To show the temperature sensitivity, calculate T_{max} for a 10 per cent higher feed concentration.

$$X = (1.1)(1.63) = 1.79$$

$$Y \cong 1.3$$

The average temperature rise is about 25 per cent greater than before, which is not a very high sensitivity. As shown by Fig. 15-6, an adiabatic rise of only nine times the critical temperature drop means a fairly stable reactor, compared with cases when $T_{ad} - T_0$ is 15 to 30 times ΔT_c.

15-8. SPECIAL STABILITY PROBLEMS

With exothermic chemical reactions, the increase in rate with temperature can be considered a positive feedback effect which tends to make the system unstable. Positive feedback also occurs in autocatalytic systems and can lead to sustained oscillations at constant reactor temperature (22). Fermentation is an example of such a process, since the growth rate is proportional to the number of cells present ($dN/dt = kN$). As shown by Finn and Wilson (23), a continuous fermenter is stable only if the retention time V/F exactly equals the growth constant k. A slightly higher flow rate removes more cells than are produced, and the cell concentration falls toward zero. If the growth constant increases with decreasing cell concentration because of changes in pH, continuous cycling results.

An indirect type of positive feedback occurs when the product from an adiabatic reactor is used to preheat the feed. A slight increase in reaction rate leads to an increase in the exit temperature, and after this hotter product flows through the exchanger, the feed is hotter, which increases the rate still further. Orcutt and Lamb (24) analyzed the operation of an ammonia reactor and showed how the regions of stable and unstable operation could be predicted. A similar study of an exchanger-reactor was made by Douglas, Orcutt, and Berthiaume (25).

Recycling part of the exit stream to the reactor inlet is a direct type of feedback which can be either positive or negative depending on the reaction kinetics. If the reactor is isothermal, recycling the unreacted material is usually a positive feedback but the reactor is still stable. An increase in reactant concentration in the feed means an increase in the reactant concentration in the product, but the gain factor is less than 1 for ordinary kinetics (first-, second-, or fractional-order). Therefore a slight decrease in reaction rate leads to a gradual increase in concentration at the reactor inlet until the total material reacted per hour is again equal to the fresh feed rate.

With an adiabatic reactor, the temperature and conversion can be very sensitive to changes in recycle rate or recycle composition. A commercial reactor which cycled continuously because of a recycle stream was analyzed by Batke, Franks, and James (26). A similar case is treated in Example 15-4.

Example 15-4

The liquid-phase reaction $A + B \rightarrow C$ is to be carried out with a large excess of B and with a solid catalyst in an adiabatic packed-bed reactor. The fresh feed contains 0.3 mole A/ft^3, and the adiabatic temperature rise would be 360°F. To limit the temperature rise 2 lb of cooled product is recycled for each pound of fresh feed. The combined streams enter the reactor at 100°F and have a superficial velocity of 30 ft/hr. The packed height is 10 ft. Will the reactor be stable under these conditions?

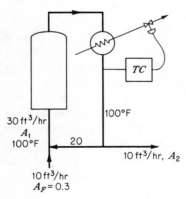

FIGURE E15-4a Diagram for Example 15-4.

$$r = k(A) \frac{\text{moles } A}{(\text{hr})(\text{ft}^3 \text{ bed})}$$

$$k = 0.75 \text{ at } 100°F$$

$$E = 20,000 \text{ Btu/lb mole}$$

$$\rho = 50 \text{ lb/ft}^3$$

$$c_p = 0.5$$

The first step is to calculate the conversions for various feed concentrations.

$$F \, dA = -k \, dV \, A \qquad A = \text{moles of } A/\text{ft}^3$$

$$\int \frac{-dA}{kA} = \frac{V}{F}$$

The integral is evaluated graphically or numerically, the increase in rate constant with temperature being allowed for.

$$T - 100 = 360°F \frac{A_1 - A}{0.30}$$

$$\ln \frac{k}{k_0} = \frac{20,000}{R} \frac{T - 100}{560(T + 460)}$$

A few calculations for $A_1 = 0.10$ are given below.

A	T	$\dfrac{k}{k_0}$	$\dfrac{k_0}{kA}$	$k_0 \displaystyle\int \dfrac{-dA}{kA} = \dfrac{k_0 V}{F}$
0.10	100	1	10	
0.0917	110	1.37	7.95	0.075
0.0833	120	1.86	6.45	0.135
0.075	130	2.49	5.35	0.184
0.0667	140	3.31	4.52	0.225
0.0583	150	4.36	3.94	0.260

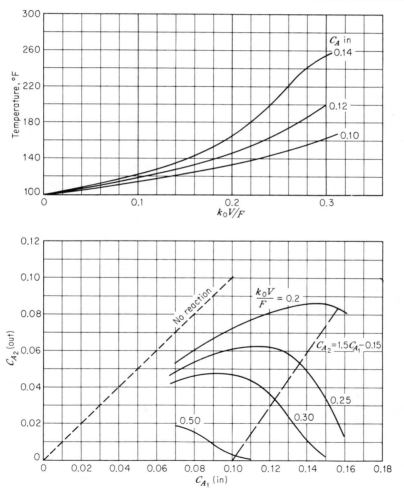

FIGURE 15-8   Temperature profiles and conversions for an adiabatic reactor (Example 15-4).

By interpolation, the values of A_2 for specific values of $k_0 V/F$ were obtained. Figure 15-8 shows that the exit concentration goes through a maximum with increasing inlet concentration if $k_0(V/F)$ is large enough. This occurs because the average reactor temperature increases with increasing feed concentration. When the temperature profile near the reactor exit is quite steep, a 1 per cent increase in feed concentration can actually cause a decrease of more than 1 per cent in the exit concentration.

For a bed length of 10 ft and $k_0 = 0.75$

$$\frac{k_0 V}{F} = \frac{(0.75)(10)}{30} = 0.25$$

A material balance on A gives

$$A_1 = \tfrac{2}{3}A_2 + \tfrac{1}{3}A_F = \tfrac{2}{3}A_2 + 0.1$$

or

$$A_2 = 1.5A_1 - 0.15$$

The material balance and the kinetic equations are satisfied at $A_1 = 0.136$, $A_2 = 0.053$. However, the reactor gain at this point is $K = -1.0$ (change in A_2/ change in A_1), and stability may be a problem.

The reactor transfer function is the gain multiplied by the time delay. The heat capacity of the solid, which has a slight stabilizing effect, is neglected. The external piping and the heat exchanger contribute a further time delay, which is lumped with the reactor delay. Equivalent block diagrams are shown in Fig. E15-4b. The sign change makes the second diagram similar to that for negative-feedback proportional control. If the overall gain, $\tfrac{2}{3}(-K)$, exceeds 1 at 180° phase lag, the process is unstable. Since $-K$ is 1.0 the overall gain at normal conditions is $\tfrac{2}{3}$ at all frequencies and so the reactor is barely stable. Slight disturbances would lead to fluctuations in reactor temperature and conversion which would take several cycles to damp out.

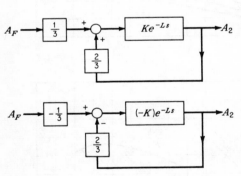

FIGURE E15-4b Block diagram for Example 15-4.

An increase in feed concentration from 0.3 to 0.36 would make the overall gain greater than 1, and the process would cycle with a period of $2L$ min.

Various methods could be used to increase the stability of the reactor.

A stirred tank could be added to the recycle line to help damp concentration fluctuations. A tank with a time constant equal to L would reduce $K_{overall}$ by about 50 per cent at the critical frequency.

A higher recycle temperature could be used when the reactor exit concentration is low, to compensate for the low concentration in the blended feed. To accomplish this, a low gain could be used on the temperature controller, or part of the recycle stream could be bypassed around the exchanger.   The evaluation of this proposal is left to the student.

A moderate change in bed length or catalyst activity (changing T_0 or changing the catalyst) would not help much, since the gain is still close to 1 at $k_0 V/F = 0.30$. However, if $k_0 V/F$ could be raised to 0.50, the conversion would be 95 to 100 per cent for all possible feed concentrations and the reactor would be stable.

A fourth alternative is to raise the recycle rate.   At 2.7/1, $k_0 V/F = 0.2$, and the calculated operating point is $A_1 = 0.144$, $A_2 = 0.086$.   The reactor gain is almost zero at this point.   However, the conversion of fresh feed is only 71 per cent compared with 82 per cent for the original case.   After a separation step, additional reactant would probably be returned to the reactor unit, raising the feed concentration to 35 to 40 per cent, which would shift the operating point and make the reactor gain about -0.2 to -0.4.

15-9. COMPOSITION CONTROL

After steps have been taken to ensure good temperature control, the control of reactor composition or degree of conversion can be considered. The following discussion deals mainly with continuous stirred reactors; plug-flow reactors are considered in the section on time-delay processes in Chap. 9. For most reactors the product concentration is not automatically controlled but is held nearly constant by having a controlled reactor temperature, controlled feed streams, and a constant holdup in the reactor. Often there is no point in controlling the conversion at a given level, since the maximum possible conversion is desired.

Justification for automatic composition control arises if the reaction rate fluctuates (because of changing catalyst activity, temperature fluctuations, or changing feed purity) and if the resulting fluctuations in product concentration or quality are costly. As one example, the product concentration may go through a maximum with increasing reactant conversion because the product decomposes to form by-products. For this case the reactor should be controlled to give the economic optimum product concentration, which would usually be somewhat less than the maximum value. For certain polymerization reactions, a relatively low conversion is desired (50 per cent or less), to avoid excessive chain branching or to get a desired distribution of molecular weights.

The success of composition control schemes depends to a large degree on the measurement device. It would be desirable to have direct control of the yield, the by-product concentrations, or the product quality, but these variables can rarely be measured both rapidly and accurately. Usually some physical property such as density, refractive index, or viscosity is used as a rough measure of the product concentration or the degree of conversion. The set point of the "composition" controller is then adjusted from time to time to allow for slow shifts in product distribution which affect the measured bulk property or the economic optimum conversion. As faster and more

reliable analytical instruments become available, there will be more justification for direct continuous control of yield or quality.

The reactor composition can be controlled by regulating the reactant
feed rate, the holdup of liquid or gas, the temperature, or the catalyst
concentration, to name the most common variables. For chain reactions
such as polymerizations, the reaction can be controlled by adding small
amounts of chain stoppers or chain transfer agents. The dynamic response
of the control system depends on which input variable is regulated, since
the time constant for concentration changes is not the same as for changes
in level, temperature, or other variables.

If the reactant feed rate is controlled to maintain a given product
concentration, there are two major time constants in the reactor transfer
function, as shown below. In the following analysis a large excess of
solvent is assumed so that changes in reactant feed rate do not change the
holdup time in the reactor.

Consider the reaction $A \rightarrow B$

For A,
$$V \left(\frac{dA}{dt} \right) = FA_F - FA - V \left(\frac{dr}{dA} \right) A$$

$$r = \text{moles}/(\text{hr})(\text{ft}^3) \tag{15-17}$$

The transformed equation is

$$A \left[Vs + F + V \left(\frac{dr}{dA} \right) \right] = FA_F$$

$$A \left[\frac{Vs}{F + V(dr/dA)} + 1 \right] = \frac{F}{F + V(dr/dA)} A_F \tag{15-18}$$

$$\frac{A}{A_F} = \frac{k_1}{T_1 s + 1}$$

For B,
$$V \left(\frac{dB}{dt} \right) = -FB + V \left(\frac{dr}{dA} \right) A \qquad B_F \text{ assumed } 0$$

$$\left(\frac{V}{F} s + 1 \right) B = \frac{V}{F} \left(\frac{dr}{dA} \right) A \tag{15-19}$$

Combining with the transfer function A/A_F gives

$$\frac{B}{A_F} = \frac{k_p}{(T_1 s + 1)(T_2 s + 1)} \tag{15-20}$$

where $T_1 = \dfrac{V}{F + V(dr/dA)}$

$T_2 = \dfrac{V}{F}$

As was shown in Chap. 3 [Eq. (3-21)], the time constant T_1 is $\dfrac{V/F}{1 + kV/F}$
for a first-order reaction, or $0.5V/F$ at 50 per cent conversion.   If the two
reactor time constants are 2 hr and 1 hr, a simple proportional control
system will have a period of 2 to 3 hr when adjusted for a reasonable damping
coefficient ($\zeta = 0.2$ to -0.3) [see Eq. (4-16)].   Note that controlling the
concentration of reactant in the reactor leads to only one time constant in
the reactor transfer function, and the control system would be much faster
than the system based on the product concentration.   The product concentration is likely to be more important than the reactant concentration,
but neither method ensures a constant conversion or yield of the reactant fed.

When the feed rate is fixed, the per cent conversion can be controlled
by changing the catalyst concentration or by making slight changes in the
set point of the temperature controller.   These cases are considered in
Example 15-5.   Note that there are two time constants in the transfer
functions, one coming from the lag in reaching a new temperature or catalyst
concentration and the other from the lag in conversion following a change
in reaction rate.   The conversion could also be controlled by changing the
level in the tank, which changes the holdup time.   A possible disadvantage
of this scheme is that changes in level change the effective heat-transfer area
and the agitation pattern.

Example 15-5

A liquid-phase reaction is carried out continuously in a 2,000-gal stirred tank
with a holdup time of 1.2 hr.   The reaction rate varies with the square root of the
catalyst concentration and the first power of the reactant concentration.   The
reaction is inhibited by unidentified impurities in the feed, and the catalyst flow
must be regulated to maintain the desired 60 per cent conversion of the reactant.
Conversion is measured by a refractometer, and the only measurement lag is a
50-sec time delay in the sampling system.   The heat of reaction is small, and the
temperature is easily held to within 0.5°C by controlling the jacket temperature.

a. How effective would a conventional feedback control system be in compensating for changes in feed purity?

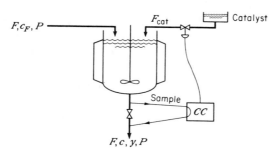

FIGURE E15-5*a*   Diagram for Example 15-5.

b. Would a faster response be obtained by changing the reaction temperature rather than the catalyst concentration?

$$r = k(y)^{1/2}cV = k_1cV$$

$$k = f(P)$$

$$c = \text{reactant concentration}$$

$$y = \text{catalyst concentration}$$

$$P = \text{impurity concentration}$$

The material balance for the reactant is written with c_F, c, y, and P denoting changes from the normal values.

$$V\left(\frac{dc}{dt}\right) = Fc_F - Fc - \left(\frac{\partial r}{\partial c}\right)c - \left(\frac{\partial r}{\partial y}\right)y - \left(\frac{\partial r}{\partial P}\right)P$$

The transformed equation is

$$c(T_1s + 1) = K_Fc_F + K_yy + K_PP$$

where
$$T_1 = \frac{V}{F + k_1V}$$

Since the normal conversion is 60 per cent

$$\frac{k_1V}{F} = \frac{x}{1 - x} = \frac{0.6}{0.4} = 1.5$$

$$T_1 = \frac{1.2}{2.5} = 0.48 \text{ hr}$$

The material balance for the catalyst is

$$V\left(\frac{dy}{dt}\right) = \Delta F_{cat} - \bar{F}y$$

Note that ΔF_{cat} is the change in the catalyst flow and F is the total flow leaving the reactor ($F_{cat} \ll F$)..

$$(Vs + F)y = \Delta F_{cat}$$

$$y = \frac{\Delta F_{cat}}{F(T_2s + 1)}$$

where
$$T_2 = \frac{V}{F} = 1.2 \text{ hr}$$

The change in catalyst concentration can be put on a percentage basis. Since $\bar{y} = \bar{F}_{cat}/F$,

$$\frac{y}{\bar{y}} = \left(\frac{\Delta F_{cat}}{\bar{F}_{cat}}\right)\frac{1}{1.2s + 1}$$

The same result could be obtained by using the change in catalyst concentration in the combined feed, $y_f = \Delta F_y/F$, as the independent variable. The transfer function for impurities is the same as for catalyst and so the impurity level in the feed is shown as a load change entering before the two main time constants.

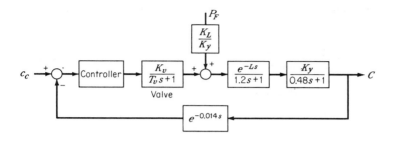

FIGURE E15-5b Block diagram for Example 15-5.

The valve time constant would be only a few seconds unless there is a very long transmission line, and so it will be neglected. The time delay caused by imperfect mixing in the tank could be 10 to 100 sec, depending on agitation conditions

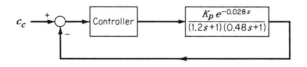

FIGURE E15-5c Block diagram for Example 15-5.

(see Chap. 16). A value of 50 sec is assumed, and the overall transfer function for stability analysis has a total delay of 100 sec.

ω, rad/hr	ϕ, deg	A.R.
4	147	0.09
4.5	152	0.076
5	156	0.063
10	179	0.0167

The usual rules for choosing the gain give quite different results for this example. Half the maximum overall gain is 30, whereas a gain of 12 is needed for 30° phase margin. Though half the maximum gain gives the minimum error integral for step changes in load (see Chap. 9), a lower gain is recommended for this case, to get a larger damping coefficient and better response to periodic disturbances. An overall gain of 15 is chosen to give about 25° phase margin. With the time delay neglected, the system frequency will be about

$$\sqrt{\frac{K+1}{T_1 T_2}} = 5.2 \text{ rad/hr}$$

The controller gain needed for an overall gain of 15 is calculated from the valve and process gains.

$$K_v = \frac{5\% \text{ change in catalyst flow}}{\% \text{ change in valve signal}} \qquad \text{typical value for an equal-percentage valve}$$

$$K_y = \frac{\% \text{ change in concentration}}{\% \text{ change in catalyst flow}}$$

K_y can be obtained by the following line of reasoning: Full scale on the refractometer is assumed to correspond to a conversion range of 40 to 70 per cent. A 2 per cent change in y changes k_1 by 1 per cent because of the 0.5 exponent. At about 60 per cent conversion, a 1 per cent change in k_1 changes the conversion 0.24 per cent, since

$$dx = \frac{dk_1 V/F}{(1 + kV/F)^2} = \frac{(0.01)(1.5)}{2.5^2}$$

A 0.24 per cent change in conversion corresponds to 0.8 per cent of the recorder scale.

$$K_y = \frac{0.8}{2.0} = 0.4$$

$$K = K_v K_y K_c = 5(0.4)K_c = 2.0K_c$$

For $K = 15$, $K_c = 7.5$

a. With only proportional control and a gain of 7.5, the errors resulting from load changes will be reduced by a factor of $1 + K = 16$, and the recovery curves will show a period of $2\pi/5.2 = 1.2$ hr. Reset action may not be needed, but derivative action would reduce the period as much as twofold.

Suppose that the composition signal is used to adjust the jacket inlet temperature, and suppose the time constant of the jacket itself is quite small. The temperature response of the kettle is given by Eq. (3-29).

$$\left[\left(\frac{V\rho c_p}{F\rho c_p + UA} \right) s + 1 \right] \theta_2 = \left(\frac{UA}{UA + F\rho c_P} \right) \theta_j$$

If the ratio $UA/F\rho c_p = 5$,

$$T_3 = \frac{V/F}{1 + 5} = \frac{1.2}{6} = 0.2 \text{ hr}$$

The transfer function relating reactant concentration and temperature has a time constant of 0.48 hr, the same as for changes in catalyst concentration. With time constants of 0.48 and 0.2 hr and a time delay of 0.028 hr (the mixing delay might be less important for this case than for catalyst control), an overall gain of 10 gives a phase margin of 25° at a frequency of 9.5 rad/hr.

b. Control of reactor temperature would almost halve the period of oscillation, but the peak error after a load change would be greater because the overall recommended gain would be only two-thirds as great. Still, the error integral following a load change would probably be smaller with this scheme.

PROBLEMS

1a. Show why Eq. (15-11) is usually a sufficient criterion for the stability of a fluidized catalytic reactor.

b. Consider the reactor stability for the following conditions.   The reactor

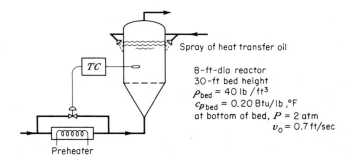

Spray of heat transfer oil

8-ft-dia reactor
30-ft bed height
$\rho_{bed} = 40$ lb $/ft^3$
$c_{p\,bed} = 0.20$ Btu/lb ,°F
at bottom of bed, $P = 2$ atm
$v_0 = 0.7$ ft/sec

Preheater

is to be controlled by varying the preheat temperature of the feed.   What controller gain is recommended?

Reaction $A + B \rightarrow C$ $\Delta H = -45,000$ Btu/mole

At 270°F, $r = \dfrac{0.026 p_A p_B}{1 + 10 p_A}$ moles A/(hr)(lb catalyst)

At 300°F, $r = \dfrac{0.043 p_A p_B}{1 + 8 p_A}$ p_A, p_B in atm

Normal conditions   270°F reactor temperature

                            60% conversion of A

                            170°F feed temperature

2. A liquid-phase oxidation $A + \frac{1}{2}O_2 \rightarrow B$ is carried out continuously in a 2,000-gal stirred reactor with air blown through the solution.   Normal reactor conditions are 50°C, 2 atm pressure above the liquid, a liquid residence time of 20 min, and 40 per cent conversion of the A in the feed.   The reaction is first-order to both oxygen and A.   To keep the moles of A reacted per hour constant, the exit gas is analyzed and the air rate decreased if the oxygen consumption is too great. The oxygen analysis loop has a time delay of 2 min and a time constant of 1 min.

Draw a block diagram for the control system, and estimate the critical frequency and maximum overall gain.

3. The reaction $2A \rightarrow B + C$ is carried out in a 2-in. tube packed with $\frac{1}{8}$ in. spherical catalyst pellets. Because A is strongly adsorbed, the reaction is almost zero-order. What is the maximum safe feed temperature if the peak temperature must be kept below $530°F$? What jacket temperature should be used if the resistance of the wall and the jacket film are negligible?

$$G = 600 \text{ lb/(hr)(ft}^2)$$

$$c_p = 0.3 \text{ Btu/(lb)(°F)} = 20 \text{ Btu/(mole)(°F)}$$

$$k_e = 0.3$$

$$h_w = 16$$

$$\Delta H = -20,000 \text{ Btu/mole of } A$$

Reaction rate

$$0.20 \text{ mole/(hr)(ft}^3 \text{ of bed) at } 470°F$$

$$0.41 \text{ mole/(hr)(ft}^3 \text{ of bed) at } 500°F$$

$$0.78 \text{ mole/(hr)(ft}^3 \text{ of bed) at } 530°F$$

4. Use a stepwise calculation to show what would happen if the reactor of Example 15-1 were operated at $139°F$ with a constant jacket temperature of $113°F$. Assume that the conversion is initially 50 per cent and that the reactor temperature rises $1°F$ in a 1-min period.

5. Under what conditions would the stability criterion of Eq. (15-8) be satisfied without satisfying the steady-state criterion of Eq. (15-9)? Give a numerical example where this situation exists.

6. The transfer function for an unstable reactor is $\theta/\theta_j = -0.6/(-5s + 1)$ $(T = -5 \text{ min})$. The other elements in the control loop have time constants of 7 min and 2 min. Show by use of Nyquist plots or the Routh-Hurwitz criterion whether or not the system would be stable with proportional control, and, if so, at what range of gains.

7. A vinyl polymerization is carried out in a stirred reactor to 25 to 35 per cent conversion. The exact conversion is not critical, but the average molecular weight of the polymer must be carefully controlled to maintain product quality. Draw a block diagram of a control system based on the following scheme, and estimate the proper controller parameters, assuming reasonable numerical values for unspecified parameters. The molecular weight is determined from the viscosity and adjusted by adding chain stopper. The viscosity must be corrected for changes in conversion, which is determined from the solution density.

8. A continuous-flow stirred-tank reactor 6 ft in diameter and 12 ft high is cooled by a film of water flowing down the outside. The reactor is filled to a depth of 10 ft, and the heat generated is 1,200,000 Btu/hr at the normal conditions of $60°C$ reactor temperature, 50 per cent conversion, and 2 hr holdup time. The reaction rate doubles for a temperature rise of $18°C$. The solution has a viscosity of 4 cp, a density of 70 lb/ft^3, and a heat capacity of 0.7.

   a. What temperature and flow rate of cooling water should be used to ensure stable operation?

   b. Discuss the dynamics of heat removal by a falling film of water.

REFERENCES

1. Heerden, C. van: Autothermic Processes, *Ind. Eng. Chem.*, **45**:1242 (1953).
2. Wilson, K. B.: Calculations and Analysis of Longitudinal Temperature Gradients in Tubular Reactors, *Trans. Inst. Chem. Engrs.*, **24**:77 (1946).
3. Boynton, D. E., W. B. Nichols, and H. M. Spurlin: How to Tame Dangerous Chemical Reactions, *Ind. Eng. Chem.*, **51**:489 (1959).
4. Harriott, P.: Designing Temperature-stable Reactors, *Chem. Eng.*, **68**(10):165, (11):81 (1961).
5. Cannon, K. J., and K. G. Denbigh: Studies on Gas-Solid Reactions—Causes of Thermal Instability, *Chem. Eng. Sci.*, **6**:155 (1957).
6. Harriott, P.: An Example of Instability in a Packed-bed Reactor, *Am. Inst. Chem. Eng. J.*, **8**:562 (1962).
7. Bilous, O., and N. R. Amundson: Chemical Reactor Stability and Sensitivity, *Am. Inst. Chem. Eng. J.*, **1**:513 (1955).
8. Aris, R., and N. R. Amundson: An Analysis of Chemical Reactor Stability and Control, *Chem. Eng. Sci.*, **7**:121 (1958).
9. Aris, R., and N. R. Amundson: Stability of Some Chemical Systems under Control, *Chem. Eng. Progr.*, **53**:227 (1957).
10. Foss, A. S.: Chemical Reaction System Dynamics, *Chem. Eng. Progr.*, **54**(9):39 (September, 1959).
11. Weber, T.: Control of a Continuous-flow Agitated-tank Reactor, Ph.D. thesis, Cornell University, 1963.
12. Bilous, O., H. D. Block, and E. L. Piret: Control of Continuous-flow Chemical Reactors, *Am. Inst. Chem. Eng. J.*, **3**:248 (1957).
13. Nemanic, D. J., J. W. Tierney, R. Aris, and N. R. Amundson: An Analysis of Chemical Reactor Stability and Control—IV. Mixed Derivative and Proportional Control, *Chem. Eng. Sci.*, **11**:199 (1959).
14. Bilous, O., and N. R. Amundson: Effect of Parameters on Sensitivity of Empty Tubular Reactors, *Am. Inst. Chem. Eng. J.*, **2**:117 (1956).
15. Barkelew, C. H.: Stability of Chemical Reactors, *Chem. Eng. Progr. Symposium Ser. No. 25*, **55** (1959).
16. Beek, J.: Relationship between Pellet Size and Performance of Catalysts, *Am. Inst. Chem. Eng. J.*, **7**:337 (1961).
17. Yoshida, F., D. Ramaswami, and O. A. Hougen: Temperatures and Partial Pressures at the Surfaces of Catalyst Particles, *Am. Inst. Chem. Eng. J.*, **8**:5 (1962).
18. Bondi, A., R. S. Miller, and W. G. Schlaffer: Rapid Deactivation of Fresh Cracking Catalyst, *Ind. Eng. Chem., Process Design Develop.*, **1**:196 (1962).
19. Argo, W. B., and J. M. Smith: Heat Transfer in Packed Beds, *Chem. Eng. Progr.*, **49**:443 (1953).
20. Beek, J.: Design of Packed Catalytic Reactors, "Advances in Chemical Engineering," vol. III, p. 203, Academic Press Inc., New York, 1962.
21. Beek, J., and E. Singer: A Procedure for Scaling-up a Catalytic Reactor, *Chem. Eng. Progr.*, **47**:534 (1951).
22. Lotka, A. J.: Undamped Oscillations Derived from the Law of Mass Action, *J. ACS*, **42**:1595 (1920).

23. Finn, R. K., and P. E. Wilson: Population Dynamics of a Continuous Propagator for Microorganisms, *Agr. and Food Chem.*, **2**(2):66 (1954).

24. Orcutt, J. C., and D. E. Lamb: Stability of a Fixed Bed Catalytic Reactor System with Feed-effluent Heat Exchange, "Automatic and Remote Control," vol. 4, p. 274, Butterworth Scientific Publications, London, 1960.

25. Douglas, J. M., J. C. Orcutt, and P. W. Berthiaume: Design and Control of Feed-effluent Exchanger-Reactor Systems, *Ind. Eng. Chem.*, *Fundamentals*, **1**:253 (1962).

26. Batke, T. L., R. G. Franks, and E. W. James: Analog Computer Simulation of a Chemical Reactor, *ISA J.*, **4**(1):14 (January, 1957).

16 CONTROL OF pH AND CONTROL OF BLENDING PROCESSES

The last section of Chap. 15 deals with composition control of chemical reactors for cases where the rate of reaction influences the reactor dynamics. However, if the reaction is extremely fast, the reaction rate has little influence on the dynamics of the system; the composition of the product depends mainly on the relative amounts of reagents added and on the effectiveness of mixing or blending these reagents. Most ionic reactions can be considered instantaneous, and thus control of pH is usually just a mixing and sampling problem, similar to the problem of blending gasolines or mixing any nonreacting miscible fluids. It is this problem of controlling the exact composition of a mixture that is considered here. The control of pH is emphasized because the nonlinear process characteristics make it particularly interesting. Most of the examples and problems deal with stirred tanks, because we know something about their performance. Pipeline mixers or in-line blenders might be preferred for some applications, but there is little published information on these devices.

16-1. RESPONSE OF AGITATED TANKS

In many control problems a stirred tank has been assumed to be perfectly mixed in deriving transfer functions for changes in concentration or temperature. For a blending tank, this assumption leads to a first-order equation and a time constant equal to V/F, the holdup time. However, there has to be some delay before a change in feed concentration is noticed at the tank exit or at any other spot where a sampling probe might be located. The delay depends on the size of the tank, the fluid velocities, and, to some extent, on the locations of the inlet pipe and the sample tap. The delay could be reduced by sampling close to the inlet pipe, but the result would not be representative of the tank contents: the sample is usually taken from the exit line or at a point well away from the tank inlet. The exact dynamics for a change in inlet conditions must be quite complex, but fortunately the frequency response can be represented quite well (1, 2) by a time constant equal to V/F and a time delay, which is called the "mixing delay" to distinguish it from the mixing times defined in other studies. The mixing delay, though much smaller than V/F, is often the second largest lag in pH or blending control systems and thus has a large influence on the system performance.

MIXING-DELAY DATA

Kramers (1) was the first to use frequency-response tests (the exit pH was measured) to determine the mixing delay for a continuous-flow stirred tank. For a 12-in. tank with a holdup time of 10 min and a 4-in. propeller operating at 300 rpm, the mixing delay was 2.5 sec, roughly the same as the visual mixing time. In a similar study Colucci (2) determined mixing delays for a 3-ft baffled tank with a flat paddle stirrer. A typical Bode diagram is shown in Fig. 16-1; the amplitude ratios are based on calculated concentrations, since the measured pH values give a distorted curve. The mixing

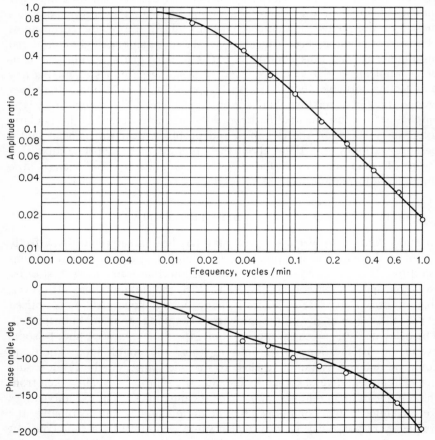

FIGURE 16-1 Frequency response for a 3-ft tank with a 10-in. paddle at 150 rpm (2). Solid lines are theory for $V/F = 8.67$ min and $L = 0.31$ min.

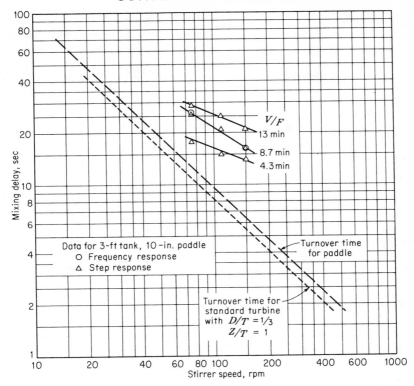

FIGURE 16-2   Mixing delay for a continuous-flow stirred tank.   [*Data of Colucci* (2).]

delay was obtained by fitting the phase-lag curve and correcting for the small lags of the sample line and electrode system.

The mixing delays obtained by Colucci are shown in Fig. 16-2. For his system, the mixing delay was about the same as the time for an initial response to a step change in feed concentration, and so some additional step-response data are included in Fig. 16-2. The mixing delay varies inversely with a fractional power of the stirrer speed and decreases with decreasing holdup time. Similar effects of speed and holdup time on the time for complete mixing were found by MacDonald and Piret (3) and by van de Vusse (4). Published mixing-time data can be correlated by using the reciprocals of the batch mixing time and the flow mixing time (4), and if more data were available, a similar correlation could probably be developed for the mixing delays.

The mixing delays shown in Fig. 16-2 are two to four times the turnover time based on the theoretical pumping capacity of the paddle.

$$\text{Turnover time} = \frac{\text{volume}}{\pi^2 D^2 w N} \qquad (16\text{-}1)$$

Smaller mixing delays would be expected with a standard six-blade turbine, because the power dissipated in the turbulent region is 3.6-fold greater than for a paddle of the same diameter (5) and the flow is perhaps 1.5-fold greater, $(3.6)^{1/3}$. The mixing delay would generally be greater for larger tanks, because the turnover time varies with $1/N$ for geometrically similar tanks and N varies with $D^{-2/3}$ for scaling up at constant power per unit volume. A mixing delay equal to the turnover time has been used in designing large blending vessels (6), but there was no proof that this value was correct.

MIXING-TIME STUDIES

There have been many other studies of mixing in stirred tanks, and while these are pertinent to our problem, the mixing times reported are not expected to be the same as the effective time delays for process control calculations. Consider what happens when a pulse of acid is added to a batch of water in a tank and the concentration is measured by a probe in the tank. As sketched in Fig. 16-3, there would be a pure time delay of t_1 sec before any acid reached the probe. The concentration would then rise, generally overshoot the final value, and show some oscillations before becoming constant at time t_4.

The mixing time could be defined (7) as the time for 90 per cent response, or t_2 in Fig. 16-3, as the time t_3 for the concentration fluctuations to become less than certain value (8), or as the time for uniformity to be reached, which would be t_4. Some authors (9, 10) have added an equivalent amount of acid to a tank containing base and measured the time for complete disappearance of color streaks, which would correspond to t_4. By intuitive reasoning, the effective mixing delay should be somewhat greater than t_1 and less than t_4, corresponding to some sort of average concentration at

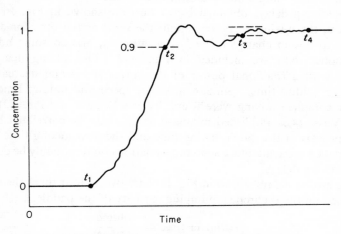

FIGURE 16-3   Response of a stirred tank to a pulse input.

the probe. The delay would not be strongly influenced by the persistence of minor inhomogeneities in the tank.

For highly turbulent systems, the times for complete mixing are probably between 1.1 and 2 times the effective mixing delays. Published mixing times for batch systems were 4 to 10 times the turnover time with a flat paddle (4) and about 2 to 4 times the turnover time with a turbine (10). More experimental studies are needed to relate mixing time and mixing delays, particularly for tanks where the times are large.

Example 16-1

An aqueous stream is diluted continuously in a 500-gal tank equipped with a 16-in. turbine. A portion of the exit stream is sent to a continuous density controller which adjusts the flow of water to the tank. The total flow is about 100 gpm, and there is a 10-sec delay in the measurement line. Estimate the ultimate period and maximum gain of the control system if the turbine speed is 90 rpm.

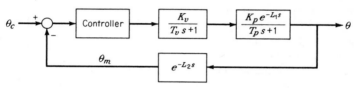

FIGURE E16-1 Diagram for Example 16-1.

The block diagram for the control system is shown in Fig. E16-1. The process time constant T_p is V/F, or 5 min. To estimate the mixing delay L_1, the turnover time is calculated from Eq. (16-1).

$$V = 500 \text{ gal} = 67 \text{ ft}^3$$

$$D = 1.33 \text{ ft}$$

$$w = 0.2D \text{ for standard turbine}$$

$$\text{Turnover time} = \frac{67}{\pi^2(0.2)(1.33)^3(1.5)} = 9.6 \text{ sec}$$

Based on the data in Fig. 16-2 and the prediction that a turbine has about 50 per cent greater pumping capacity than a paddle, the mixing delay is estimated to be 1.0 to 1.5 times the turnover time, or about 12 sec. The valve time constant, which would be only 1 to 2 sec for a short line, is neglected.

With a 5-min time constant and a total delay of 22 sec, the time constant will contribute almost 90° phase lag at the critical frequency.

$$\omega_c L \cong \frac{90}{57.3} = 1.57$$

$$\omega_c = \frac{1.57}{22} \times 60 = 4.3 \text{ rad/min}$$

Check:
$$\omega T_p = (4.3)(5) = 21$$
$$\phi_{T_p} = -88°$$
$$P_u = \frac{2\pi}{\omega_c} = 1.5 \text{ min}$$
$$\text{A.R.} = \frac{1}{\omega T} = \frac{1}{21}$$
$$K_{\max} = 21$$

If the mixing delay were 20 sec, the ultimate period would be $^{30}\!/_{22} \times 1.5 = 2.1$ min and the maximum gain about 30.

16-2. CONTROL OF pH

Systems for pH control can be divided into two types, those where pH control is incidental to carrying out the main reaction and those where the equipment serves only to mix a reagent and a process stream to give a certain pH.   Fermentation is an example of the first type, and because of the large holdup of solution and slow disappearance of reagent, pH control is fairly easy and on-off control is usually adequate.

An example of the second and more difficult type of problem is the continuous neutralization of a process or waste stream.   If the control point is at or close to the equivalence point, adding a slight excess of one reactant makes a large relative change in the final concentration of that reactant.   Therefore a 1 to 2 per cent change in controlled flow may correspond to a large change in pH, perhaps one to four units.   If the neutralization is carried out in a stirred tank, the size of the tank is a design variable; increasing the tank size provides more capacity to damp out load changes but also increases the mixing delay and the cost of the system.

The nonlinear change in pH with reagent flow is another factor that makes close control of neutralization difficult.   A change in set point can change the process gain and optimum controller gain severalfold and perhaps lead to unstable oscillations.   Finally, systems used for pH control of waste streams often have to cope with severe load changes resulting from changes in concentrations and changes in flow.   Large load changes affect the process gain and may even alter the time constants of the system, which makes it especially difficult to design a control system that is moderately fast yet stable for all conditions.

PROCESS CHARACTERISTIC CURVES

The nonlinear behavior of pH systems is shown by the typical titration curves of Fig. 16-4.   These process characteristic curves show whether control at a given pH is likely to be routine or very difficult.   If the curve is very steep at the control point, the process is sensitive to slight changes in reagent flow or process flow, and large excursions are likely with a simple

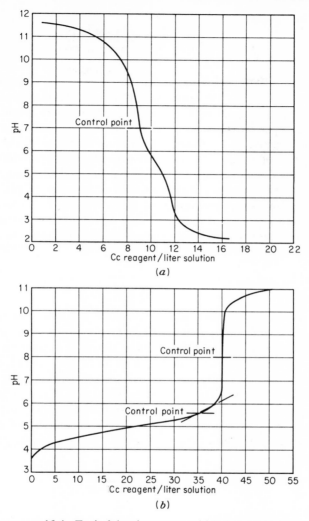

FIGURE 16-4 Typical titration curves. (a) Industrial process
water containing carbonates and organic
matter titrated with dilute H_2SO_4. [*From
Chaplin* (11), *courtesy of Instruments Publishing
Co., Inc.*] (b) Weak acid titrated with strong
base.

control system.   If the slope is changing rapidly, the controller settings will
have to be considerably lower than those based on a linear analysis at the
set point to allow for changes in process gain.   For the curve in Fig.
16-4a, the slope at the control point is about 2.4 pH units/(cm³ of reagent)
(liter of solution).   Based on a pH range of 5 and a normal reagent flow of

9, the process gain is

$$K_p = \frac{2.4/5}{1/9} = 4.3 \frac{\% \text{ change in pH}}{\% \text{ change in flow}}$$

This value of K_p is not very large, and a controller gain of 1 or greater could probably be used, since the mixing tank would provide some damping. However, the process gain of 4.3 is correct only for very small changes in flow. For a 5 per cent change in reagent flow, the actual changes in pH would be an increase of 1.3 units or a decrease of 0.7 unit, corresponding to average gains of 5.2 or 2.8. To make sure of stable operation, a process gain of about 6 should be used in calculating the controller gain.

A more difficult control problem would be to keep the pH at about 8 for the neutralization process shown in Fig. 16-4b, on the assumption that the reaction is to be carried out in a small stirred tank. The slope at the control point corresponds to a process gain of about 60 based on a pH range of 5.

$$K_p = \frac{8/5}{1/40} = 64$$

With such a high process gain, a low controller gain would be required for stable operation, and this gain might be below the lowest value available on standard controllers. Using a pH range of 10 on the measuring instrument would permit the controller gain to be doubled (or various attenuators could be used), but this would not solve the main problem of large excursions following load changes. The maximum deviations after sudden changes in load depend on the overall closed-loop gain and on the process gain for load changes, K_L. For load changes near the beginning of the control loop, the peak error is about 1.5 times the steady-state error for proportional control [Eq. (5-25)].

$$\frac{\text{Max error}}{\text{Load change}} \cong \frac{1.5K_L}{1 + K} \tag{16-2}$$

A 1 per cent change in flow or concentration of the process stream would have about the same effect as a 1 per cent change in reagent flow; so K_L would be about the same as K_P, or 64 for this example. With an overall gain of 10 for the closed loop, a 1 per cent change in load would therefore cause a peak error of about 10 per cent in the pH. Of course the linearization of the pH curve may introduce considerable error; for a better estimate of the pH change after an upset, the concentration of excess acid or base could be calculated by a stepwise procedure or obtained from an analog with nonlinear elements.

One way of increasing the stability of the control system is to increase the size of the reactor or mixing tank. The increased capacity provides additional damping of pH fluctuations, which means that a higher controller gain can be used and the peak error reduced.

If the other lags remained constant, doubling the size of the tank would almost double the maximum gain; the tank holdup is usually much larger than the other lags in the system, and there would be only a slight decrease in the critical frequency. Unfortunately, increasing the tank size usually increases the mixing delay, which reduces the potential advantage of a large tank. Also, it may be more expensive to use a large tank to get adequate damping than to use a small tank and a more accurate control system, perhaps one involving two controllers. Most of the reagent could be added to the inlet pipe or to a small preliminary tank and the main controller used to regulate the small flow needed to complete the reaction. Other schemes involving cascade control or split flow have also been used (11, 12). Before considering some specific examples, the dynamic behavior of all elements in the control loop must be known. Agitated tanks and control valves have already been discussed, and the response of pH electrodes is treated briefly in the next section.

RESPONSE OF pH ELECTRODES

A pH meter is expected to show a nonlinear response to step or sinusoidal changes in solution concentrations, since the pH depends on the logarithm of the concentration. However, even after changing the pH readings to equivalent concentrations, there are many peculiarities in the response. The response is up to severalfold faster in buffered than in unbuffered solutions (13, 14). A faster response was also reported for high and low pH than for nearly neutral solutions (13). However, others have found a slower response in going from water to acid than in going from acid to water (14). Values of the effective time constant, based on a 63 per cent concentration response, have ranged from 1 to 30 sec. The use of effective time constants is the most convenient way of comparing the responses and is probably good enough for control calculations as long as the electrode lag is one of the minor lags.

The major factors affecting the electrode lag are the thickness of the boundary layer around the electrode, the concentrations and diffusivities of the various ions, and the capacity (for ions) of the glass in the electrode. If the capacity of the glass were negligible, the effective time constant could be obtained from the capacity of the liquid layer around the electrode and the resistance to diffusion. On assuming a stagnant layer of thickness z, the time for 63 per cent response at the electrode surface is obtained from the solutions for diffusion in slabs.

$$t_{63\%} = \frac{0.5z^2}{D} \tag{16-3}$$

where $D =$ molecular diffusivity

For a velocity of 30 cm/sec past a 1-cm cylinder, the effective film thickness for mass transfer is about 40 μ for a diffusivity of 2×10^{-5} cm²/sec. This thickness corresponds to a time constant of about 0.4 sec, which is the right order of magnitude for buffered solutions but much higher than the value for nearly pure water. Apparently the capacity of the glass is greater than that of the thin layer of external liquid. The response in buffered solutions is rapid in spite of the glass capacity, since buffer molecules or ions near the electrode can dissociate to provide some of the extra hydrogen ions required for the new equilibrium. Thus only a fraction of the extra hydrogen ions have to diffuse through the entire boundary layer.

There are two time constants associated with the standard flow cells for industrial pH control. They are a holdup time of V/F if the cell is well mixed and a diffusion time constant that depends on the velocity past the electrode and the ion concentrations. Tests by Gusti and Hougen (14) with a large flow cell showed that both these lags decreased with increasing flow rate, as expected. With a special low-holdup cell, the response was more nearly first-order, but too fast to show clearly the effect of flow rate. The results of Disteche and Dubuisson (15) are even better evidence that the electrode time lag depends on diffusion in the boundary layer. By using a sudden jet of solution to sweep away a drop wetting the electrode, the time lag was reduced to 30 μsec, which was attributed to the electrical resistance of the glass.

To explain and predict the effects of pH level, direction of pH change, and buffer concentration of the electrode response, it will be necessary to consider the changes in the glass itself as well as the complex equations for counterdiffusion of ions in the liquid. This subject certainly deserves more attention from a scientific standpoint, though there may be little need for extensive studies for engineering purposes. With buffered solutions, the electrode response can be taken as 1 to 2 sec, which is generally negligible in pH control systems. For unbuffered solutions a time constant of 10 to 20 sec can be used for prediction purposes. This value could be lowered to a few seconds if necessary by using a high-velocity flow cell.

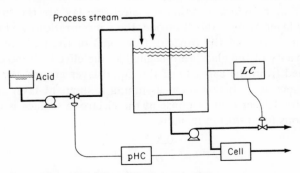

FIGURE E16-2   Diagram for Example 16-2.

16-3. EXAMPLES OF pH CONTROL SYSTEMS

Example 16-2

The continuous neutralization of an industrial process stream is carried out in a 6,000-gal tank with manual control of the acid reagent.   The process characteristic is given in Fig. 16-4a.   Estimate the maximum controller gain and the critical frequency for a proposed control system using a pH meter in a bypass line and a proportional-reset controller.   Would the control system keep the pH at 7 ± 0.5 even after sudden changes of 5 per cent in the process flow, which represent the most severe load changes encountered?

$$V = 6{,}000 \text{ gal}$$

$$F = 400 \text{ gpm}$$

$$N = 75\text{-rpm flat paddle stirrer}$$

$$\frac{D}{T} = 0.28$$

$$\text{Bypass flow} = 2 \text{ gpm}$$

$$\text{Sampling delay} = 6 \text{ sec in main line}$$

$$= 6 \text{ sec in bypass line}$$

$$\text{Cell holdup} = 0.1 \text{ gal}$$

The valve time constant is estimated to be 0.1 min, and the electrode lag is taken as 0.05 min.   The lags in the control loop are

$$T_1 = \frac{6{,}000}{400} = 15 \text{ min}$$

$$T_v = 0.1 \text{ min}$$

$$T_E = 0.05\text{-min electrode lag} \qquad T_C = \frac{0.1}{2} = 0.05\text{-min cell mixing lag}$$

$$L_1 = 0.5\text{-min mixing delay in tank} \qquad \text{from Fig. 16-2}$$

$$L_2 = 0.2\text{-min total sampling delay}$$

With one large time constant and several small ones, the critical frequency can be estimated by combining the small time constants with the time delays and assuming that the large time constant contributes 90° phase lag.

For $T_1 = 15$ min

$$\Sigma L \cong 0.9 \text{ min} \qquad 0.7\text{-min true delay} + 0.1 + 0.05 + 0.05$$

$$\omega_c \cong \frac{90}{(57.3)(0.9)} = 1.75 \text{ rad/min}$$

Check: At $\omega_c = 1.8$,

		ϕ, deg
$\omega T_1 = 27$		87
$\omega T_v = 0.18$		10
$\omega T_E = 0.09$		5
$\omega T_C = 0.09$		5
$\omega(L_1 + L_2) = 1.26$		72
		179

At 1.8 rad/min,   A.R. $= \frac{1}{27}$ $K_{max} = 27$

The process gain at the control point is 4.3, but a value of 6 is used to allow for nonlinearities (see page 350).   An equal percentage valve is used with a gain of 2.

$$K_v = \frac{2\% \text{ change in flow}}{\% \text{ change in pressure}}$$

$$K_p = \frac{6\% \text{ change in pH}}{\% \text{ change in flow}} \quad \text{on scale of 5-10}$$

$$K_{c,max} = \frac{K_{max}}{K_v K_p} = \frac{27}{2 \times 6} = 2.2 \frac{\% \text{ change in flow}}{\% \text{ change in pH}}$$

$$K_{c,\text{recommended}} = 1.1$$

Using Eq. (16-2) to predict the peak error gives

$$\frac{\Delta pH}{5} \cong 1.5 \frac{K_L}{1 + K} \times \frac{\Delta_{\text{flow}}}{\text{av flow}}$$

$$\frac{K_L}{1 + K} = \frac{6}{1 + (1.1)(2)(6)} = 0.42$$

For a 5 per cent change in flow

$$\Delta pH = 5(1.5)(0.42)(0.05) = 0.16$$

The above calculations show that the system will give satisfactory control and that an even smaller tank might have been used.   Of course, there are many uncertainties in the calculations, and the numerical values of gain, frequency, and error may be off by 50 to 100 per cent.   The mixing delay in the tank is the most uncertain lag in the control loop, but the lack of definite information on load changes is probably as big a handicap in predicting the performance of industrial systems.

Example 16-3

A waste stream containing weak acid is to be neutralized with a strong base in a continuous system with the pH controlled at 8.0.   The acid characteristic is shown in Fig. 16-4b.   Would the simple control system used in Example 16-2 keep the pH always between 7 and 9 even for sudden 5 per cent changes in load?   The flow rate is 400 gpm, the same as for Example 16-2.

a. Try the same tank size, 6,000 gal.

$$\omega_c = 1.8 \text{ rad/min}$$
$$K_{\max} = 27$$
$$K_p = 64 \qquad \text{at pH} = 8$$
$$K_{c,\max} = \frac{K_{\max}}{K_v K_p} = \frac{27}{2 \times 64} = 0.21$$
$$K_{c,\text{recommended}} = 0.1$$

Controllers with gains of 0.1 are available, but the peak error for this system would exceed one pH unit for a 5 per cent load change.

$$\frac{K_L}{1 + K} = \frac{64}{1 + 0.1(2)(64)} = 4.6$$
$$\Delta \text{pH} = 5(1.5)(4.6)(0.05) = 1.7$$

The peak error could be reduced to one unit by using a tank two to three times as large. Although the mixing delay would be slightly greater, the greater damping in the tank would permit the controller gain to be increased about 1.5- to 2-fold.

b. Instead of using a larger tank, the control could be improved by reducing the smaller time constants and time delays. The pH electrodes could be installed in the tank or in the outlet line to eliminate the sampling delay. A higher agitator speed could be used to decrease the mixing lag and a valve positioner used to reduce the valve lag.

Original system	*Revised system*	
$V = 6,000 \text{ gal}$	$V = 6,000 \text{ gal}$	
$T_1 = 15 \text{ min}$	$T_1 = 15 \text{ min}$	
$T_v = 0.1 \text{ min}$	$T_v = 0.05 \text{ min}$	with positioner
$T_E = 0.05 \text{ min}$	$T_E = 0.05 \text{ min}$	diffusion lag
$T_C = 0.05 \text{ min}$	$T_C = 0$	electrodes in tank
$L_1 = 0.5 \text{ min}$	$L_1 = 0.3 \text{ min}$	speed raised to 200 rpm
$L_2 = 0.2 \text{ min}$	$L_2 = 0$	
$\omega = 1.8 \text{ rad/min}$	$\omega = 4.0$	
$K_{\max} = 27$	$K_{\max} = 60$	
$K_{c,\text{recommended}} = 1.1$	$K_{c,\text{recommended}} = 2.2$	

With twice the controller gain, the peak error would be about half as great, and the revised system would be satisfactory. However, systems using smaller tanks and two controllers might do the job just as well or better.

c. Consider the use of two 600-gal tanks in series for the neutralization. The first tank would have a control point of 5.6, corresponding to 90 per cent reaction, and the second tank would have a control point of 8.0.

The two control loops would have identical characteristics except for the process gain and controller gain. The mixing time is estimated to be 0.2 min, based on a stirrer speed of 150 rpm.   The holdup time is 1.5 min, and the valve lag and electrode lag are taken as 0.05 min.

For each loop, $\qquad \omega_c \geq \dfrac{90}{(57.3)(0.30)} = 5.3$ rad/min

Exact calculation shows $180°$ phase lag at $\omega = 5.7$ rad/min.

$$K_{max} \cong \omega_c T_1 = 8.5$$

Use $K = 4.2$.   For the first system

$$K_p = \frac{\frac{1}{5}}{\frac{9}{36}} = 0.8 \frac{\%\ \text{change in pH}}{\%\ \text{change in flow}}$$

$$K_{c_1} = \frac{K}{K_v K_p} = \frac{4.2}{2 \times 0.8} = 2.6$$

For the second system, $K_p = 6.4$, 10 per cent of the former value, because the reagent flow is 0.1 as great.

$$K_{c_2} = \frac{4.2}{2 \times 6.4} = 0.33$$

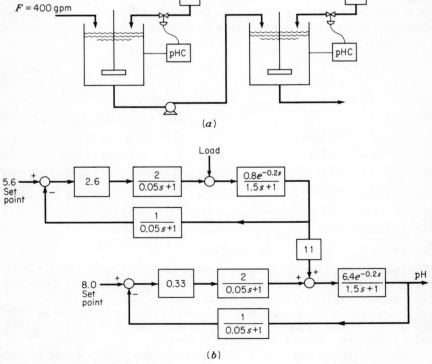

(a)

(b)

FIGURE 16-5   Diagram for Example 16-3c.

To show exactly how the combined systems would respond to a load change would require analog-computer simulation of the system in Fig. 16-5b. The gain of 11 is introduced between the two loops because a 1 per cent change in pH at pH = 5.6 is equivalent to a 1.25 per cent change in primary reagent flow or an 11 per cent change in secondary flow.

$$K_{L_2} = \frac{1}{0.8} \times \frac{36 \text{ cm}^3 \text{ primary flow}}{4 \text{ cm}^3 \text{ secondary flow}} = 11$$

Rough calculations indicate that a 5 per cent step change would result in a peak error of about one-twentieth pH unit at the first tank and about one-half pH unit at the second tank, which is within the desired limits.

Example 16-4

An acid waste is neutralized by adding lime in a small tank, with 0.5 min holdup, and allowing the slurry to react completely in a long trough, with 3 min holdup time. What control systems might be used?

Satisfactory control could probably not be achieved by measuring the pH at the trough exit and using a proportional controller to regulate the addition of lime. The maximum overall gain would be 1 because of the relatively large dead

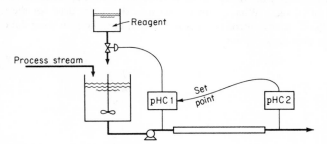

FIGURE E16-4 Diagram for Example 16-4.

time, and fluctuations in the acid content of the waste would go almost uncorrected. Somewhat better results could be obtained by using an intermittent control action, i.e., making an adjustment and waiting to see the effect before making further adjustments (see Chap. 9).

A more promising scheme is to use a cascade system, with the primary controller adjusting the set point of a pH controller at the tank exit (1). If complete reaction occurs in 3 min, a significant fraction (perhaps half) has reacted in 0.5 min and changes in strength of the waste stream would be noticed in the mixing tank.

Other control schemes, such as the use of split-flow and feedforward systems, are discussed in the literature (1, 11, 12). A complex system tested by analog simulation is described by Field (16). One of the interesting features of the system was the use of antilog converters so that the controllers received a signal proportional to ionic concentration.

PROBLEMS

1. For control of neutralization reactions, can the maximum gain be increased appreciably by diluting the reagent whose flow is regulated? Check your reasoning by calculating the process gain for the following cases:

a. Neutralization of 0.001 N HNO$_3$ with 0.1 N NaOH, main flow $= 1,000$ gpm, control point pH $= 7$

b. Same as (*a*), but using 0.01 N NaOH

2*a.* Derive the transfer function relating exit concentration to inlet concentration for a first-order reaction carried out in a perfectly mixed tank. What happens as the kinetic constant approaches infinity? What is the physical interpretation of this effect?

b. Considering the answers to (*a*), is it consistent to use a time constant of V/F for pH control when the reactions are almost instantaneous?

3. The flow of ammonia to an ammonium sulfate crystallizer is regulated by a ratio controller operating on the acid-flow signal and the ammonia-flow signal. Because of changing reagent quality, a pH controller is used to adjust the set point of the ratio controller. The pH electrodes are in an external sample line, which introduces a 20-sec time delay. The reactor has a holdup time of 10 min and a mixing lag of 30 sec. What controller settings should be used for the three-mode pH controller? The control point is pH $= 3.5$, and the following data from Ref. 17 show the process sensitivity.

% excess acid	pH of slurry
0.1	4.3
0.2	3.9
0.3	3.6
0.4	3.4
0.5	3.2

4. Compare the performance of the following systems for continuous pH control:

a. $F = 10,000$ gpm

$V = 1,200$ gal, $N = 50$ for turbine with $D/T = \frac{1}{3}$

pH electrodes at tank exit

b. $F = 10,000$ gpm

2 tanks in series, $V_1 = V_2 = 600$ gal

Same power consumption as (*a*)

pH electrodes at exit of second tank

Use the acid characteristic of Fig. 16-4*a*.

5. Sketch the response of a continuous-flow stirred tank to a step change in feed concentration and to a pulse change. Identify the times that correspond to $t_1, t_2, t_3,$ and t_4 in Fig. 16-3.

6. A 1,000-gal tank is used to mix fluids A and B in the ratio 1/2.4. The total flow rate is about 150 gpm, and the estimated mixing delay is 0.3 min. There is a

5-sec delay in the sample line, and the control valve, which regulates the flow of A, has a time constant of 2 sec.

    a. What setting would you recommend for a two-mode controller?

    b. Would derivative action be recommended for the controller?

    c. What would be the peak error following a 10 per cent decrease in the flow of B?

    7. A simplified method of predicting the effectiveness of pH control systems is given by Greer (18). For a step change in acid feed rate to a stirred tank, the peak error is calculated by using the equation for a first-order response and a "lag time" t.

$$\frac{C - C_0}{C_1 - C_0} = 1 - e^{-tF/V}$$

Show that this equation underestimates the error if the true time delay is substituted for t. About what value should be used for t?

    8. Calculate the step response of a pH electrode for the following changes, assuming that all the resistance and capacity are in a liquid film with a time constant of 1.0 sec. What would be the apparent time constant for each case, based on 63 per cent response in pH?

    a. pH 6 to pH 3 (unbuffered solutions)

    b. pH 3 to pH 6 (unbuffered solutions)

    c. pH 4 to pH 3 (unbuffered solutions)

REFERENCES

1. Kramers, H.: The Dynamic Behavior of Processes, in Particular for pH Control, *Trans. Soc. Instr. Technol.*, **8**:144 (1956).
2. Colucci, F. W., Jr.: pH Control of a Continuous Neutralization in a Stirred Tank, M.S. thesis, Cornell University, 1963.
3. MacDonald, R. W., and E. L. Piret: Agitation Requirements for Continuous Flow Stirred Tank Reactor Systems, *Chem. Eng. Progr.*, **47**:363 (1951).
4. van de Vusse, J. G.: Mixing by Agitation of Miscible Liquids, *Chem. Eng. Sci.*, **4**:178, 209 (1955).
5. Rushton, J. H., E. W. Costich, and H. J. Everett: Power Characteristics of Mixing Impellers, *Chem. Eng. Progr.*, **46**:395, 467 (1950).
6. Velguth, F. W.: Determination of Minimum Capacities for Control Applications, *ISA J.*, **1**(2):33 (February, 1954).
7. Harriott, P.: Mass Transfer to Particles Suspended in Agitated Tanks, *Am. Inst. Chem. Eng. J.*, **8**:93 (1962).
8. Kramers, H., G. M. Baars, and W. H. Knoll: A Comparative Study on the Rate of Mixing in Stirred Tanks, *Chem. Eng. Sci.*, **2**:35 (1953).
9. Fox, E. A., and V. E. Gex: Single Phase Blending of Liquids, *Am. Inst. Chem. Eng. J.*, **2**:539 (1956).
10. Norwood, K. W., and A. B. Metzner: Flow Patterns and Mixing Rates in Agitated Vessels, *Am. Inst. Chem. Eng. J.*, **6**:432 (1960).
11. Chaplin, A. L.: Applications of Industrial pH Control, The Instruments Publishing Co., Pittsburgh, 1950; also in *Instruments*, **22, 23** (1949, 1950).

12. Colver Nutting, D.: The Industrial Application of pH Measurement and Control, *Instrument Practice*, **8**:50, 123, 221, 327, 416 (1954).

13. Geerlings, M. W.: Dynamic Behaviour of pH-Glass Electrodes and of Neutralization Processes, in "Plant and Process Dynamic Characteristics," Butterworth Scientific Publications, London, 1957.

14. Gusti, A. L., Jr., and J. O. Hougen: Dynamics of pH Electrodes, *Control Eng.*, **8**(4):136 (April, 1961).

15. Disteche, A., and M. Dubuisson: Transient Response of the pH Glass Electrode, *Rev. Sci. Instr.*, **25**:869 (1954).

16. Field, W. B.: Design of a pH Control System by Analog Simulation, *ISA J.*, **6**(1):42 (January, 1959).

17. Campbell, C. G., R. L. Scaife, and T. C. Landwehr: pH Controls for Nitrate and Sulfate Processing, *ISA J.*, **5**(7):53 (1958).

18. Greer, W. N.: The Measurement and Automatic Control of pH, *Tech. Publ.* EN-96 (1), Leeds and Northrup Co., Philadelphia, 1962.

APPENDIX 1 NYQUIST STABILITY CRITERION

The stability of a closed-loop control system can be determined by plotting the open-loop frequency response on polar graph paper. The polar plot also shows how close a system is to instability and is a guide in evaluating possible changes in the system. The Nyquist criterion was developed by H. Nyquist in 1932 (Ref. 7, Chap. 1), and the derivation is given in most texts on servomechanisms. The results are presented here without proof.

The Nyquist diagram is a polar plot of the frequency response for the open-loop transfer function $G(s)$. The gain is plotted as the radius vector, and the phase shift is plotted in degrees clockwise from the right-hand abscissa. There is a separate point for each frequency; the frequencies are indicated next to a few of the points, or an arrow is used to show the direction of increasing frequency. Sometimes a dotted line is used for negative frequencies, and the results from $\omega = -\infty$ to $\omega = +\infty$ form a closed curve. For most cases, instability of the closed loop occurs if the plot encircles the (-1) point in a clockwise direction. This occurs if the plot for positive frequencies passes to the left of the (-1) point, which just means that the overall gain is greater than 1 at 180° phase lag.

When the open-loop system is unstable, a more general rule must be used. The closed-loop system is stable if the number of denominator roots of $G(s)$ with positive real parts is equal to the number of counterclockwise encirclements of the (-1) point.

$$N = P \qquad \text{for stability}$$

$$P = \text{no. denominator roots of } G(s) \text{ with positive real parts}$$

$$N = \text{no. counterclockwise encirclements of } (-1) \text{ point}$$

The value of N can be obtained from just the curve for positive frequencies by following the method presented by Nixon.† A radial line is drawn from the (-1) point, and wherever this line cuts the curve, an arrow is placed to show the direction of increasing frequency. The number of counterclockwise arrows minus the number of clockwise arrows is equal to N. The value of P can be obtained by inspection if the open-loop transfer function is in factored form; if not, the Routh criterion can be used. A few examples of stable and unstable systems are shown in Fig. A1-1.

Example A1-1

The open-loop transfer function for a three-capacity process is

$$G = 2/[(10s + 1)(5s + 1)(2s + 1)]$$

† F. E. Nixon, "Principles of Automatic Controls," p. 115, Prentice-Hall, Inc., Englewood Cliffs, N.J., 1953.

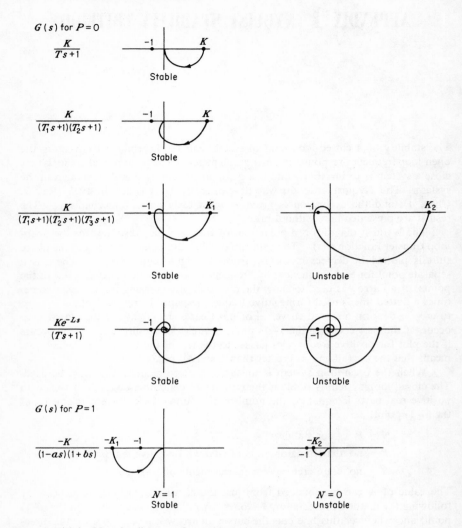

FIGURE A1-1   Nyquist diagrams for stable and unstable systems.

where the time constants are in minutes. (*a*) If a proportional controller with a gain of 3.0 is used, will the system be stable? (*b*) If reset action is added and a reset time of 5 min is used, will the system be stable for a controller gain of 3.0?

  a. The open-loop response of the process plus controller is plotted in Fig. A1-2. Since the curve does not encircle the (-1) point, the closed-loop system will be stable. (The Bode diagram for this process is Fig. 5-11.)

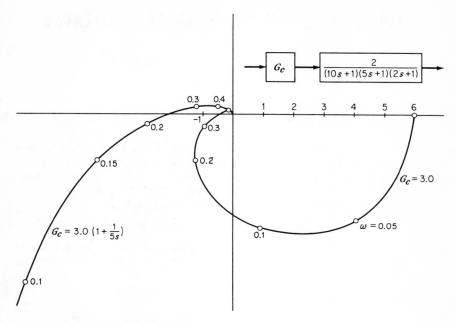

FIGURE A1-2   Nyquist diagrams for Example A1-1.

b. Adding reset action to the controller adds 90° phase lag at low frequencies, 45° phase lag at $\omega = 0.2$ ($\omega T_R = 1$), and 0° at very high frequencies.   Also the controller gain exceeds K_c, as shown in Fig. 6-2.   The open-loop plot now passes to the left of the (-1) point and would encircle it if the line for negative frequencies were added.   The closed-loop system is therefore unstable.

APPENDIX **2** ROUTH'S STABILITY CRITERION

A control system is unstable if the denominator of the closed-loop transfer function has any roots that are real and positive or have positive real parts. The Routh test† is a simple and exact method for determining whether or not positive roots exist, though the method does not show how close the system is to instability.

Consider a system with the following characteristic equation:

$$B_n s^n + B_{n-1} s^{n-1} + B_{n-2} s^{n-2} + \cdots + B_0 = 0$$

If any of the coefficients are negative or zero, the system is unstable (B_n is assumed to be positive). If all the coefficients are positive, they are arranged in an array as follows.

$$
\begin{array}{llll}
B_n & B_{n-2} & B_{n-4} & B_{n-6} \quad \cdots \\
B_{n-1} & B_{n-3} & B_{n-5} & B_{n-7} \quad \cdots \\
a_1 & a_2 & a_3 \quad \cdots \\
b_1 & b_2 & b_3 \quad \cdots \\
c_1 & c_2 \quad \cdots \\
d_1 & d_2 \quad \cdots
\end{array}
$$

where

$$a_1 = \frac{B_{n-1}B_{n-2} - B_n B_{n-3}}{B_{n-1}} \qquad a_2 = \frac{B_{n-1}B_{n-4} - B_n B_{n-5}}{B_{n-1}}$$

$$b_1 = \frac{a_1 B_{n-3} - B_{n-1} a_2}{a_1} \qquad b_2 = \frac{a_1 B_{n-5} - B_{n-1} a_3}{a_1}$$

$$c_1 = \frac{b_1 a_2 - a_1 b_2}{b_1} \qquad c_2 = \frac{b_1 a_3 - a_1 b_3}{b_1}$$

The number of terms in each row decreases as the array is developed, and the process is continued until only zeros are obtained for additional coefficients.

The number of changes in sign in the left-hand column is the number of roots with positive real parts.

Example A2-1

Consider the stability of a system with the following closed-loop transfer function:

$$\frac{G}{1 + G} = \frac{10}{4s^3 + 8s^2 + 5s + 11} \qquad G = \frac{10}{(2s + 1)^2(s + 1)}$$

† F. J., Routh, "Dynamics of a System of Rigid Bodies," 3d ed., Macmillan & Co., Ltd., London, 1877.

The coefficient array for the denominator is

$$
\begin{array}{cc}
4 & 5 \\
8 & 11 \\
-\tfrac{1}{2} & \\
11 &
\end{array}
$$

Since there are two sign changes in the left-hand column, there are two roots with positive real parts and the system is unstable.

Example A2-2

Use the Routh criterion to examine the closed-loop stability of the system in Example A1-1.

a.
$$ G = \frac{(3)(2)}{(10s + 1)(5s + 1)(2s + 1)} $$

The denominator of the closed-loop transfer function $G/(1 + G)$ is

$$ (10s + 1)(5s + 1)(2s + 1) + 6 = 100s^3 + 80s^2 + 17s + 6 $$

The coefficient array is

$$
\begin{array}{cc}
100 & 17 \\
80 & 6 \\
{}^{760}\!/_{80} & \\
6 &
\end{array}
$$

Since there are no sign changes in the left-hand column, the system is stable.

b.
$$ G = 3\left(1 + \frac{1}{5s}\right)\frac{2}{(10s + 1)(5s + 1)(2s + 1)} = \frac{6(5s + 1)}{5s(100s^3 + 80s^2 + 17s + 1)} $$

The denominator of $G/(1 + G)$ is

$$ 500s^4 + 400s^3 + 85s^2 + 35s + 6 $$

The coefficient array is

$$
\begin{array}{ccc}
500 & 85 & 6 \\
400 & 35 & \\
41.2 & 6 & \\
-23.3 & & \\
6 & &
\end{array}
$$

The two sign changes in the left-hand column indicate that the closed-loop system is unstable.

NAME INDEX

SUBJECT INDEX